Cenozoic Seas

The View from Eastern North America

Cenozoic Seas

The View from Eastern North America

Edward J. Petuch
Florida Atlantic University

Photography by Mardie Drolshagen Banks

CRC Press
Taylor & Francis Group
Boca Raton London New York

CRC Press is an imprint of the
Taylor & Francis Group, an **informa** business

CRC Press
Taylor & Francis Group
6000 Broken Sound Parkway NW, Suite 300
Boca Raton, FL 33487-2742

First issued in paperback 2019

ISBN-13: 978-0-8493-1632-6 (hbk)
ISBN-13: 978-0-367-39466-0 (pbk)

Library of Congress Card Number 2003065367

Library of Congress Cataloging-in-Publication Data

Petuch, Edward J.
Cenozoic seas : the view from eastern North America / Edward J. Petuch ; photograpy by Mardie Drolshagen Banks
p. cm.
Includes bibliographical references and index.
ISBN 0-8493-1632-4 (alk. paper)
1. Paleoceanography—Cenozoic. 2. Paleoceanography—Altantic Coast (North America) 3. Paleoecology—Cenozoic. 4. Paleoecology—Altantic Coast (North America) 5. Paleontology—Cenozoic. 6. Marine animals, Fossil—Atlantic Coast (North America) I. Title.

QE39.5.P25P45 2003
560′.178—dc22 2003065367

Visit the CRC Press Web site at www.crcpress.com

Visit the Taylor & Francis Web site at
http://www.taylorandfrancis.com

and the CRC Press Web site at
http://www.crcpress.com

To my wife, Linda Joyce Petuch, and to my
children, Eric, Brian, and Jennifer

Foreword

The extraordinarily rich fossil record of the Atlantic and Gulf coastal plains of the United States has long attracted paleontologists. The last thirty-five million years of Earth history are wonderfully, if not always perfectly, chronicled by a succession of fossil assemblages extending from New Jersey to Florida. Having this record available is one thing, but interpreting it is quite another. In order to make sense of the changing geography, the comings and goings of ecosystems and species, and the controls that climate exercises on living things, we need a descriptive framework in which all the fossil-bearing formations and their contents are neatly classified in units of time and space. This is what Petuch provides for the eastern United States in this book. Petuch lays the groundwork for those who want to study evolution and extinction within specific communities. It will now be possible to carry out comparative studies within ecosystems — coral-dominated reefs, lagoon-bottom associations dominated by scallops or by suspension-feeding turritellid gastropods, and communities of mangroves and estuaries, and to evaluate how different community types were affected by changes in geography and climate. In short, the descriptive foundation allows us to view events in the history of life in the specific environmental and biotic contexts in which evolution, invasion, and extinction take place.

There are no apples and oranges in the fossil record, but the present book will stimulate the kind of research in which metaphorical apples and oranges will no longer be confused. We shall all benefit from the fruits of Petuch's pioneering effort.

Geerat J. Vermeij, Ph.D.

University of California-Davis

Preface

The coastal waters of Recent eastern North America, from the Gulf of St. Lawrence to the Florida Keys and the eastern Gulf of Mexico, contain some of the richest marine faunas found anywhere on Earth. As a marine biologist, I was always fascinated by the seemingly endless array of organisms and communities that exists along this area; from the Arctic faunas of the north, to the cool-temperate ecosystems of the mid-Atlantic United States, south to the warm-temperate faunas of the Gulf of Mexico and the southeastern United States and the tropical coral reef environments of southern Florida. While studying these modern faunas as an undergraduate and graduate student, I was impressed by the sharply-defined biogeographical patterns of the coastal communities but was also puzzled by the presence of seemingly anomalous faunal components in many of these ecosystems. A whole new view of this modern marine world was revealed to me during my doctoral and postdoctoral research, when I began to study the Cenozoic fossil record of the eastern North American coast. Slowly and painstakingly, as I collected data in fossil beds from Maryland to Florida, the questions engendered by these anomalous modern faunal components and biogeographical patterns began to be answered. Instead of viewing coastal marine communities in a present-day three-dimensional spatial system, I was now able to see them in a four-dimensional context, with a time component that spanned over 35 million years. The sheer magnificence and beauty of the history of these organisms and their communities immediately converted me to a new academic discipline, coastal paleoceanography. This book is a result of that conversion.

Having specialized in marine malacology, I always view modern and ancient oceans in the context of molluscan systematics and ecology. While investigating the Cenozoic fossil beds of the East Coast, I was delighted to find that the vast majority of well-preserved organisms were mollusks; mostly bivalves and gastropods. This overwhelming abundance of mollusks is actually a boon for paleoceanographers and paleoecologists, as both classes have been shown to give great insight into ancient marine physical parameters. Since most marine mollusks are physiologically tightly constrained by water temperature, depth, substrate, and salinities, their presence or absence can be used to reconstruct the physiographic and biogeographic patterns of ancient seas. In this book, I describe the physical and biological oceanography of seven separate eastern North American paleoseas, covering a time frame that extends from the early Oligocene to the Holocene. Depending on sea level fluctuations, the basins of these paleoseas periodically flooded and dried, resulting in the formation of 34 separate subseas. Fascinatingly, each of these subseas contained its own set of endemic organisms and communities, some being among the most bizarre and beautiful that have ever evolved on our planet. As a way of classifying the biotic and ecological patterns of these subseas, I have named and described over 120 separate marine communities. Unlike descriptions of modern marine ecosystems, where the community is named after the dominant primary producer, these paleocommunities are named after the most prominent fossil organism. Since most primary producers do not fossilize, these index paleospecies are used as proxies or indicators of the entire community. As an artifact of the preservation of the fossil record, my catalog of paleocommunities is highly skewed toward the hard-skeletoned mollusca and scleractinia. These community descriptions are meant to be only brief, thumbnail sketches that emphasize the largest and most prominent organisms in as many trophic levels as possible.

While mapping the distribution of the fossil communities over time, I was able to gain insight into the approximate configurations and sizes of the paleoseas, subseas, and coastlines. To my amazement, many of these lost coastal seas extended far inland and covered large areas of what are now the mid-Atlantic and southeastern United States and sprawling urban centers. Putting things in perspective, it is fascinating to imagine that Richmond, Virginia was once the shoreline of a large, scallop-filled sea, that Miami, Fort Lauderdale, and Naples, Florida were once the edge of a giant coral reef complex similar to Kwajalein or Eniwetok Atolls, or that Beluga Whales once swam in Lake Champlain. What is equally fascinating is the constant turnover of marine communities in the seven paleoseas, with well-established faunas being almost instantaneously decimated and quickly replaced by newly-evolved faunas. This flickering spectrum of constantly-changing communities and species is one of the most beautiful images that emerges from the history of these ancient marine worlds. It is also one of the more horrific, as it reflects the environmental catastrophies and sea level fluctuations that occurred on such a regular basis throughout the Cenozoic. I have tried to capture some of these images in this book, primarily to shed light onto cyclical patterns of evolution, extinction, and community replacement in the American seas. Hopefully, this information will be of predictive value, allowing insights into present and future patterns of global warming and cooling. It should be remembered that we too are part of the biotic continuum and that the modern coastal areas of eastern North America will someday be the paleoseas of a future world.

Edward J. Petuch
2 April 2003

Acknowledgments

Without the help of many persons and organizations during the past 25 years of research and fieldwork, this book would never have been produced. To them I am greatly indebted. For allowing me to conduct field work on their property, I thank the following: Mr. Enrique Tomeu (Palm Beach Aggregates, Inc., Loxahatchee); Mr. D.L. Brantley (Brantley Shell Pit, Arcadia); the Fanjul Family (Florida Crystals, Inc., Everglades Agricultural Area quarries); Mr. Howard A. Griffin (Griffin Brothers, Inc. quarries, Ft. Lauderdale); Mr. Hugh Cannon (Quality Aggregates, Inc., Sarasota); the Richardson Family (APAC pit, Sarasota); Mr. Pete Miniear (Dickerson Florida, Inc. Indrio Mine, Ft. Pierce, Florida) and the Dickerson Group, Inc.; Messrs. Robert Roach, Jose Dimas, Richard Rhodes, and Mrs. Iris Shellhorn (Florida Rock Industries, Naples Quarry; formerly Mule Pen Quarry, Naples); Mr. Ronald Capeletti (Capeletti Brothers quarries); Messrs. Edwin and Stanley Rucks and the Rucks Family (Rucks Pit, Fort Drum); Messrs. Ralph Chamness and I.K. Gilmore (Lee Creek Mine, Texasgulf Chemicals, Inc.); Mr. Allen Ridgdill (Ridgdill Quarry, Moore Haven); Mrs. Robin Weeks (Weeks Pit, Lakeport); Mr. Fred Smith (Smith Pit, Okeechobee); Ms. Amy Plummer (Plum Point, Maryland); and the Mecca Family (Mecca Orange Groves quarry, North Palm Beach). For the donation of valuable research specimens (many photographed in this book), I thank the following: Mr. Eric Kendrew, Mr. Brian Schnirel, Mr. Jonathan Arline and Mrs. Joanna Arline, Mr. Steven Rial, Mrs. Phyllis Diegel, Mr. Richard Duerr, Mr. Larry and Mrs. Judy Haley, Mr. Dale Stream, the late Mr. Charles and Mrs. Violet Hertweck, Mrs. Meta Jones, Mrs. Mary Mansfield, Mr. Steve and Mrs. Roxane Wilson, Mr. Richard and Mrs. Diane Pennington, Mr. Gregory Herbert, Dr. Gregory Dietl, the late Mr. Vladimir Eismont, Mrs. Susan Khan, Mr. Frank and Mrs. Becky Hynes, Mrs. Marilyn Barkley, Mr. Robert Schmidt, Mr. Joseph Turner, Mr. Donald Asher, Ms. Jamie Smith, Mr. James Houbrick, Mr. Joseph Buchek, Ms. Irene White, Mr. Norman Riker, Mr. Theodore Davis, Mr. Jack Spengler, Mr. Edward Volek, Mr. George Salony, Dr. Anton Oleinik, Mr. Edwin Rucks, Mrs. Eleanor Marr, Mr. Anthony Cinelli, Mr. Edward Mattingly, Mr. Frank Boer, Mr. Josiah Strauss, Mr. Rudolph Pascucci, Mr. Anthony Cinelli, Mr. Herbert and Mrs. Fonda Waldron, and Mrs. Susan Stephens. Special thanks goes to the Division of Research and the College of Science (Vice President Larry Lemanski and Dean Nathan Dean), Florida Atlantic University, for supporting the photographic and technical aspects of the book's production. Special thanks also goes to four talented Florida Atlantic University staff members: Ms. Stefanie Resciniti, for cover design and text layout; Mr. Thomas Corcoran, for constructing the paleoseas tables (Tables 1, 2); to Mr. Tobin Hindle, for helping to design the paleoseas maps; to Ms. Mardie D. Banks, for photographing all the specimens, setting up the digital files, and for helping me design the plates. Dr. Charles Roberts, Florida Atlantic University, developed the GIS programming techniques for simulated satellite photographs of paleolandforms and created the images of Pliocene and Pleistocene Florida used in this book. Special thanks, also, to my old post-doc mentor, Dr. G.J. Vermeij, University of California-Davis, and Dr. Anton Oleinik, Florida Atlantic University, for reviewing the manuscript and for offering numerous helpful suggestions.

Biography

Edward J. Petuch was born in Bethesda, Maryland in 1949. Being raised in a Navy family, he spent many of his childhood years collecting living and fossil shells in such varied localities as California, Puerto Rico, Chesapeake Bay, and Wisconsin. His early interests in paleontology and marine biology eventually led to B.A. and M.S. degrees in Zoology from the University of Wisconsin-Milwaukee. While in Wisconsin, his thesis work concentrated on the molluscan biogeography of coastal West Africa. There he traveled extensively in Morocco, Western Sahara, Senegal, Gambia, Sierra Leone, and the Cameroons. Continuing his education, Petuch studied marine biogeography under Gilbert Voss at the Rosenstiel School of Marine and Atmospheric Science, University of Miami. During that time, his dissertation work involved intensive collecting and working on shrimp boats in Colombia, Venezuela, Barbados, the Grenadines, and Brazil. After receiving his Ph.D. in Oceanography in 1980, Petuch undertook two years of postdoctoral research on paleoecology with Geerat Vermeij at the University of Maryland. There, he also held a research associateship with the Department of Paleobiology, National Museum of Natural History, Smithsonian Institution and conducted intensive field work on the Plio-Pleistocene fossil beds of Florida and the Miocene of Maryland.

Petuch has also collected fossil and living mollusks in Australia, Papua-New Guinea, Fijis, French Polynesia, Japan, the Mediterranean coasts of North Africa and Europe, the Bahamas, Mexico, Belize, Nicaragua, and Uruguay. This research has led to the publication of almost 100 papers. His seven previous books, *Atlas of Living Olive Shells of the World* (with Dennis Sargent), *New Caribbean Molluscan Faunas, Neogene History of Tropical American Mollusks, Field Guide to the Ecphoras, Edge of the Fossil Sea, Atlas of Florida Fossil Shells*, and *Coastal Paleoceanography of Eastern North America* are well-known reference texts within the malacological and paleontological communities. Presently, Petuch is a Professor of Geology at Florida Atlantic University in Boca Raton, Florida. He resides in Lake Worth, Florida with his wife Linda and three children, Eric, Brian, and Jennifer. When not collecting or studying mollusks, Petuch leads an active career as a musician and member of the university-affiliated Cuvier Trio, playing the recorders and harpsichord and specializing in Baroque and Renaissance music.

Cenozoic Seas: The View from Eastern North America

Table of Contents

Chapter 1. Paleoseas of Cenozoic Eastern North America

The Atlantic Coastal Plain and Gulf Coastal Plain regions of the eastern United States have long been known to contain vast, thick Cenozoic marine deposits. Being on the trailing edge of the North American Plate, the Gulf and Atlantic Coastal Plains have been almost continuously subject to structural downwarping, making them particularly sensitive to sea level fluctuations. This has allowed for the preservation of a virtually complete eustatic history of the area, a feature that is generally lost along leading plate edges. The sediments that accumulated along these downwarped, flooded continental margins also contain some of the richest and best preserved marine fossil beds found anywhere on Earth, enabling unprecedented insights into both the ancient marine ecosystems and the paleoceanography of the regions.

Unlike other trailing plate edges elsewhere, several large, open-ended basins formed along the Gulf and Atlantic Coastal Plains during the late Cenozoic (Oligocene-Pleistocene). These pericontinental basins were separated from each other by wide structural arches that had formed either during the Appalachian Orogeny or during some subsequent local reactivation or faulting event. Although traditionally referred to by geologists as "embayments" (Gibson, 1983; Ward, 1992), these pericontinental systems were far more complex than simple bays and represented true seas in the strict oceanographic sense. Like all seas, they were bodies of salt water, they were structurally bound on at least three sides, they occupied geologically discrete basins, they each contained their own distinct configurations of currents and water masses, and they contained their own distinctive endemic organisms and ecosystems. In this book, these late Cenozoic "embayments" will be accorded full paleosea status.

Depending on overall depth, geomorphological configuration, and rate of sedimentary infilling from deltas, each paleosea exhibited its own pattern of eustatic fluctuations and number of marine transgressive intervals. In the time interval from the early Oligocene to the late Pleistocene, some paleoseas had as many as ten large transgressive intervals, while others had as few as four. Some of the longer-lived paleoseas existed throughout this entire time, while others existed for only one or two epochs. In accordance with the standard paleoceanographic nomenclature used for epicontinental seas (Sloss, 1963), I here refer to the unconformity-bound transgressive intervals of marginal paleoseas as "subseas." These pericontinental subseas differed from epicontinental subseas in that they were influenced primarily by eustatic fluctuations and not by tectonics or orogeny.

As presently understood, the eastern North American paleoseas are divided into two distinct basinal groups, encompassing the Gulf basins and the Atlantic basins (Ward, 1985). Sedimentary deposition within these two groups differs greatly, with the Atlantic basins characteristically having thin marine transgressive units that are separated by numerous regressive unconformities and with the Gulf basins having much thicker, faster-growing units that are separated by fewer unconformities. These differences in sedimentary deposition reflect the erosional patterns of the Appalachian Mountains during the Neogene, when the bulk of the sediments was being deposited in the south and when giant deltaic environments existed along Alabama, northern Florida, and Georgia. In this book, each separate marine sedimentary unit is considered to be a sea floor, or series of sea floors, within an individual subsea.

Geochronology and Geography of the Eastern American Paleoseas

By the beginning of the Oligocene, the eastern Gulf basinal systems contained only two paleoseas; the northern *Choctaw Sea* and the southern *Okeechobean Sea*. The Atlantic basinal systems were more complex and contained four paleoseas; the southern *Charleston Sea*, the central *Albemarle Sea* and *Salisbury Sea*, and the northern *Raritan Sea*. Included within the scope of this book, but not occupying a true structural basin, is a seventh paleosea, the far northern *Champlain Sea*. Formed by flooding due to isostatic depression caused by the great Wisconsinan continental ice sheet, the *Champlain Sea* was the only paleosea that was not eustatically controlled. Detailed overviews of the geomorphology, oceanography, and ecology of these seas and their subseas are given in the following chapters. The geochronology of the five main paleoseas, and their subsea intervals, is shown on Tables 1 and 2.

The Choctaw Sea

Named for the Choctawhatchee ("Choctaw River") of northern Florida, this northeastern Gulf sea occupied a large area of the present-day Florida Panhandle, including the Choctawhatchee, Apalachicola, and Ochlockonee River deltas, and extended as far west as Okaloosa County and as far east as Taylor County. In its greatest development, the Choctaw Sea also extended inland as far north as Bainbridge, Georgia (Figures 1, 3, and 5). Existing from the Rupelian Oligocene to the latest Piacenzian Pliocene, the Choctaw Sea encompassed six separate subseas. Based upon their marine faunas, these subseas can be further subdivided into two groups; the early subseas (Rupelian Oligocene to Serravallian Miocene; see Chapters 3 and 4) and the late subseas (Tortonian Miocene to Piacenzian Pliocene; see Chapters 5 and 6). Of the early subseas, the oldest, the *Bainbridge Subsea* (named for the fossil beds at Bainbridge, Georgia), existed during the early part of the Oligocene, from the Rupelian to the earliest Chattian Ages, and spanned the Suwannee Strait between Georgia and Florida (Figure 1). The Choctaw Sea basin dried out for most of the Chattian Age, but reflooded in the latest Chattian as the much smaller *Chattahoochee Subsea* (named for the Chattahoochee Formation of the Florida Panhandle). During Chattahoochee time, the Suwannee Strait closed off completely and Georgia connected to Florida as a single peninsula. By the early Aquitanian Miocene, this subsea disappeared and the basin remained dry until the beginning of the Burdigalian Miocene. At this time, the *Chipola Subsea* (named for the Chipola Formation of the Florida Panhandle and previously referred to as the "Chipola Sea" (Petuch, 1997)) flooded the area and persisted until the end of the Langhian Miocene (Figure 3).

After a short regressive period during the earliest Serravallian Miocene, the basin was reflooded during the early Serravallian Miocene to produce the first of the late subseas, the *Walton Subsea* (named for the fossil beds in Walton County, Florida Panhandle). The early Tortonian Age of the Miocene saw another regressive dry time within the Choctaw Sea basin. By the mid-Tortonian, however, a fifth transgressive flooding event produced the *Alaqua Subsea* (named for the fossil beds along Alaqua Creek, Walton County, Florida Panhandle). This subsea persisted until the mid-Messinian Miocene, when a major regressive period occurred. The basin remained dry throughout the late Messinian and the entire Zanclean Pliocene, only to be flooded in the Piacenzian Pliocene for a sixth and final time. This subsea, the *Jackson Subsea* (named for the Jackson Bluff Formation of the Florida Panhandle), was previously referred to as the "Jackson Sea" (Petuch, 1997) and existed only until the late Piacenzian Pliocene (Figure 5). For detailed discussions of the individual geo-

Epoch	Age/Stage	Choctaw Sea	Okeechobean Sea	Charleston Sea	Albemarle Sea	Salisbury Sea
Miocene	Messinian	Alaqua	Charlotte		Rappahannock	
	Tortonian	Walton	Polk			St. Mary's
	Serravallian			Coosawhatchee		Patuxent
	Langhian	Chipola	Arcadia		Pamlico	Calvert
	Burdigalian					
	Aquitanian	Chattahoochee	Tampa	Edisto	Silverdale	Old Church
Oligocene	Chattian					
	Rupelian	Bainbridge	Dade	Ashley	River Bend	?

Table 1

Table 1. Oligocene and Miocene chronology of the five main eastern North American paleoseas, showing the chronological distributions of their subseas.

Epoch	Age/Stage	Choctaw Sea	Okeechobean Sea	Charleston Sea	Albemarle Sea	Salisbury Sea
Pleistocene	Wisconsinan					
	Sangamonian		Lake Worth			
	Illinoian					
	Yarmouthian		Belle Glade			
	Kansan					
	Aftonian		Loxahatchee			
	Nebraskan					
	Calabrian		Caloosahatchee	Waccamaw	Croatan	
Pliocene	Piacenzian	Jackson	Tamiami	Duplin	Yorktown	
	Zanclean		Murdock	Santee	Williamsburg	

Table 2

Table 2. Pliocene and Pleistocene chronology of the five main eastern North American paleoseas, showing the chronological distributions of their subseas.

morphologies and ecosystems of the Choctaw subseas, see Chapters 3, 4, 5, and 6.

The Okeechobean Sea

The other Gulf basin, the *Okeechobean Sea* (named for Lake Okeechobee, Florida; see Petuch, 1993; 1997), occupied the southern tip of Florida and extended from present-day Tampa, southward across the entire Everglades Basin, northward into the Kissimmee River Valley, and eastward to present-day Martin County (Figures 1, 3, 5, and 7). The longest-lived of all the eastern American paleoseas, the Okeechobean Sea existed from the early Oligocene to the late Pleistocene and encompassed eleven separate subseas. Like the Choctaw Sea, the Okeechobean Sea subseas can be subdivided into two groups, based upon their marine faunas; in this case, the early subseas (Rupelian Oligocene to Messinian Miocene; see Chapter 5) and the late subseas (Zanclean Pliocene to Sangamonian Pleistocene; see Chapters 6, 7, 8, 9). The oldest of the early subseas, the *Dade Subsea* (named for the fossil reefs in Dade County, Florida), represented the incipiency of the Okeechobean Sea and was contemporaneous with the Bainbridge Subsea of the Choctaw Sea (Rupelian-earliest Chattian Oligocene) (Figure 1). The Okeechobean Sea basin was emergent throughout most of the Chattian Age, flooding in the latest Chattian and developing into the *Tampa Subsea* (named for the fossil beds at Tampa, Florida). Contemporaneous with the Chattahoochee Subsea of the Choctaw Sea, the Tampa Subsea persisted into the early Aquitanian Miocene but dried out after that and remained emergent for the rest of the Aquitanian Age. The beginning of the Burdigalian Miocene saw the reflooding of the Okeechobean Sea basin and the establishment of the *Arcadia Subsea* (named for the Arcadia Formation of southern Florida). This subsea persisted until the Langhian-Serravallian boundary and was the exact contemporary of the Chipola Subsea. A short emergent interval during the early Serravallian Miocene gave way to a major transgressive flooding that ranged from the early Serravallian to the early Tortonian Miocene and produced the *Polk Subsea* (named for the fossil beds in Polk County, Florida). This subsea was contemporaneous with the Walton Subsea of the Choctaw Sea. After a short regressive period during the early to mid-Tortonian, the last of the early subseas, the *Charlotte Subsea* (named for the fossil beds in Charlotte County, Florida) flooded the area. This subsea was contemporaneous with both the Alaqua Subsea of the Choctaw Sea.

During the late Messinian Miocene, a major regression took place and the entire Okeechobean Sea basin was emergent. This was followed by a brief transgressional period during the early Zanclean Pliocene, producing the short-lived *Murdock Subsea* (named for the Murdock Station Formation of southwestern Florida). By late Zanclean time, the Okeechobean Sea basin was again emergent. This was soon reversed by a major transgressive event in the early Piacenzian Pliocene, establishing the large and complex *Tamiami Subsea* (named for the Tamiami Formation of southern Florida and previously referred to as the "Pinecrest Subsea" (Petuch, 1997)), an exact contemporary of the Jackson Subsea of the Choctaw Sea farther north (Figure 5). A series of rapid eustatic fluctuations took place during the late Piacenzian Age (Campbell, 1993; Petuch, 1997), with the final sea level lowering being the most severe and resulting in a brief emergent time. At the very end of the Piacenzian Age, the Okeechobean Sea basin again reflooded and produced the *Caloosahatchee Subsea* (named for the Caloosahatchee Formation of southern Florida) (Figure 7). This subsea existed across the Pliocene-Pleistocene boundary and throughout the entire Calabrian Pleistocene. During the Nebraskan Glacial Stage of the Pleistocene, and during the accompanying major sea level drop, the entire Okeechobean area was emergent. For the rest of the Pleistocene Epoch, three short-lived subseas developed within the Okeechobean Sea basin; the *Loxahatchee Subsea* (named for the fossil beds at

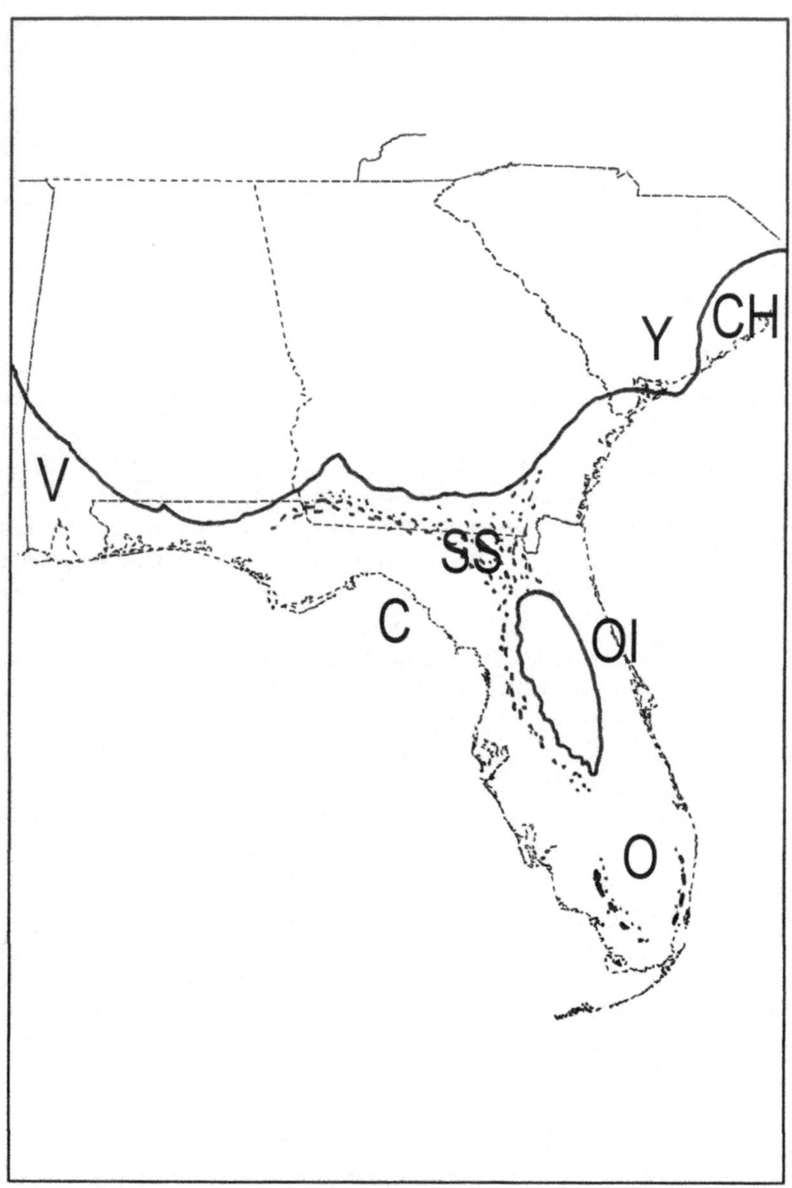

Figure 1. Approximate configuration of the southeastern North American seas during the Rupelian to earliest Chattian Oligocene, superimposed upon the outline of the Recent North American coastline. Paleoseas include: C= Choctaw Sea (Bainbridge Subsea), O= Okeechobean Sea (Dade Subsea), CH= Charleston Sea (Ashley Subsea), and V= Vicksburg Sea (not covered in this book). Prominent geomorphological features include: SS= Suwannee Strait (also referred to as the Gulf Trough) with its complex of coral reefs (speckling on map), OI= Orange Island and its complex of coral reefs (speckling on map), and Y= the Yamacraw Arch and highlands, separating the Charleston Sea from the Choctaw Sea-Suwannee Strait. During this time, most of the Okeechobean Sea was covered with deep water (100-200m), with only small reef complexes in the south.

Loxahatchee, Palm Beach County, Florida) (Aftonian Interglacial Stage), the *Belle Glade Subsea* (named for the fossil beds at Belle Glade, Palm Beach County, Florida) (Yarmouthian Interglacial Stage), and the *Lake Worth Subsea* (named for the fossil beds at Lake Worth, Palm Beach County, Florida) (Sangamonian Interglacial Stage). These three subseas were separated by regressions during the Kansan and Illinoian Glacial Stages, respectively. For detailed discussions of the individual geomorphologies and ecosystems of the Okeechobean subseas, see Chapters 3, 5, 6, 7, 8, and 9.

The Raritan Sea

The northernmost of the paleoseas of the Atlantic basinal systems, the *Raritan Sea* (named for the Raritan River and Raritan Bay of New Jersey), extended from Atlantic County, New Jersey northward to at least Cape Cod, Massachussetts (Figures 2 and 4). The Raritan Sea is also the most poorly known of the Atlantic paleoseas, as most of its sediments are deeply buried beneath Plio-Pleistocene deltas or, in the north, have been removed by Pleistocene glaciation. The paltry Raritan fossil material, collected mostly from deep well cores or from poorly preserved coastal exposures (eg. Sankaty Head, Nantucket, Massachussetts; Shattuck, 1904), shows that the sea existed only from the mid-Burdigalian Miocene to the end of the Langhian Miocene. This shortest-lived of the coastal paleoseas may also have had subsea intervals during the Rupelian Oligocene (contemporaneous with the Bainbridge Subsea of the Choctaw and Dade Subsea of the Okeechobean Sea) or during the late Chattian Oligocene-early Aquitanian Miocene (contemporaneous with the Chattahoochee Subsea of the Choctaw Sea or the Tampa Subsea of the Okeechobean Sea). Since there is, at present, no paleontological evidence for these earlier subseas, their existence is purely conjectural and they will remain un-named. Some cold water faunal components, probably derived from Raritan Sea ecosystems, are present in early Miocene paleoseas farther south and these are discussed in Chapter 4.

The Salisbury Sea

South of the Raritan Sea and extending from present-day southern New Jersey to the James River of Virginia and inland to Richmond, Virginia and west of Washington, D.C. (at its greatest development), was the largest of the Atlantic coast paleoseas, the *Salisbury Sea* (named for Salisbury, Maryland on the Delmarva Peninsula). As can be seen in Figures 2 and 4, the Salisbury Sea was separated from the Raritan Sea to the north by the Normandy Arch highlands of present-day southern New Jersey. Ranging from the late Chattian Oligocene to the mid-Tortonian Miocene, the Salisbury Sea encompassed four subseas. The oldest of these, and representing the incipiency of the Salisbury Sea, was the *Old Church Subsea* (named for the Old Church Formation of the Chesapeake Group) (Figure 2). Existing from the late Chattian Oligocene until the early Aquitanian Miocene, the Old Church Subsea was contemporaneous with the Chattahoochee and Tampa Subseas farther south. As in these southern paleoseas, the basin of the Old Church Subsea dried out during the mid-Aquitanian but remained emergent longer, until the mid-Burdigalian Miocene. At that time, the basin reflooded to produce the largest and best-developed subsea, the *Calvert Subsea* (named for the Calvert Formation of the Chesapeake Group (Figure 4). This expansive and complex subsea disappeared at the end of the Langhian Miocene and the area remained emergent until the early Serravallian Miocene. During this regressive interval, large river systems rapidly began to fill the northern end of the basin with sediments, greatly reducing the areas that would be susceptible to flooding by future sea level rises. By the mid-Serravallian Miocene, the paleosea basin again

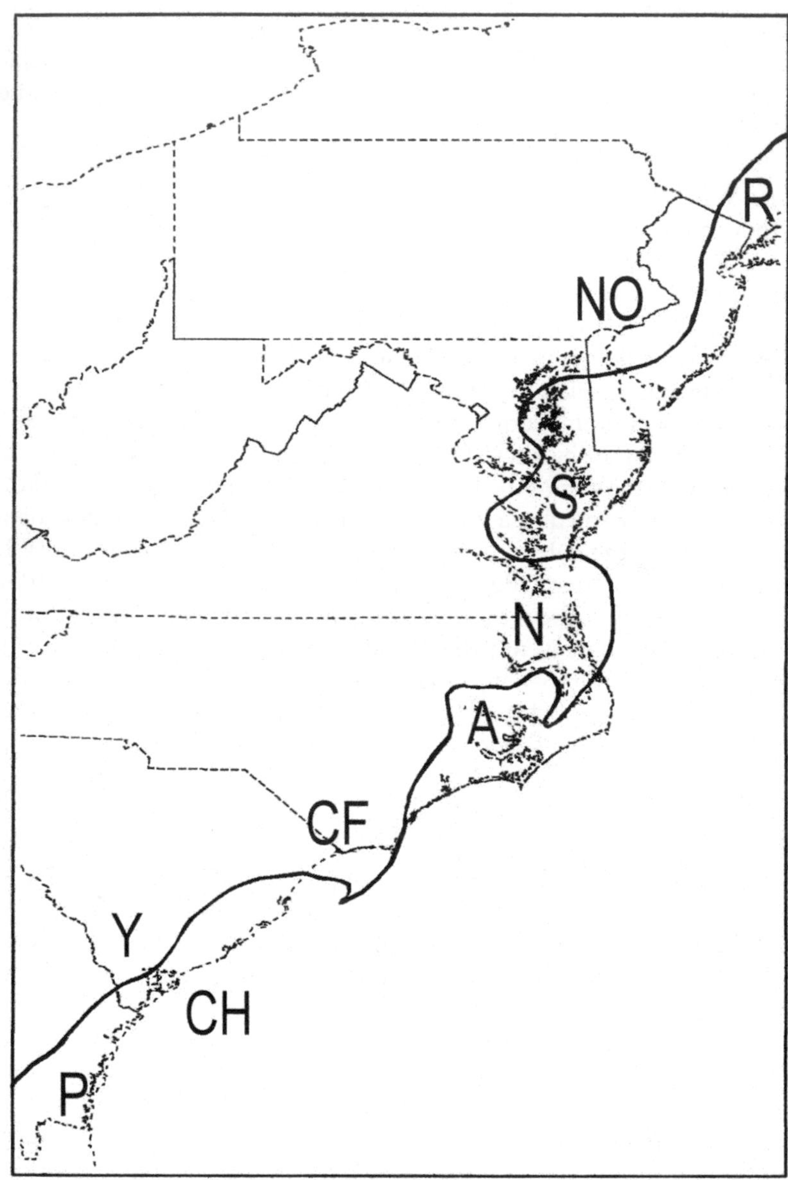

Figure 2. Approximate configuration of the eastern North American seas during the late Chattian Oligocene to early Aquitanian Miocene, superimposed upon the outline of the Recent North American coastline. Paleoseas include: R= Raritan Sea, S= Salisbury Sea (Old Church Subsea), A= Albemarle Sea (Silverdale Subsea), CH= Charleston Sea (Edisto Subsea), and P= Parachucla Embayment, Charleston Sea. Prominent geomorphological features include: NO= Normandy Arch and highlands, N= Norfolk Arch and Peninsula, CF= Cape Fear Arch and Peninsula, and Y= Yamacraw Arch and highlands.

reflooded to produce the much smaller *Patuxent Subsea* (named for the Patuxent River of Maryland). Another brief emergent period occurred during the mid-to-late Serravallian and this was followed by a reflooding during the latest Serravallian to produce the still-smaller *St. Mary's Subsea* (named for the St. Mary's Formation of the Chesapeake Group). By the mid-Tortonian Miocene, the giant delta systems of the Patuxent and St. Mary's Subseas had largely infilled the Salisbury Sea basin and the paleosea ceased to exist. For detailed discussions of the individual geomorphologies and ecosystems of the Salisbury subseas, see Chapters 4 and 5.

The Albemarle Sea

To the south of the Salisbury Sea, and separated from it by the Norfolk Arch highlands, was the *Albemarle Sea* (named for the Albemarle Sound of North Carolina) (Figures 2, 4, 6, and 8). At its greatest development, the Albemarle Sea extended from the present-day Potomac River area in the north, south to southern North Carolina, and inland as far west as near Greenville, North Carolina. The longest-lived of the Atlantic paleoseas, the Albemarle Sea ranged from the Rupelian Oligocene until the end of the Calabrian Pleistocene and encompassed seven subseas. Like the Choctaw and Okeechobean Seas, the subseas of the Albemarle Sea can be broken into two groups, based upon their marine faunas; the early subseas (Rupelian Oligocene to late Langhian Miocene) and the late subseas (late Tortonian Miocene to Calabrian Pleistocene). The oldest of the early subseas, the *River Bend Subsea* (named for the River Bend Formation of North Carolina), existed during the Rupelian to earliest Chattian Oligocene and was contemporaneous with the Bainbridge and Dade Subseas farther south. The River Bend Subsea and the Albemarle Sea basin were emergent briefly during the late Chattian only to be reflooded during the latest Chattian to produce the *Silverdale Subsea* (named for Silverdale, Onslow County, North Carolina). Contemporaneous with the Chattahoochee and Tampa Subseas to the south and with the Old Church Subsea to the north, the Silverdale Subsea existed until the early Aquitanian Miocene, when the entire basin was again emergent. By the late Burdigalian Miocene, the last but largest of the early subseas, the *Pamlico Subsea* (named for the Pamlico River of North Carolina) reflooded the area as far north as the present-day Albemarle Sound (Figure 4). The Pamlico Subsea was contemporaneous with the last part of the deposition of the Chipola and Arcadia Subseas to the south and with the Calvert Subsea and Raritan Sea to the north.

The end of the Langhian Miocene saw the beginning of a long period of emergence for most of the Albemarle Sea basin, for the entire Serravallian until the late Tortonian Miocene. At that time, the depocenter of the Albemarle basin moved northward across the Norfolk Arch to incorporate the southernmost part of the now mostly infilled Salisbury Sea. This is the only known example of "basin-capture" seen in the eastern North American paleoseas, and resulted from the erosion and removal of the Norfolk Arch highlands during the late Miocene. The first and largest of the late subseas, the *Rappahannock Subsea* (named for the Rappahannock River of Virginia), filled the new, enlarged basin until the mid-Messinian Miocene, when the area again became emergent. Only for a short time during the mid-Zanclean Pliocene did the Albemarle Sea area flood again, this time producing the *Williamsburg Subsea* (named for Williamsburg, Virginia). This short-lived subsea, contemporaneous with the Murdock Subsea farther south, disappeared in the late Zanclean and was followed by another short period of emergence. The beginning of the Piacenzian Pliocene saw a reflooding of the enlarged basin and the formation of the *Yorktown Subsea* (named for the Yorktown Formation of the Chesapeake Group and previously referred to as the "Yorktown Sea" (Petuch, 1997)) (Figure 6). This large and complex

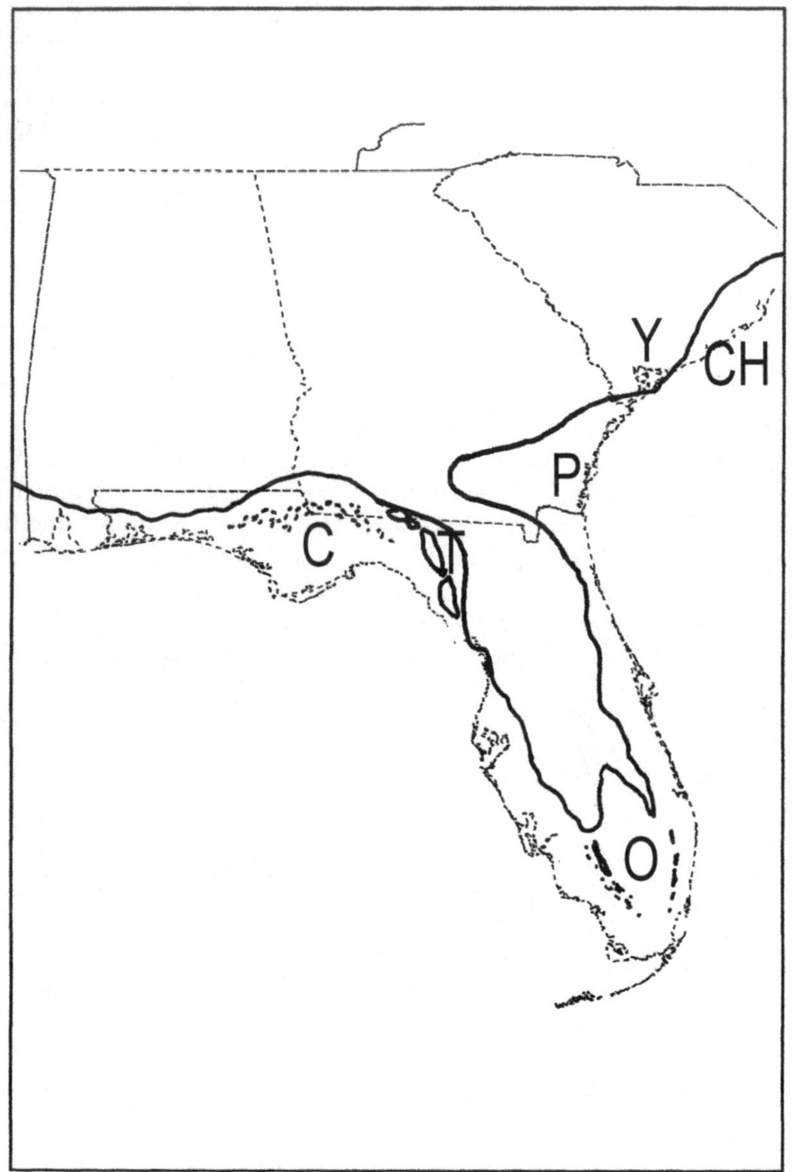

Figure 3. Approximate configuration of the southeastern North American seas during the Burdigalian to early Langhian Miocene, superimposed upon the outline of the Recent North American coastline. Paleoseas include: C= Choctaw Sea (Chipola Subsea) with its reef complexes (speckling on map), O= Okeechobean Sea (Arcadia Subsea), CH= Charleston Sea (Coosawhatchee Subsea), and P= Parachucla Embayment, Charleston Sea. Prominent geomorphological features include: T= Torreya Lagoon System and Y= Yamacraw Arch and highlands. During this time, the Okeechobean Sea was rapidly infilling to become a shallow lagoon and the southern coral reef complexes were expanding to produce full reef tracts.

subsea was contemporaneous with the Jackson and Tamiami Subseas farther south. After a brief period of emergence in the late Piacenzian, the last of the late subseas, the *Croatan Subsea* (named for the Croatan National Forest area, Craven County, North Carolina) reflooded the Albemarle basin in the latest Piacenzian (Figure 8). Contemporaneous with the Caloosahatchee Subsea farther south, the Croatan Subsea existed until the end of the Calabrian Pleistocene. By the Nebraskan Stage of the Pleistocene, the Albemarle Sea ceased to exist. For detailed discussions of the individual geomorphologies and ecosystems of the Albemarle subseas, see Chapters 3, 4, 5, 6, and 8.

The Charleston Sea

The southernmost of the Atlantic paleoseas, the *Charleston Sea* (named for Charleston, South Carolina), extended, at its maximum development, from present-day southern North Carolina south to northern Georgia and extended inland at least as far as Santee, South Carolina (Figures 1, 3, 4, 5, 6, 7, and 8). Separated from the Albemarle Sea to the north by the Cape Fear Arch highlands, the Charleston Sea was the most geomorphologically complex of the eastern North American paleoseas, comprising two separate embayments and a series of lagoon systems in the south. Like the Albemarle Sea, the Charleston Sea existed from the Rupelian Oligocene to the end of the Calabrian Pleistocene, but encompassed only six separate subseas. Also like the Albemarle Sea, the subseas of the Charleston Sea can be broken into two separate groups, based on their marine faunas; the early subseas (Rupelian Oligocene to late Langhian Miocene) and the late subseas (mid-Zanclean Pliocene to late Calabrian Pleistocene). Because of poor fossil preservation, high levels of postdepositional diagenesis, and deep burial by Pleistocene alluvia, the early subseas are still poorly known. As is presently understood, the oldest of the Charleston subseas, the *Ashley Subsea* (named for the Ashley Formation of the Cooper Group of South Carolina) (Figure 1) existed from the Rupelian to early Chattian Oligocene and was contemporaneous with the Bainbridge and Dade Subseas farther south and with the River Bend Subsea farther north. After an emergent time during the mid-Chattian, the Charleston Sea basin was flooded in the latest Chattian Oligocene to produce the *Edisto Subsea* (named for the Edisto Formation of South Carolina). Existing into the earliest Aquitanian Miocene, the Edisto Subsea was contemporaneous with the Chattahoochee and Tampa Subseas in the south and the Silverdale and Old Church Subseas in the north. The majority of the basin dried out during the mid-Aquitanian and remained emergent until the late Burdigalian Miocene. At that time, the area was again reflooded by the last of the early subseas, the long-lived *Coosawhatchee Subsea* (named for the Coosawhatchee Formation of Georgia) (Figure 3), a contemporary of the Pamlico, Calvert, and Patuxent Subseas and Raritan Sea to the north and the latest parts of the Chipola, Arcadia, and Walton Subseas to the south. By the end of the mid-Serravallian Miocene, the Charleston Sea was again emergent.

Little is known about the the paleoceanography of the Charleston Sea during the time interval spanning the Serravallian, Tortonian, and Messinian Miocene. During this time, the entire area was under the influence of extensive fluvial, deltaic, and estuarine systems that rapidly buried or removed much of the older sea floors. In the early Zanclean Pliocene, the Charleston Sea basin again reflooded to produce the first of the late subseas, the *Santee Subsea* (named for the fossil beds at Santee, Orangeburg County, South Carolina). Contemporaneous with the Murdock Subsea to the south and the Williamsburg Subsea to the north, the Santee Subsea existed only until the late Zanclean, when the entire area was again emergent. The beginning of the Piacenzian Pliocene saw a major reflooding of the area, producing the large and complex *Duplin Subsea* (named for the Duplin

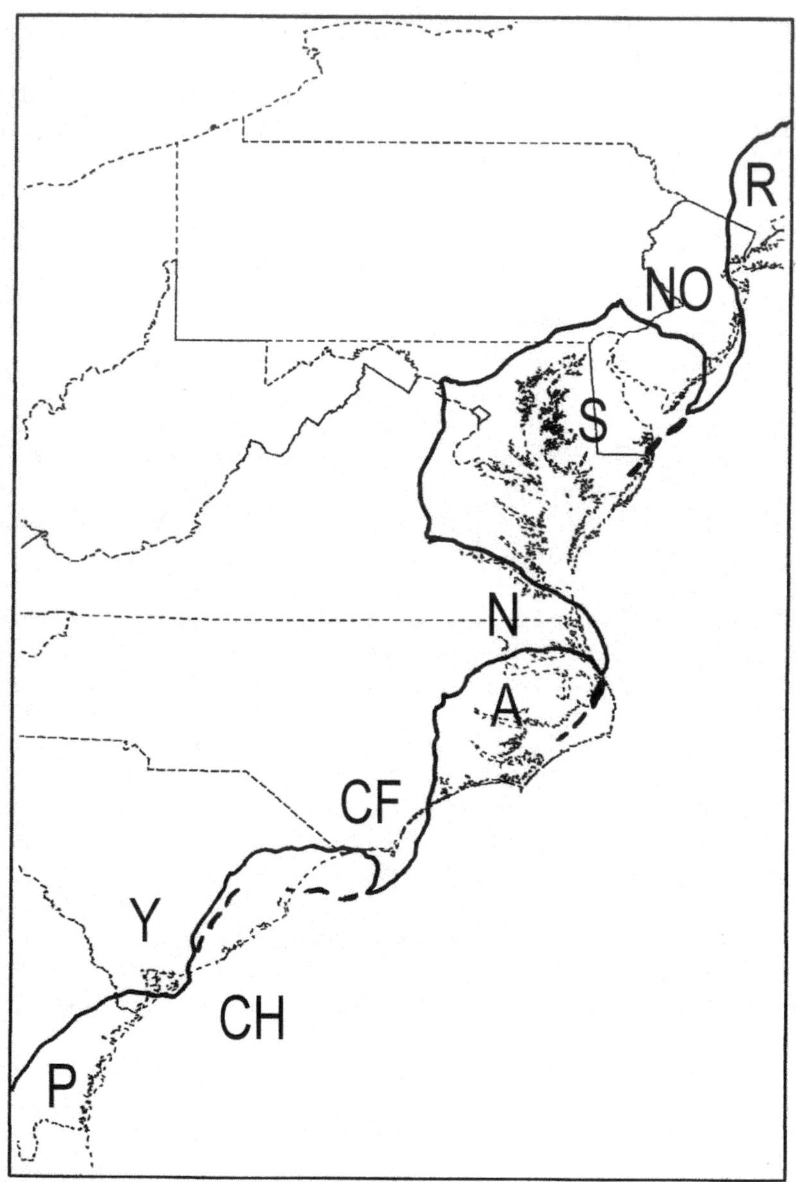

Figure 4. Approximate configuration of the eastern North American seas during the late Burdigalian to late Langhian Miocene, superimposed upon the outline of the Recent North American coastline. Paleoseas include: R= Raritan Sea, S= Salisbury Sea (Calvert Subsea), A= Albemarle Sea (Pamlico Subsea), CH= Charleston Sea (Coosawhatchee Subsea), and P= Parachucla Embayment, Charleston Sea. Prominent geomorphological features include: NO= Normandy Arch and Peninsula, N= Norfolk Arch and Peninsula, CF= Cape Fear Arch and Peninsula, and Y= Yamacraw Arch and highlands.

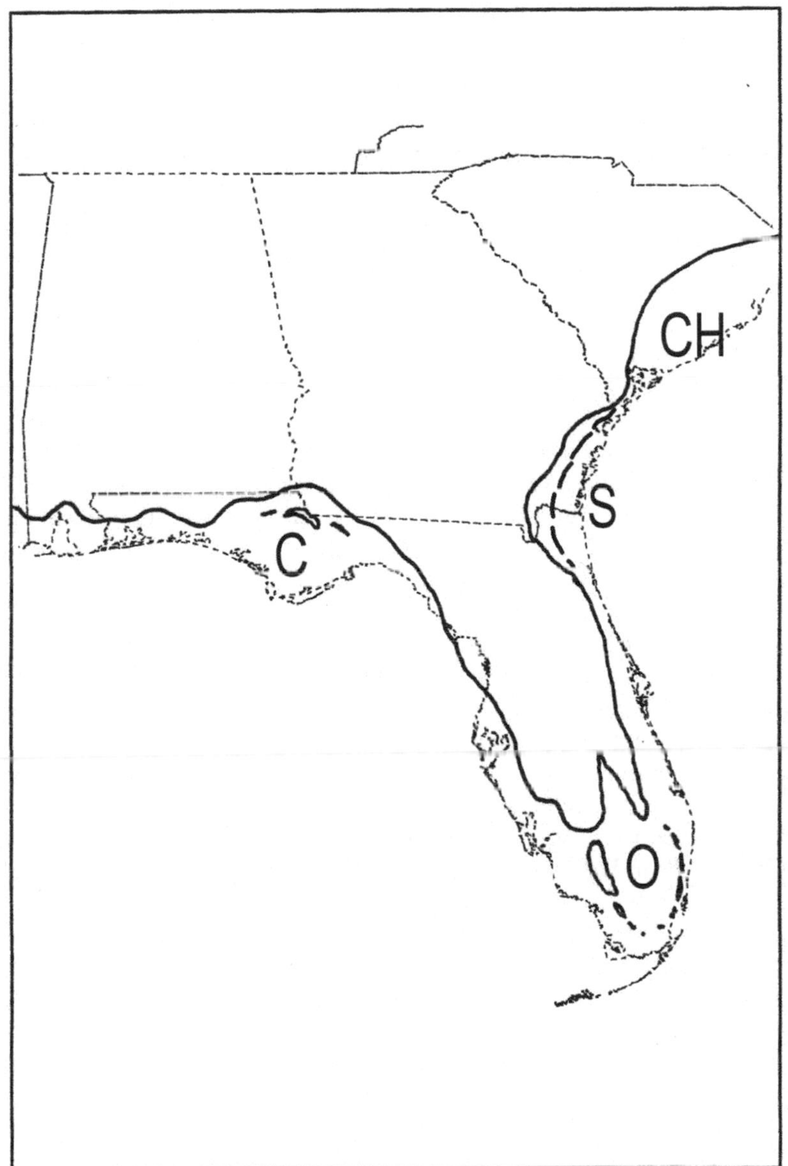

Figure 5. Approximate configuration of the southeastern North American seas during the Piacenzian Pliocene, superimposed upon the outline of the Recent North American coastline. Paleoseas include: C= Choctaw Sea (Jackson Subsea), O= Okeechobean Sea (Tamiami Subsea) and the Everglades Pseudoatoll, CH= Charleston Sea (Duplin Subsea), and S= Satilla Lagoon System, Charleston Sea.

Formation of southern North Carolina and northernmost South Carolina and previously referred to as the "Duplin Sea" (Petuch, 1997)) (Figure 6). Existing until the late Piacenzian, the Duplin Subsea was the exact contemporary of the Jackson and Tamiami Subseas to the south and the Yorktown Subsea to the north. After a brief emergent time, the basin again reflooded in the latest Piacenzian to produce the *Waccamaw Subsea* (named for the Waccamaw Formation of southern North Carolina and South Carolina) (Figure 8). As in the contemporaneous Caloosahatchee Subsea to the south and the Croatan Subsea to the north, the Waccamaw Subsea existed only until the end of the Calabrian Pleistocene. After that time, the Charleston Sea was essentially infilled and was replaced by an extensive series of coastal lagoons (the *Socastee Lagoon System*, see Chapter 9). For detailed discussions of the individual geomorphologies and ecosystems of the Charleston subseas, see Chapters 3, 6, and 8.

Deposition within the eastern North American Paleoseas

The sedimentary units found within the Gulf and Atlantic Coastal basins are the "windows" to the oceanography of the American paleoseas. The mineralogical composition, size classes, and mode of formation of their component sediments give great insight into the temperature, salinity, and bathymetry of any given area within the individual subseas. Overall, the geological formations of the subseas fall under two broad lithologic and environmental types; those produced in carbonate environments and those produced in siliciclastic environments. The carbonate strata formed in shallow, offshore, tropical environments away from river mouths while siliciclastic strata formed along the sandy shorelines of both tropical and warm temperate environments, often along river deltas. Coastal upwelling systems, with their nutrient-rich water and thick plankton blooms, also produced distinctive reducing environments that led to the formation of phosphorites in warmer water areas and diatomaceous oozes in cooler water areas.

In this book, seven main types of tropical depositional environments are recognized in the geological formations that were laid down within the eastern American paleosea basins. Primary among these are the coral reefs of the southern subseas. In some cases, as within the Okeechobean Sea (see Chapter 7), these carbonate systems represented true, zonated coral reefs, with discernable fore-reefs, reef crests, reef platforms, and back reefs. More often, the coralline facies were produced by smaller coral bioherms and "patch reefs" within carbonate lagoons. These same lagoons also housed the second and third types of tropical environments; the open carbonate sand and oolite bottoms and the extensive beds of Turtle Grass (*Thalassia* sp.) growing on carbonate mud (calcilutite) bottoms. Both of these depositional environments interfingered with the coral bioherms and zonated reefs. A fourth type of tropical environment included estuaries and their accompanying organic-rich mud flats, derived from fluvial deposition of clays and mud-sized particles. Typically, large "reefs" of tropical oysters (genera *Hyotissa* and *Gigantostrea*) also grew and accumulated within these estuaries and produced the fifth type of tropical depositional environment. In close proximity to the mud flats and oyster reefs, and growing successionally over them, was the sixth type of environment, the mangrove jungles (most probably species of Red Mangroves, *Rhizophora*, and Black Mangroves, *Avicennia*) and their accompanying peaty sands. The seventh tropical environmental type, upwelling systems, usually occurred in deeper offshore areas, and can be differentiated from temperate upwelling areas by the greater species-richness and diversity of the associated ecosystems. These environments of deposition and their associated tropical marine communities and ecosystems are discussed in Chapters 3, 4, 5, 7, 8, and 9.

The warm temperate paleoseas discussed in this book also contained seven main types of depositional environments. Two of the most prevalent types were open, quartz sand bot-

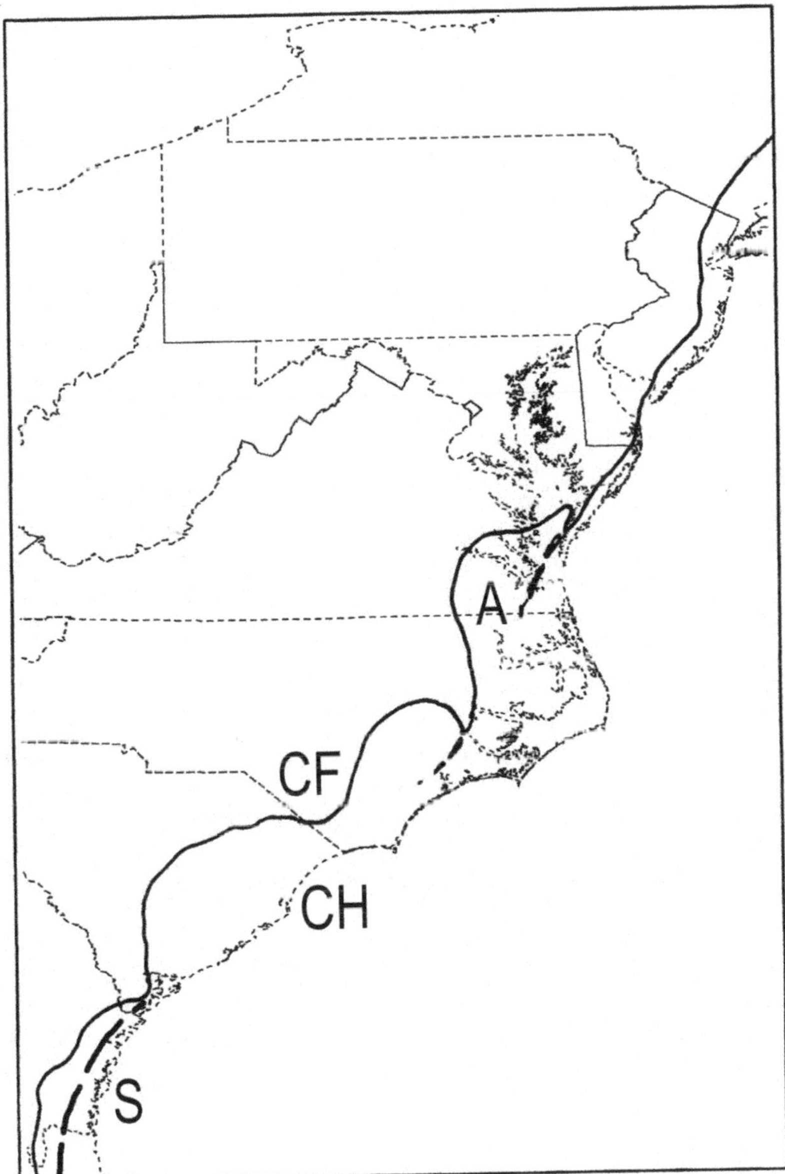

Figure 6. Approximate configuration of the eastern North American seas during the Piacenzian Pliocene, superimposed upon the outline of the Recent North American coastline. Paleoseas include: A= Albemarle Sea (Yorktown Subsea), CH= Charleston Sea (Duplin Subsea), and S= Satilla Lagoon System, Charleston Sea. Prominent geomorphological features include: CF= Cape Fear Arch and highlands.

toms in shallow water areas behind barrier islands and quartz sand bottoms in deep lagoons and sounds. These often housed immense beds of scallops and turritellid gastropods. In close proximity to the shallow water sand bottom areas was the third type of temperate environment, the Eel Grass (*Zostera*) beds. These sea grass beds grew extensively on muddy bottoms in quiet, shallow bays and lagoons. Temperate estuaries represented the fourth type of environment, where mud flats and mud banks predominated. As in the tropical estuaries, a fifth type of depositional environment, the oyster reefs, also grew within these estuarine lagoons. Unlike the tropical estuaries, these oyster reefs were dominated by the genera *Ostrea*, *Crassostrea*, and *Conradostrea*. The sixth environmental type, river deltas, typically laid down thick sequences of clays and mud, most often unfossiliferous. The seventh environmental type, upwelling systems, occurred in deeper, offshore waters and differed from the tropical analogues in having a low species-richness in the associated ecosystems. Most often, these temperate upwelling areas produced thick beds of "Fuller's Earth" diatom frustule sediments. The warm temperate-to-temperate depositional environments and their associated marine communities and ecosystems are discussed in Chapters 3, 4, 5, 6, 8, and 9.

The fourteen main tropical and warm temperate depositional environments, and their numerous variations and combinations, have produced a large number of lithostratigraphic groups, formations, and members within the paleosea basins covered in this book. These are outlined here by paleosea and subsea. The following listing of formations and members is a compilation of the correlation charts and works of Campbell (1993), Gibson (1983), Petuch (1994, 1997), Scott (1988), Ward (1985, 1992, 1993), and Zullo and Harris (1992).

Choctaw Sea

-*Bainbridge Subsea*; **Marianna Formation, Suwannee Formation, Flint River Formation**
-*Chattahoochee Subsea*; **Chattahoochee Formation, St. Mark's Formation, Penney Farms Formation**
-*Chipola Subsea*; **Chipola Formation, Torreya Formation, Mark's Head Formation, Oak Grove Formation**
-*Walton Subsea*; **Shoal River Formation, Yellow River Formation**
-*Alaqua Subsea*; **Red Bay Formation**
-*Jackson Subsea*; **Jackson Bluff Formation**

The Chipola, Oak Grove, Shoal River, Yellow River, Red Bay, and Jackson Bluff Formations comprise the *Alum Bluff Group* (based upon their high percentages of carbonates). The St. Mark's, Penney Farms, Torreya, and Mark's Head Formations are marine or estuarine components of the *Hawthorn Group* (based upon their high phosphorite content).

Okeechobean Sea

-*Dade Subsea*; **Suwannee Formation** (undescribed coral reef facies)
-*Tampa Subsea*; **Tampa Formation** (considered a member of the Arcadia Formation by Scott, 1988, but here retained at full formational status)
-*Arcadia Subsea*; **Arcadia Formation**
-*Polk Subsea*; **Peace River Formation**
-*Charlotte Subsea*; **Bayshore Formation**
-*Murdock Subsea*; **Murdock Station Formation, Wabasso Beds** equivalent
-*Tamiami Subsea*; **Tamiami Formation**, with lower **Buckingham Member** (Unit 10 at Sarasota; previously referred to as the Buckingham Formation), middle **Pinecrest Member** (Units 9-5 at Sarasota; previously referred to as the "Pinecrest Beds or Formation," see Olsson, 1964), and upper **Fruitville Member** (Units 4-2 at Sarasota; name here demoted

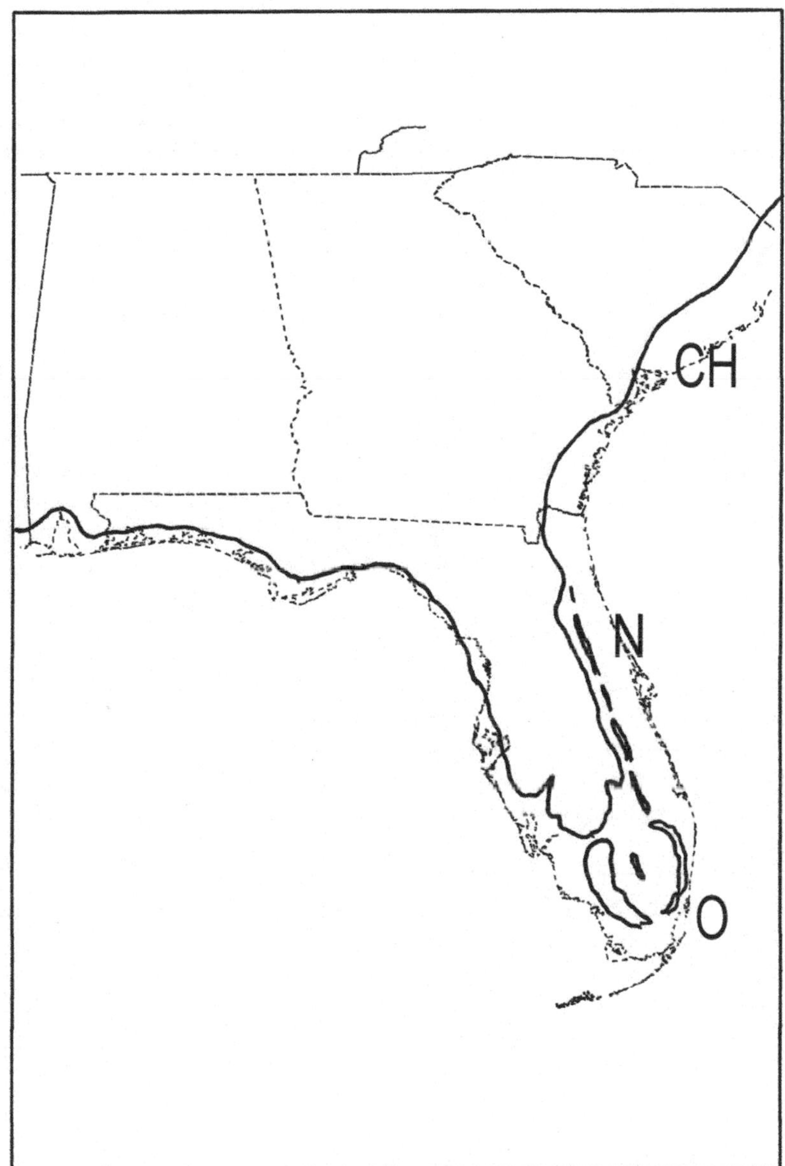

Figure 7. Approximate configuration of the southeastern North American seas during the latest Piacenzian Pliocene to Calabrian Pleistocene, superimposed upon the outline of the Recent North American coastline. Paleoseas include: CH= Charleston Sea (Waccamaw Subsea), N= Nashua Lagoon System, and O= Okeechobean Sea (Caloosahatchee Subsea).

from full formational status to member status; see Waldrop and Wilson, 1990), the **Ochopee Member** (limestone member equivalent to the Pinecrest Member), and the **Golden Gate Member** (coral limestone member equivalent to the Fruitville Member; see Missimer, 1992) (for the standard reference Sarasota Units, see Petuch, 1982)

-*Caloosahatchee Subsea;* **Caloosahatchee Formation**, with un-named lower member (Unit 1 at Sarasota, see Lyons, 1992), middle **Ft. Denaud Member** and **Bee Branch Member**, and upper **Ayer's Landing Member** (see DuBar, 1958), and Nashua Formation

-*Loxahatchee Subsea;* **Bermont Formation** (lower units), with **Holey Land Member** and **"Loxahatchee Unit"** (lower member) (see Petuch, 1994)

-*Belle Glade Subsea;* **Bermont Formation** at stratotype and **Belle Glade Beds** (upper units) (see Petuch, 1994)

-*Lake Worth Subsea;* **Ft. Thompson Formation**, with lower **Okaloacoochee Member** and upper **Coffee Mill Hammock Member, Key Largo Formation, Miami Formation, Anastasia Formation**

Based upon their high percentages of carbonates and carbonate clasts (especially massive reef corals and macromollusks), the Tamiami, Caloosahatchee, Bermont, Ft. Thompson, Key Largo, Miami, and Anastasia Formations are included together here in the *Okeechobee Group* (informally proposed by Scott, 1992 as the Okeechobee Formation and here elevated to group status). The phosphorite-rich Bayshore and Murdock Station Formations, originally proposed as the lowest members of the Tamiami Formation (Hunter, 1968), are here removed from the Okeechobee Group and are placed in the Hawthorn Group.

Charleston Sea

-*Ashley Subsea;* **Ashley Formation, Cooper Group**

-*Edisto Subsea;* **Edisto Formation, Parachucla Formation**

-*Coosawhatchee Subsea;* **Coosawhatchee Formation, Charlton Formation** (considered a member of the Coosawhatchee Formation by Scott, 1988, but here retained at full formational status)

-*Santee Subsea;* **Goose Creek Formation** (lower part, limestone)

-*Duplin Subsea;* **Duplin Formation, Raysor Formation, Goose Creek Formation** (upper beds), **Bear Bluff Formation, Orange Bluff Beds** (Satilla Beds)

-*Waccamaw Subsea;* **Waccamaw Formation**

-*Socastee Lagoon System;* **Canepatch Formation, Flanners Beach Formation, Okefenokee Formation, Socastee Formation, Wicomico Formation, Talbot Formation, Princess Anne Formation, Pamlico Formation**

Albemarle Sea

-*River Bend Subsea;* **River Bend Formation**

-*Silverdale Subsea;* **Belgrade Formation**, with lower **Pollocksville Member** and upper **Haywood Landing Member**

-*Pamlico Subsea;* **Pungo River Formation**, with lower **Bonnerton Member** and upper **Belhaven Member**

-*Rappahannock Subsea;* **Eastover Formation**, with lower **Claremont Manor Member** and upper **Cobham Bay Member**

-*Williamsburg Subsea;* **Yorktown Formation** (lower beds), with **Sunken Meadow Member**, (Yorktown Zone 1 of Mansfield, 1928)

-*Yorktown Subsea;* **Yorktown Formation** (upper beds), with middle **Rushmere** and **Mogarts Beach Members**, and upper **Moore House Member**, (Yorktown Zone 2 of Mansfield, 1928), and **Chowan River Formation**, with lower **Colerain Landing Member** and upper

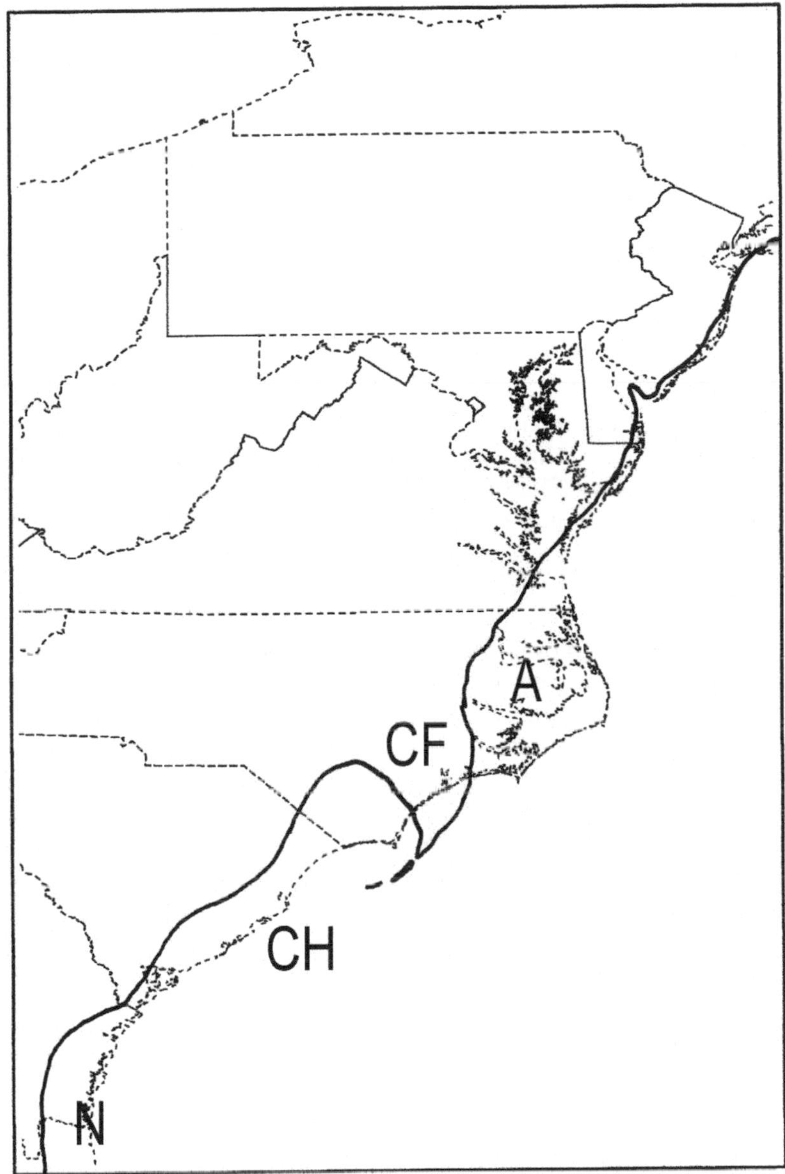

Figure 8. Approximate configuration of the eastern North American seas during the latest Piacenzian Pliocene to Calabrian Pleistocene, superimposed upon the outline of the Recent North American coastline. Paleoseas include: A= Albemarle Sea (Croatan Subsea), CH= Charleston Sea (Waccamaw Subsea), N= Nashua Lagoon System. Prominent geomorphological features include: CF= Cape Fear Arch and Peninsula.

Edenhouse Member
-*Croatan Subsea;* **James City Formation**
-*Socastee Lagoon System;* **Norfolk Formation**

For correlations between Florida (Okeechobean Sea) and Virginia (Albemarle Sea), the Rushmere Member is equivalent to the Buckingham Member of the Tamiami Formation, the Mogarts Beach Member is equivalent to the Pinecrest and Ochopee Members of the Tamiami Formation, and the Moore House Member is equivalent to the Fruitville and Golden Gate Members of the Tamiami Formation. The Chowan River Formation is equivalent to the uppermost beds of the Tamiami Formation (particularly the Kissimmee Tamiami beds) and the James City Formation is equivalent to the Ft. Denaud, Bee Branch, and Ayer's Landing Members of the Caloosahatchee Formation and the Nashua Formation.

Salisbury Sea

-*Old Church Subsea;* **Old Church Formation** (Shattuck Zone 1)
-*Calvert Subsea;* **Calvert Formation**, with lower **Fairhaven Member** (Shattuck Zones 2-9), middle **Plum Point Member** (Shattuck Zones 10-11), and upper **Calvert Beach Member** (Shattuck Zones 12-15), **Kirkwood Formation**
-*Patuxent Subsea;* **Choptank Formation**, with lower **Drumcliff Member** (Shattuck Zone 16), middle **St. Leonard Member** (Shattuck Zone 17), and upper **Boston Cliffs Member** (Shattuck Zones 18-19)
-*St. Mary's Subsea;* **St. Mary's Formation**, with lower **Conoy Member** (Shattuck Zone 20), middle **Little Cove Point Member** (Shattuck Zones 21-23), upper **Windmill Point Member** (Shattuck Zone 24), and "**Chancellor's Point Member**" ("Zone 24A," see Petuch, 1997)

The Shattuck Zones (1-24; see Shattuck, 1904) are traditional standard references for Maryland Miocene stratigraphy and they are used here along with the lithostratigraphic units of Ward (1992). The Old Church, Calvert, Choptank, and St. Mary's Formations of the Salisbury Sea and the Eastover, Yorktown, Chowan River, and James City Formations of the Albemarle Sea are all included together as the *Chesapeake Group.*

Raritan Sea; **Raritan Formation, Sankaty Beds**

Champlain Sea; **Riviere du Loup Formation, Champlain Formation, Prescott Beds**

Chapter 2. Biogeography of the Eastern American Paleoseas

The marine biogeography of western Atlantic fossil organisms, encompassing their large-scale distributions along the Caribbean Basin and North and South American coasts, is the single most powerful tool for deciphering paleoceanographic parameters. In the case of the neritic environments covered in the following chapters, these biogeographical patterns reflect the interaction of only three parameters: temperature, salinity, and substrate type. Since most marine organisms are physiologically restricted to relatively narrow temperature ranges, their latitudinal distribution in the fossil record is a precise reflection of paleotemperature gradients and oceanographic climate. Likewise, osmoregulatory ability limits most organisms to narrow salinity ranges, reflecting the physiographic characteristics of ancient water masses. As will be seen in the marine communities discussed in this book, most benthonic organisms are also ecologically and morphologically tied to very specific substrate types and these preferences give insight into past depositional environments. Altogether, these ecological parameters and constraints define discrete geographical units whose latitudinal limits are set by the ranges of endemic index organisms.

Marine biogeographical units can be considered to be single, giant biological entities or "superorganisms" evolving above the organismal and community levels. In this book, I have adopted a classification scheme that uses the concept of the *province* as the primary biogeographical unit. All provinces are based upon the application of the "50% Rule" to a single large taxonomic group, usually a phylum or a class (Valentine, 1973; Briggs, 1974, 1995). Here, I have chosen the gastropod and bivalved mollusks, together, as the defining group because of their good preservation and abundance in the geological record and because of their sharply defined ecological preferences. Within this framework, two adjacent geographical areas can be considered separate molluscan provinces if at least 50% of the species-level taxa are endemic to each area. This mutual exclusivity includes the faunas of all possible analogous biotopes within the two provinces (Petuch, 1988, 1997). Areas within the boundaries of a single province that have at least 30% endemism at the species level are referred to as *subprovinces*. As in ecotones (Odum, 1971), some subprovinces represent areas of faunal overlap between two provinces. These ecotone analogues are referred to as *provinciatones* and they characteristically contain their own suites of provinciatonal endemics. Most provinces, and their subprovince and provinciatone subdivisions, contain large numbers of endemic genera and unique communities. As demonstrated by Valentine (1973) and Briggs (1995), the provinces of other marine organisms, such as echinoderms, corals, and fishes, essentially conform to the geographical limits of molluscan faunal provinces.

Water temperature is the primary oceanographic parameter that defines the provincial arrangement of the eastern North American seas and subseas. With the exception of the Champlain and Raritan Seas, the five main paleoseas (Choctaw, Okeechobean, Charleston, Albemarle, and Salisbury Seas) can be classified as having had tropical, subtropical, or warm-temperate marine climates. In many cases, the higher latitude provinces were mixed systems, containing both subtropical and warm-temperate subprovinces. The provinces within the tropical seas, where water temperatures do not fall below 20 degrees Celsius, are referred to as *eutropical provinces*. These are low-latitude and circumequatorial in distribution and their faunas are generally derived from Eocene Tethyan Sea lineages. On the other hand, provinces within subtropical or warm-temperate seas, where water temperatures regularly fall below 20 degrees Celsius but not below 10 degrees Celsius, are referred to as *paratropical provinces*. These are distributed at higher latitudes and always form a mirror-image pair (north and south) on either side of the eutropical province. Paratropical provinces typically contain mixed faunas, with Tethyan-derived tropical components coexisting with boreal or antiboreal-derived cold water components. Provinciatonal subprovinces form along the boundaries of eutropical and paratropical provinces and contain geographically-restricted taxa that cannot exist at either higher or lower latitudes.

Environmentally, eutropical marine provinces and subprovinces are demarcated by the presence of coral reefs and their accompanying carbonate systems, by Turtle Grass (*Thalassia*) beds in lagoonal situations, and by mangrove forests in estuaries. Each of these main environmental types has its own characteristic associated molluscan fauna, with the same genera and families being found in tropical areas around the world. In the fossil record, eutropical areas can be delineated by the presence of several key gastropod families, subfamilies, and genera. Some of the key families and subfamilies include: Potamididae (all genera, associated with mud flats and estuaries), Strombidae (all genera, associated with lagoons and coral reef areas), Xenophoridae (all genera, associated with carbonate lagoons or deep water areas), Modulidae (all genera, associated with Turtle Grass beds and carbonate lagoons), Melongenidae (all genera, associated with lagoons, mud flats, and some deep water areas), Vasinae (family Turbinellidae, all genera, associated with coral reefs and carbonate lagoons), Volutinae and Lyriinae (both family Volutidae, all genera, associated with coral reefs and carbonate lagoons), and Peristerniinae (family Fasciolariidae, all genera, associated with coral reefs, carbonate lagoons, and some deep water areas). Some of the key eutropical index genera include: *Cassis* (Cassidae), *Turbinella* (Turbinellidae), (in the Americas) *Muracypraea, Luria (Basilitrona), Pseudozonaria,* and *Pustularia* (all Cypraeidae), *Jenneria, Pseudocyphoma, Cyphoma,* and *Marginocypraea* (all Ovulidae), *Myurella, Panaterebra,* and *Myurellina* (all Terebridae), *Protoconus, Eugeniconus, Virgiconus, Hermes,* and *Artemidiconus* (all Conidae), and *Taenioturbo, Marmarostoma, Lithopoma,* and *Astralium* (all Turbinidae). Examples of all of these genera and families are illustrated, in ecological and chronological context, in the following chapters.

Paratropical marine provinces and subprovinces typically lack well-developed carbonate environments, and the associated coral reefs, mangrove forests, and *Thalassia* beds, but contain some cold-tolerant, eurythermal offshoots of eutropical families. In the Recent Americas a classic example of a paratropical province is seen in the Californian Province, which lacks coral reefs, mangroves, and *Thalassia* beds but where the tropical families Conidae (genus *Chelyconus*), Cypraeidae (genus *Neobernaya*), Potamididae (genus *Cerithidea*), and Bursidae (genus *Bursa*) each are represented by a single species. These cold-tolerant derivatives coexist with classic cold water genera such as *Ophiodermella* and *Megasurcula* (Turridae), *Euspira* (Naticidae), and *Ceratostoma* (Muricidae). The key eutropical families Strombidae, Modulidae, Vasinae (Turbinellidae), Lyriinae (Volutidae), and Peristerniinae (Fasciolariidae) (all of which have examples living farther south, in the Panamic Province, along southern Baja California) are absent in the Californian Province. As pointed out by Vermeij (1978), the cryophilic offshoots of tropical families are usually generalist feeders with weak shell architecture, demonstrating that their invasion into higher and lower latitudes may be the result of avoidance of competition and predation. Similar paratropical patterns are seen in the Miocene Salisbury Sea, where a few species of tropical groups such as the Conidae (*Gradiconus*), Olividae (*Oliva (Strephona)*), and Xenophoridae (*Xenophora*) coexisted with classic cold water genera such as *Euspira*

(Naticidae), *Admete* (Cancellariidae), and *Oenopota* (Turridae). Some old, long-lived paratropical provinces have developed large species radiations of tropically derived groups, producing a distinct tropical appearance in a nontropical area. As can be seen in the Southwestern Australia, Tasmanian, and Southeastern Australian Provinces of Recent Australia and the Southwestern African and Agulhas Provinces of Recent South Africa (Briggs, 1995), species swarms of the Turbinidae, Conidae, Cypraeidae, Vasinae (Turbinellidae), and Lyriinae (Volutidae) have evolved in cold water areas that are devoid of both coral reefs and stenothermal tropical families such as the Strombidae and Modulidae.

Recent advances in the accuracy of stratigraphic correlations, geochronologic dating, and paleoenvironmental interpretations (examples discussed, in context, in the following chapters) have allowed for the first high-resolution view of the biogeographic patterns in the eastern American paleoseas. These discoveries have shown that the original "Miocene Caribbean Province" of Woodring (1974) and my Miocene-to-early Pleistocene "Caloosahatchian and Gatunian Provinces" and Miocene "Transmarian Province" (Petuch, 1982, 1988, and 1997) were essentially correct but were too simplistic and chronologically too broad in scope. This was particularly true at the subprovince level, where previous inaccuracies in dating led to chronologically disparate faunas being included together in the same chronosubprovince. When these new data are incorporated into the provinicial schemes proposed by Woodring and myself, a more detailed and complex biogeographical pattern presents itself, both spatially and temporally. In the time period spanning the early Oligocene to the Recent, twelve separate molluscan faunal provinces and at least forty-six subprovinces are now known to have existed within the western Atlantic and Eastern Pacific Regions. These are arranged chronologically and discussed in the following sections.

Marine Provinces of Rupelian to earliest Chattian Oligocene Time

By the beginning of Oligocene time, three eutropical and paratropical molluscan faunal provinces had formed within the western Atlantic. These included the northern *Proto-Transmarian Province*, the central *Antiguan Province*, and the southern *Pernambucan Province* (Figure 9). At this time, the Isthmus of Panama had not yet formed and the Atlantic and Pacific Oceans were connected by a wide strait that extended from Honduras south to northern Colombia (the Bolivar Strait). The tropical waters of the North Equatorial Current of the Atlantic Ocean flowed westward through the Bolivar Strait and bifurcated along the western coast of Central and northern South America. From there, one branch of the bifurcated current flowed northward to at least southern California while the other branch flowed southward to northern Peru. Another Atlantic warm water current, the Arawak Current (Petuch, 1997), flowed northward along the West Indian Arc, through the Gulf of Mexico region as a loop current, and northward to at least Recent Cape Fear, North Carolina. The North Equatorial Current also bifurcated and flowed southward along the Brazilian coast to at least Recent Cabo Frio. These currents produced the tropical, subtropical, and warm-temperate marine climates found in all three provinces.

The Proto-Transmarian Province

Named for its ancestry to the Miocene Transmarian Province, this northernmost early Oligocene paratropical province extended from South Carolina northward, conjecturally, to the Gulf of St. Lawrence (Figure 9) and encompassed the Ashley Subsea of the Charleston Sea, the River Bend Subsea of the Albemarle Sea, and the pre-Old Church Salisbury Sea basin. Since no well-preserved fossil beds dating from this time are known

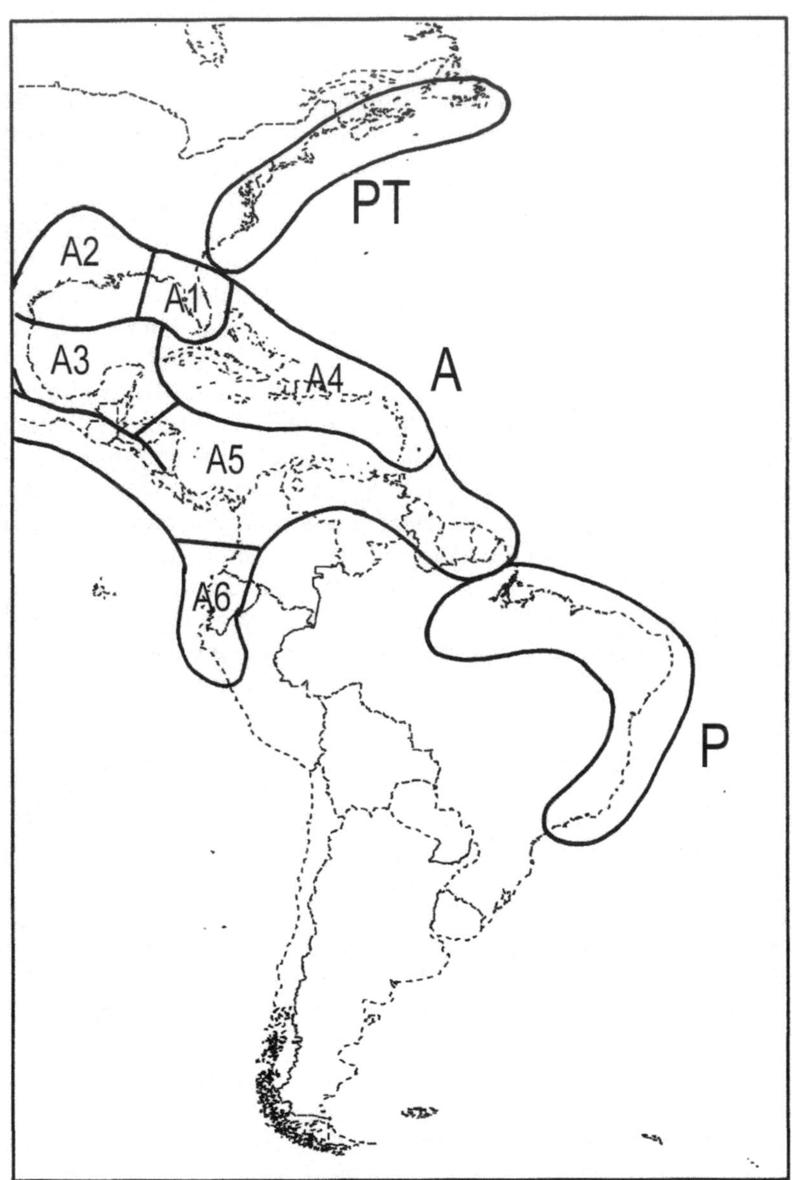

Figure 9. Approximate configuration of the marine molluscan provinces and subprovinces of the western Atlantic and Eastern Pacific during Rupelian to earliest Chattian Oligocene time, superimposed upon the outline of the Recent Americas. Biogeographical units include: PT= Proto-Transmarian Province, A= Antiguan Province, A1= Hernandoan Subprovince, A2= Vicksburgian Subprovince, A3= Alazanian Subprovince, A4= Guanican Subprovince, A5= Bohioan Subprovince, A6= Mancoran Subprovince, P= Pernambucan Province.

from most of this area (Ward, 1992), little is known about the basic faunal structure and subprovincial arrangement. The molluscan fauna of the southern parts of the province, in the Charleston and Albemarle Seas (in the Ashley and River Bend Formations; see Rossbach and Carter, 1991), shows some affinity to the Vicksburg Sea fauna, sharing several species such as the personiid *Distorsio crassidens* and the ficid *Ficus mississippiensis*. The Proto-Transmarian fauna also contained some important endemic species such as the oldest true *Ecphora* (sensu stricto) species (*E. wheeleri*) (Ocenebrinae, Muricidae), the oldest true *Scaphella* species (*S. saintjeani*), and an endemic *Calvertitella* (*mortoni* complex) species radiation (Turritellidae). The genera *Ecphora* (s.s.) and *Calvertitella* were important components of the subsequent Transmarian Province molluscan fauna. The cold water Raritan Current, flowing southward as a countercurrent (Petuch, 1997), most likely created a temperate marine climate in the northern part of the province (the incipient Raritan and Salisbury Seas).

The Antiguan Province

Named for the early Oligocene fossil beds on Antigua Island, Lesser Antilles, which typify the paleoprovince (the Antigua Formation), the Antiguan Province extended from South Carolina, throughout the Vicksburg Sea of the Mississippi Embayment, along Central America, through the Bolivar Strait, north to central Mexico, southwest to northern Peru, and southeast to northern Brazil (Figure 9). The Antiguan provincial limits are defined by a large number of widespread tropical organisms, including the corals *Antiguastrea cellulosa* (Montastreidae) (Mississippi, Alabama, Florida, Antigua, Aruba, and Panama), *Acropora saludensis* and *Actinacis alabamensis* (both Acroporidae) (Alabama, Florida, Antigua, and Panama), and *Astrocoenia decaturensis* (Astrocoeniidae) (Georgia, Cuba, and Antigua), the bivalved mollusks *Hyotissa antiguensis* (Gryphaeidae) (Cuba, Antigua, and Panama) and *Clementia peruviana* (Clementiinae-Veneridae) (Panama, Antigua, Colombia, and Peru), and the gastropod mollusks *Ampullinopsis spenceri* (Ampullospiridae) (Florida, Cuba, Antigua, Panama, Colombia, and Peru), *Ficus mississippiensis* (Ficidae) (Mississippi, Florida, and Panama), *Torcula hubbardi* (Turritellidae) (Cuba, Panama, and Peru), and the gastropod genera *Orthaulax* (Strombidae) (Alabama to Brazil), *Falsilyria* (Volutinae-Volutidae) (Florida to Brazil), and *Hemisinus* (Thiaridae) (Cuba to Brazil). Based upon regionally endemic species and species complexes, six subprovinces are now recognized as having existed within the Antiguan Province (Figure 9). These include: the *Hernandoan Subprovince* (southeastern Georgia, Florida, and southern Alabama), the *Vicksburgian Subprovince* (the Mississippi Embayment area and eastern Texas), the *Alazanian Subprovince* (northern Mexico to Honduras), the *Guanican Subprovince* (Cuba, Hispaniola, Puerto Rico, and the Lesser Antilles), the *Bohioan Subprovince* (Suriname, Venezuela, northern Colombia, Panama, Costa Rica, and northward to the Gulf of California), and the *Mancoran Subprovince* (southwestern Colombia, Ecuador, and northern Peru). These subprovinces are described in the following sections and illustrations of some of the index fossils are shown in Chapter 3.

The Hernandoan Subprovince

Named for Hernando County, Florida, site of rich fossil beds containing the typical fauna, this subprovince encompassed the areas of the Bainbridge Subsea, the Salt Mountain reef beds of Alabama, and the Suwannee Strait of northern Florida and southern Georgia (see Chapter 3). Being eutropical in marine climate, the Hernandoan Subprovince contained the largest and best-developed coral reef complexes on the Oligocene eastern North American mainland. Because of the varied tropical habitats associated with these coral reefs, the Hernandoan area housed a highly endemic molluscan fauna, both at the species and genus level. Some of the endemic genera include: *Prismacerithium* and *Cestumcerithium* (Cerithiidae), *Pyrazisinus* (Potamididae, the oldest-known species), and *Suwanneescapha* (Cylichnidae). Although sharing many species with the neighboring Vicksburgian Subprovince, such as *Torcula mississippiensis* (Turritellidae), *Distorsio crassidens* (Personiidae), *Sinum mississippiensis* (Naticidae), *Ficus mississippiensis* (Ficidae), *Clavolithes vicksburgensis* (Turbinellidae), and *Phalium caelatura* (Cassidae), the Hernandoan Subprovince contained whole suites of endemic, carbonate-loving species such as *Telescopium hernandoense* (and at least three other *Telescopium* species, Potamididae), *Cassis flintensis* (Cassidae), *Orthaulax hernandoensis* (Strombidae), *Falsilyria mansfieldi* and *F. kendrewi* (both Volutinae-Volutidae), *Spinifulgur gemmulatum* (Busyconidae), *Astralium polkensis* (Turbinidae), *Cypraeorbis kendrewi* (Cypraeidae), *Vasum suwanneensis* (Vasinae-Turbinellidae), *Hermes kendrewi* (Conidae) and *Turbinella suwannensis* (Turbinellinae-Turbinellidae) (see Dall, 1915, Mansfield, 1937, and Petuch, 1997). Of special interest on the Hernandoan reefs were the only-known American species of the tridacnid bivalve genus *Hippopus* (*H. hernandoensis*) and the last-living American example of the seraphsid gastropod genus *Terebellum* (*T. hernandoensis*). In the Recent, both of these genera are restricted to the tropical Indo-Pacific region and their presence on the Hernandoan reefs, along with *Telescopium*, gives a decidedly South Pacific feel to the area.

The Vicksburgian Subprovince

Named for the Vicksburg Group of Formations of Mississippi, which contain the typical fauna, this northernmost Antiguan subprovince was paratropical in marine climate and lacked extensive carbonate environments and coral reefs. Massive sedimentary infilling of the Mississippi Embayment by river deltas was also taking place at this time, and the absence of key carbonate environment taxa may be more related to substrate and facies differences than to temperature. Although containing most of the classic tropical gastropod families such as the Conidae (*Gradiconus sauridens*), Lyriinae-Volutidae (*Lyria mississippiensis*), Xenophoridae (*Xenophora humilis*), Melongenidae (*Myristica crassicornuta*), and Cypraeidae (*Cypraeorbis* cf. *ventripotens*), the Vicksburgian Subprovince lacked other key tropical families such as the Strombidae (with *Orthaulax* being strikingly absent), Volutinae-Volutidae (with the wide-ranging Antiguan genus *Falsilyria* being absent), Turbinidae, Modulidae, and Potamididae (for the best documentation, see MacNeil and Dockery, 1984). On the other hand, the Vicksburgian area housed a large number of characteristic endemic gastropod genera, some of which include *Echinofulgur* (*E. branneri* complex, Melongenidae), *Reticulacella* (Scaphellinae-Volutidae), and *Spinifulgur* (*S. spiniger* complex, Busyconidae). Of special interest were several endemic species complexes, especially the *Pleurofusia servata* complex (Turridae) and the *Sulcocypraea lintea* complex (Ovulidae).

The Alazanian Subprovince

Named for the Alazan Formation of La Ceiba, Buena Vista River, Veracruz State, Mexico, which contains the typical fauna, this subprovince is still poorly known and unstudied. The type formation, and only extensive collecting site, contains a relatively impoverished fauna but enough evidence is present to demonstrate that this area was distinct from the subprovinces to the north and south (Cooke, 1928). Although sharing several gastropod species with the Vicksburgian and Hernandoan Subprovinces, such as the previously mentioned *Torcula mississippiensis*, *Distorsio crassidens*, *Phalium caelatura*, and *Sinum mississippiensis*, and the turrid *Pleurofusia servata* and the naticid *Polinices byramensis*

(Naticidae), the Alazanian Subprovince contained several unusual and distinct endemic gastropod and bivalve species such as *Ancilla alazana* (Olividae), *Amusium alazanum* (Pectinidae), the bizarre *Thatcheria*-like turrid "*Pseudotoma*" *alazana* (representing an undescribed endemic genus), and species radiations of the turrid genera *Scobinella*, *Gemmula*, *Glyptotoma*, and *Paraborsonia*. Of special interest was the presence of the endemic melanellid gastropod genus *Protonema* (*P. bartschi*).

The Guanican Subprovince

Named for the Guanica Formation of Puerto Rico, which contains the typical fauna, this subprovince is also poorly known and relatively unstudied. Although encompassing the area of the entire Antilles Arc, preservation of the Guanican Oligocene faunas is generally poor in all formations, usually being leached or present only as molds. Of special interest within the Guanican Subprovince was the strombid gastropod genus *Orthaulax*, which underwent a large and characteristic endemic species radiation, producing at least six different species. Some of these, such as *O. aguadillensis* and *O. bermudezi* are the largest of their genus, being two or three times bigger than their northern Hernandoan congeners. Also of special interest was the presence of the large, *Campanile*-like endemic potamidid gastropod genus *Portoricia* (particularly *P. laricum*), which probably occupied the same ecological niche on mud flats as did the Hernandoan *Telescopium* species complex (see Maury, 1920). Similarly, the wide-ranging Antiguan thiarid gastropod genus *Hemisinus* (as the *H. atriformis* species complex) underwent an endemic species radiation within estuarine areas of the Guanican Subprovince. The Guanican area also contained a distinctive species complex of the turritellid gastropod genus *Torcula* (centered around *T. collazica*), and these appear to have dominated deeper lagoonal areas along the shorelines.

The Bohioan Subprovince

Named for the Bohio Formation of Panama, which contains the typical fauna (see Woodring, 1959), this was the only transisthmian subprovince, extending from Suriname to Colombia (Olsson and Richards, 1961), spanning the then-submerged Isthmus of Panama, and ranging along the Pacific coast from northwestern Colombia northward to the Mexican coast. Throughout the entire range, the molluscan fauna was remarkably similar and was characterized by such unusual endemic genera as *Glyptostyla* (*G. panamensis* complex, Turbinellidae), *Gonysycon* (*G. epomis* complex, Ficidae), and *Longiverena* (Thiaridae). As a eutropical subprovince, the Bohioan contained all the key gastropod families, particularly the Cypraeidae, with a species swarm of the genus *Cypraeorbis* (centered around *C. venezuelana*) and with the oldest-known American *Pustularia* (*P. mejasensis*), Turbinellidae (the *Turbinella buccina* complex, the ancestors of the Miocene *T. validus* complex), Strombidae (with the genus *Orthaulax* being found only in the southern part, along Panama, Costa Rica, Colombia, and Venezuela), and Conidae. Of special interest was the presence of the oldest-known member of the ranellid genus *Septa* (*S. ogygium*), a species complex of the Neozealandic turrid genus *Zemacies*, and a radiation of the fasciolariid genus *Mancorus* (*M. grabaui* complex). The widespread Antiguan Province gastropod index species *Torcula meroensis* (Turritellidae) and *Ampullinopsis spenceri* (Ampullospiridae) were especially common and prominent organisms throughout the entire subprovincial region.

The northwestern boundary of the Bohioan Subprovince is still conjectural, as fossiliferous Oligocene formations are largely absent from areas between the Isthmus of Tehuantepec and Oregon. Squires and Demetrion (1992) showed that the central coastline of Baja California Sur, Mexico contained a eutropical marine fauna during the Eocene, with reef complexes of *Actinacis, Goniopora, Stylosmilia, Colpophyllia, Montastrea, Stylophora,* and *Antillia* corals. These Eocene Baja California reefs supported a host of eutropical mollusks, including the gastropods *Bernaya* (Cypraeidae), *Paraseraphs* (Seraphsidae), *Ectinochilus* (Strombidae), *Xenophora* (Xenophoridae), *Conasprella* (Conidae), *Lyria* (Lyriinae-Volutidae), *Arene* (Liotiidae), *Architectonica* (Architectonicidae), and *Olivella* (Olividae), and the bivalves *Spondylus* (Spondylidae) and *Lima* (Limidae). Many of these genera survived the late Eocene-early Oligocene extinction event (see Chapter 10) and may have existed along the same coastline as they did in the middle Eocene, possibly as far north as southern California. This conjectural northernmost edge of the Bohioan Subprovince may have represented a separate, still un-named, subprovince, and one that contained eutropical faunas that gave rise to the rich coralline environments of the Pliocene Imperial Sea (see p. 52).

The Mancoran Subprovince

Named for the late Eocene-Oligocene Mancora Group of formations of northern Peru, which contains the typical fauna, this paratropical subprovince exhibited a rich and highly endemic fauna, with characteristic species radiations in several key tropical gastropod families. Primary among these were the *Harrisianella* (*H. peruviana* complex) and *Hannatoma* (*H. tumbezia* complex) of the Cerithiidae and the endemic genus *Diplocyma* (Potamididae). Other gastropod radiations were seen in the endemic genera *Peruluta* (*P. mancorensis* complex, Zidoniidae?-Volutidae) and *Perufícus* (*P. charanalensis* complex, Turbinellidae) and in the genera *Mancorus* (*M. burtii* complex, Fasciolariidae) and *Sulcobuccinum* (*S. mancorensis* complex, Pseudolividae) (both shared with the Bohioan Subprovince). The widespread Antiguan turritellid *Torcula meroensis* was abundant in this subprovince and formed extensive beds in lagoonal areas. Other key tropical genera were also present and these included *Gradiconus* (Conidae), *Turbinella* (*T. peruviana*, Turbinellidae), and *Cancellomorum* (*C. chiraense*, Harpidae). The large number of endemic genera and species complexes found in the Mancoran area suggests that the subprovince may have been a provinciatone between the eutropical Bohioan Subprovince and the temperate Proto-Peruvian Province.

The Pernambucan Province

Named for Pernambuco State, Brazil, site of the fossil beds of the Paleocene-early Eocene Maria Farinha Formation (White, 1887; possibly with species lineages surviving into the Oligocene) which typify the fauna of the paleoprovince, the Pernambucan Province extended from present-day Suriname to at least Cabo Frio, Rio de Janeiro, Brazil (Figure 9). Since Brazilian fossiliferous exposures dating from this time are rare and generally poorly preserved, little is known about the composition of the Pernambucan molluscan fauna. Because of this, any arrangement of subprovinces cannot be established at this time. Enough data exists, however, to show that the area constituted a separate province with tropical affinities and that it contained high levels of endemism at both the genus and species level. Some of the more important Pernambucan gastropod endemics include a species complex of an undescribed genus of *Prismacerithium*-like cerithiids (centered around "*Cerithium*" *freitasi*), an undescribed genus of large, ornate *Tympanotonus*-like potamidids (centered around "*Turritella*" *acuticarinata*), and a species complex of *Simnia*-like ovulids (centered around "*Simnia*" *gilliana*) (White, 1887; Magalhaes and Mezzalira, 1953). Besides the families Cerithiidae, Potamididae, and Ovulidae, other tropical influences included *Mazzalina ruginosa* (Fasciolariidae), *Gradiconus conditorius* (Conidae), and *Xenophora brasiliensis* (Xenophoridae). Of special interest were several cold water compo-

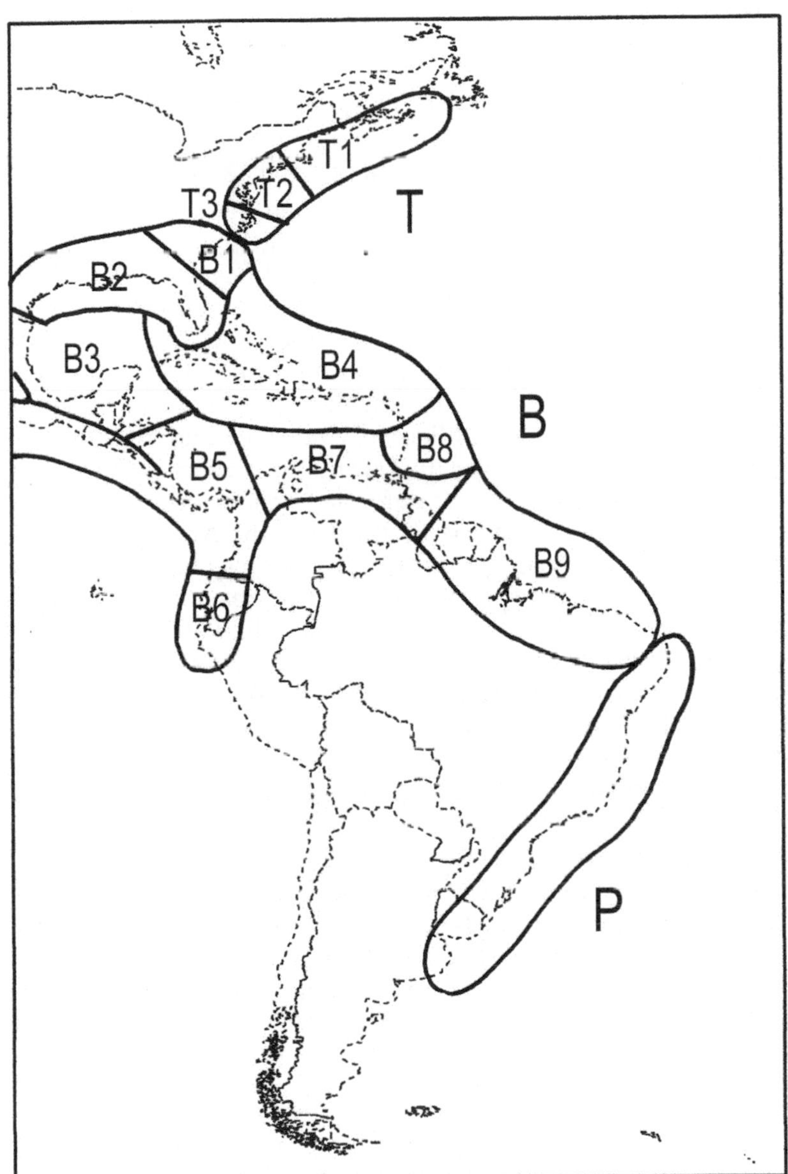

Figure 10. Approximate configuration of the the marine molluscan provinces and subprovinces of the western Atlantic and Eastern Pacific during late Chattian Oligocene to late Tortonian Miocene time, superimposed upon the outline of the Recent Americas. Biogeographical units include: T= Transmarian Province, T1= Sankatian Subprovince, T2= Calvertian Subprovince, T3= Pungoian Subprovince, B= Baitoan Province, B1= Onslowian Subprovince, B2= Chipolan Subprovince, B3= Agueguexquitean Subprovince, B4= Anguillan Subprovince, B5= Culebran Subprovince, B6= Subibajan Subprovince, B7= Cantaurean Subprovince, B8= Carriacouan Subprovince, B9= Piraban Subprovince, P= Platensian Province. The Vaquerosian Subprovince of the Baitoan Province, along the Mexican and southern Californian coasts, is not shown.

nents, including the muricid *Poirieria progne*, an undescribed genus of *Penion*-like buccinids (centered around "*Serrifusus*" *mariae*), and the turritellid *Mesalia netoana*. The presence of these antiboreal taxa demonstrates that the Pernambucan Province was at least partially paratropical.

Marine Provinces of latest Chattian Oligocene to late Tortonian Miocene Time

The mid-Chattian Oligocene saw a worldwide cooling event and the accompanying sea level drops, extinctions of many index groups, and disruptions of marine provincial structures (see Chapter 10). By the beginning of late Chattian time, the marine climates had warmed and the provinces of the western Atlantic and Eastern Pacific had reconfigured themselves. New groups of index organisms evolved and their ranges demarcated a new set of provinces and subprovinces. These included the northern *Transmarian Province*, the central *Baitoan Province*, and the southern *Platensian Province* (Figure 10). The paratropical Transmarian and Platensian Provinces and the eutropical Baitoan Province all retained their structures until the late Tortonian Miocene, when they underwent another extinction and reconfiguration period (see Chapter 10). During the existence of these three provinces, the Isthmus of Panama was still open and unformed and the oceanic currents were essentially the same as in the older Oligocene. At this time, the Arawak Current continued to convey warm waters all the way north to the Pamlico Subsea of the Albemarle Sea. The cool water Raritan Current continued southward down the coast as far as the Salisbury Sea, mixing with the warm Arawak waters and producing the Patuxent Gyre within the Old Church, Calvert, Patuxent, and St. Mary's Subseas (Petuch, 1997). The North Equatorial Current continued to bifurcate after passing through the Bolivar Strait and conveyed warm water all the way north to southern California and south to northern Peru. These provinces, and their subprovincial and provinciatonal subdivisions, are discussed in the following sections and many of their key index organisms are illustrated in Chapters 3, 4, and 5.

The Transmarian Province

Named for the St. Mary's River, St. Mary's County, Maryland (Petuch, 1988) (site of the classic St. Mary's Formation fossil beds which typify the fauna), this province extended from Nova Scotia southward to North Carolina and encompassed the Raritan, Salisbury, and Albemarle Seas (Figure 10). Based upon the ranges and percentages of restricted endemic taxa, the Transmarian Province can be further subdivided into three subprovinces: the northern *Sankatian Subprovince* (extending from Nova Scotia to the Normandy Peninsula of New Jersey), the *Calvertian Subprovince* (extending from southern New Jersey south to the Norfolk Peninsula of Virginia), and the *Pungoian Subprovince* (extending from southern Virginia south to southern North Carolina). The boundaries of the Transmarian Province were sharply demarcated by the ranges of a large number of distinctive endemic gastropod genera, some of which included the naticid *Poliniciella*, the ocenebrine muricids *Trisecphora* (Plate 1, A) and *Chesathais* (Plate 1, C), the trophonine muricids *Chesatrophon* (Plate 1, I), *Scalaspira*, *Patuxentrophon*, and *Lirosoma*, the fusinine fasciolariids *Conradconfusus*, *Pseudaptyxis*, and *Mariafusus* (Plate 1, B), the turritellids *Mariacolpus* (Plate 1, H) and *Calvertitella* (by Burdigalian time extinct south of North Carolina), and the turrids *Calverturris* (Plate 1, K), *Transmariaturris* (Plate 1, J), *Chesasyrinx* (Plate 1, F), *Chesaclava* (Plate 1, G), Nodisuculina, *Hemipleurotoma*, *Sediliopsis*, and *Mariaturricula* (Plate 1, L), the cancellariids *Marianarona* (Plate 1, E), *Cancellariella*, and *Mariasveltia*, the buccinid *Bulliopsis*, and the terebrid *Laevihastula* (Plate 1, D) (Petuch, 1993). Examples of many of these endemic taxa are shown in Chapters 4 and 5. Because of the influence of the cold water Raritan

Plate 1. Gastropod index genera of the Transmarian Province. A= *Trisecphora* Petuch, 1989 (*T. tricostata* subspecies, Ocenebrinae-Muricidae); B= *Mariafusus* Petuch, 1988 (*M. marylandicus* (Martin, 1904), Fasciolariidae); C= *Chesathais* Petuch, 1988 (*C. lindae drumcliffensis* Petuch, 1989, Ocenebrinae-Muricidae); D= *Laevihastula* Petuch, 1988 (*L. simplex* (Conrad, 1830), Terebridae); E= *Marianarona* Petuch, 1988 (*M. alternata* (Conrad, 1834), Cancellariidae); F= *Chesasyrinx* Petuch, 1988 (*C. rotifera* (Conrad, 1830), Turridae); G= *Chesaclava* Petuch, 1988 (*C. dissimilis* (Conrad, 1830), Turridae); H= *Mariacolpus* Petuch, 1988 (*M.* un-named species, Calvert Formation, Turritellidae); I= *Chesatrophon* Petuch, 1988 (*C. chespeakeanus* (Martin, 1904), Trophoninae-Muricidae); J= *Transmariaturris* Petuch, 1993 (*T. calvertensis* (Martin, 1904), Turridae); K= *Calverturris* Petuch, 1993 (*C. bellacrenata* (Conrad, 1841), Turridae); L= *Mariaturricula* Petuch, 1988 (*M. biscatenaria* (Conrad, 1834), Turridae). All specimens are from the Calvert, Choptank, and St. Mary's Formations of Maryland.

Current, no coral reefs or reef-associated carbonate environments were found within the boundaries of the Transmarian Province.

The paratropical nature of the Transmarian Province is readily demonstrated by the absence of key tropical gastropod families such as the Turbinidae, Potamididae, Strombidae, Cypraeidae, Ovulidae, Melongenidae, Peristerniinae (Fasciolariidae), Lyriinae and Volutinae (Volutidae), and the Vasinae and Turbinellinae (Turbinellidae). On the other hand, a tropically-derived component did occur within the Transmarian area and this included the bivalve genera *Divalinga* (Lucinidae), *Iphigenia* (Donacidae), *Strigilla* (Tellinidae), and *Carditamera* (Carditidae) (Martin, 1904; Ward, 1998) and the gastropod genera *Torcula* (Turritellidae), *Architectonica* (Architectonicidae), *Sassia* (Personiidae), *Oliva (Strephona)* (Olividae), *Gradiconus* (Conidae), *Mitra* s.l. (possibly an undescribed genus; Mitridae), *Xenophora* (Xenophoridae), *Dentimargo* (Marginellidae), and *Stenorhytis*, *Amaea*, and *Cirsotrema* (all Epitoniidae) (Martin, 1904). A northern cold water faunal component coexisted with these tropically derived groups and included the gastropod genera *Euspira* (*E. heros* complex, Naticidae), *Admete* (Admetinae-Cancellariidae), and *Oenopota* (Turridae), and the bivalve genera *Nuculana* (Nuculanidae), *Crenella* (Mytilidae), *Hiatella (Saxicava)* (Hiatellidae), and *Astarte* (Astartidae). The predatory ecological niche occupied by the Melongenidae in tropical waters was occupied, in the paratropical Transmarian area, by an endemic radiation of the Busyconidae, including the genera *Turrifulgur*, *Coronafulgur* (new genus; see Systematic Appendix), and *Sycopsis*. Likewise, the drilling predatory niche occupied by large muricine muricids in tropical areas, such as *Phyllonotus*, *Chicoreus*, *Hexaplex*, and *Vokesimurex*, was occupied in the Transmarian Province by endemic species radiations of the ocenebrine "ecphora" muricids *Ecphora*, *Planecphora* (new genus, see Systematic Appendix), *Trisecphora*, *Chesathais*, and *Ecphorosycon*.

The Sankatian Subprovince

Named for the Sankaty Heads on Nantucket Island, Massachussetts, which contains sparse fossil beds that typify the fauna (Shattuck, 1904; Petuch, 1988, 1997), this subprovince appears to have been only marginally paratropical and encompassed the entire Raritan Sea. Since most of the Sankatian sediments have been removed by subsequent glacial ice sheets or have been buried under inaccessible thick deltaic deposits, little is known of the faunal composition and its limits are only conjectural. Deep well cores drilled along the New Jersey shore and northern Delaware (Pilsbry and Harbison, 1933; Richards and Harbison, 1942) have brought up a number of interesting species such as the only Transmarian cerithiid, *Bittium insulaemaris*, and the endemic buccinid *Bulliopsis variabilis*. As is typical of this subprovince, these endemics coexisted with members of the naticid genus *Euspira* (*E. heros* species complex), several *Crepidula* species (*C. fornicata* complex, Calyptraeidae), and large beds of oysters in the *Ostrea glauconoides* species complex (Ostreidae). Several cooler water genera that invaded the Salisbury Sea basin in the Burdigalian, such as the ocenebrine muricids *Trisecphora* and *Chesathais*, the trophonine muricids *Scalaspira* and *Lirosoma*, and the bivalve *Glossus*, all have European-North Atlantic affinities and may have originated in the Sankatian Subprovince. After Langhian time, the Raritan Sea and Sankatian Subprovince ceased to exist and the Sankatian fauna continued on within the Calvertian Subprovince as relictual taxa.

The Calvertian Subprovince

Named for the Calvert Cliffs, Calvert County, Maryland (site of the classic fossil beds of the Calvert, Choptank, and St. Mary's Formations, which typify the fauna) (Petuch, 1988), this subprovince encompassed the entire Salisbury Sea, from the Norfolk Peninsula in the south to the Millville Delta, Delmarva Archipelago, and Normandy Peninsula in the north (Figure 10). Chronologically, the Calvertian Subprovince was long-lived, spanning the times of three different subseas; the Calvert, the Patuxent, and the St. Mary's (altogether from late Burdigalian to mid-Tortonian Miocene). Several endemic index groups were representative of the subprovince over its entire time range and occurred in all three subseas and all three geological formations. These included the gastropod genera *Ecphora* s.s. (Ocenebrinae-Muricidae), *Coronafulgur* (Maryland species complex, Busyconidae), *Scaphellopsis* and *Volutifusus* (Scaphellinae-Volutidae), *Conradconfusus* (Fusininae-Fasciolariidae), *Mariacolpus* (Turritellidae), *Chesatrophon* (Trophoninae-Muricidae), and *Mariaturricula* (Turridae), and the bivalves *Glossus* (Glossidae) and *Chesapecten* (Maryland species complex, Pectinidae). Other endemic Calvertian genera and species complexes were shorter-lived, appearing in only one or two of the subseas. Some of the groups restricted to the Calvert Subsea include the gastropod genera *Calvertitella* (Turritellidae), *Calverturris* (Turridae), *Sassia* (Maryland species complex, Personiidae), *Ficus* (*F. harrisi* complex), *Cymia* (Thaididae), and *Oliva (Strephona)* (Maryland species complex, with *O. harrisi* and *O. simonsoni*). Those restricted to both the Calvert and Patuxent Subseas included *Trisecphora, Planecphora, Chesathais,* and *Ecphorosycon* (all Ocenebrinae-Muricidae) and *Transmariaturris* (Turridae). Confined to the Patuxent Subsea were the gastropod genera *Cancellariella* (Cancellariidae) and *Panamurex (Stephanosalpinx)* and *Patuxentrophon* (Muricidae).

During Serravallian time, lowered water temperatures led to a major extinction event within the Salisbury Sea (Petuch, 1993) (Phase 1 of the "Transmarian Extinction", see Chapter 10). Many of the tropical components of the Calvert Subsea fauna disappeared from the Transmarian Province and never again occurred as far north. Some of the retreating fauna included the gastropod genera *Oliva (Strephona), Architectonica, Ficus, Sassia, Erato, Cerithiopsis, Seila, Niso, Cymia, Murexiella, Persicula,* and *Myurellina,* the nautiloid cephalopod *Aturia* (Nautilidae) (Martin, 1904), and the bivalve genera *Divalinga, Strigilla,* and *Iphigenia.* By the beginning of Tortonian time, the marine climate of the St. Mary's Subsea warmed and the Calvertian Subprovince was reinvaded by tropical groups from the south. Some of these, which were restricted to the St. Mary's Subsea, included the gastropod genera *Gradiconus* (with three species, *G. deluvianus, G. sanctaemariae,* and *G. asheri*), *Mariacassis* (*M. caelata*), *Eudolium,, Dentimargo, Mitra* s.l., and *Hesperisternia.* Interestingly, the group most affected by the Serravalian extinction was the ocenebrine "ecphora" muricids, with only one large, four-ribbed *Ecphora* (sensu stricto) species surviving into St. Mary's time. Prior to that in the Calvert and Patuxent Subseas, five or six species, comprising five different genera, co-occurred at any given time. The disappearance of the three-ribbed genus *Trisecphora* and the scaly, multiple-ribbed genus *Chesathais* is particularly striking, as only four-ribbed ecphoras exist after Patuxent (Choptank Formation) time.

The Pungoian Subprovince

Named for the Pungo River Formation of North Carolina (Petuch, 1988), which contains the typical fauna, this southernmost Transmarian subprovince encompassed the entire Pamlico Subsea and extended from the Norfolk Peninsula south to the Charleston Sea basin (Figure 10). Little is known about the Pungoian faunas, since most of the formations containing their fossils are leached and badly preserved. In some geological units, such as the Bonnerton Member of the Pungo River Formation, only calcitic fossils have been preserved, with the aragonitic fossils having been dissolved away and leaving only molds. Even with an incomplete fossil record, the Pungoian Subprovince fauna can be seen to have shared a large number of gastropod genera and species with the Calvert Subsea to the north. Included were the ocenebrine muricids *Trisecphora* (*T. chamnessi*, *T. tricostata*, and *T. schmidti*), *Ecphora* (*E. wardi*), *Chesathais* (*C. whitfieldi*), and *Ecphorosycon* (*E. pamlico*), the epitoniids *Stenorhytis pachypleura*, *Cirsotrema calvertensis*, and *Amaea prunicola*, and the turritellid *Calvertitella indenta* (preserved as molds). Typical Calvert Subsea bivalves were also present, including the pectinid *Chesapecten coccymelus* and the hiatellid *Panopea parawhitfieldi*. An interesting endemic component coexisted with these northern taxa, including a species complex of the ocenebrine genus *Trisecphora* (including *T. carolinensis* and several undescribed species), the ocenebrine genus *Siphoecphora* (*S. aurora*), and the large pectinid bivalve *Pecten maclellani*. Examples of these are illustrated in Chapter 4. The Pamlico Subsea and the Pungoian Subprovince were short-lived, existing only from late Burdigalian to early Serravallian time.

The Baitoan Province

Named for the Burdigalian Miocene Baitoa Formation of the Dominican Republic, which contains the typical fauna, this paleoprovince extended from South Carolina, across the Gulf of Mexico, the Caribbean region, and northern South America to northern Brazil, and through the Bolivar Strait to southern California in the north and northern Peru in the south (Figure 10). The Baitoan provincial limits are defined by a large number of widespread eutropical species (E.Vokes, 1979), some of which include the gastropods *Plochelaea crassilabrum* (Plicolivinae-Volutidae) (Florida, Dominican Republic, and Brazil), *Vasum haitense* (Plate 2, F) (Florida, Dominican Republic, Venezuela, and Brazil), *V. pugnans* (Florida, Dominican Republic, and Venezuela), and *V. tuberculatum* (Plate 2, I) (Florida, Dominican Republic, and Venezuela) (all Vasinae-Turbinellidae), *Potamides suprasulcatus* (Plate 2, E) and *Terebralia dentilabris* (both Potamididae) (Florida, Dominican Republic, Venezuela, and Brazil), *Hemifusus antillarum* (Plate 4, E) (Hispaniola, Venezuela, and Brazil) (Melongenidae), *Triplofusus kempi* (Plate 2, G) (Fasciolariidae) (Florida, Dominican Republic, and Venezuela), *Muracypraea henekeni* (Plate 4, C and D) (Cypraeidae) (Dominican Republic, Panama, Venezuela, and Brazil), *Chicoreus dujardinoides* (Muricidae) (Florida, Dominican Republic, and Venezuela), *Protoconus consobrinus* (Mexico, Hispaniola, Panama, and Carriacou) and *Pyruconus recognitus* (Plate 4, G) (Mexico, Hispaniola, Panama, Venezuela, and Carriacou) (both Conidae) and *Panaterebra cucurrupiensis* (Plate 4, L) (Peru, Ecuador, Panama, and Venezuela) (Terebridae). Some of the widespread Baitoan gastropod genera include *Trona* (Plate 2, A and B) (Florida to Venezuela) and *Loxacypraea* (new genus, see Systematic Appendix) (Plate 3, B and C) (Florida, Panama, and southern California) (both Cypraeidae), *Protoconus* (Plate 2, J) (Conidae) (Florida to Brazil), and *Falsilyria* (Plate 2, H) (Volutinae-Volutidae) (Florida to Brazil), and the nautiloid cephalopod genus *Aturia* (Nautilidae) (Florida to Venezuela).

Based upon regionally endemic species and species complexes, ten subprovinces are now recognized as having existed within the Baitoan Province (Figure 10). These include the *Onslowian Subprovince* (South Carolina to Georgia), the *Chipolan Subprovince* (southeastern Georgia, Florida, and the coasts of Alabama, Mississippi, Louisiana, and Texas), the *Agueguexquitean Subprovince* (Mexico to Honduras), the *Anguillan Subprovince* (Cuba, Hispaniola, Jamaica, Puerto Rico, and Virgin Islands), the *Culebran Subprovince* (Honduras, Nicaragua, Costa Rica, Panama, southern Colombia, and through the Bolivar Strait to northern Ecuador and Baja California), the *Subibajan Subprovince* (Ecuador and northern Peru), the *Cantaurean Subprovince* (northern Colombia, Venezuela, Trinidad, and Guyana), the *Carriacouan Subprovince* (Lesser Antilles), the *Piraban Subprovince* (Cayenne, Suriname, and northern Brazil), and the *Vaquerosian Subprovince* (Gulf of California to central

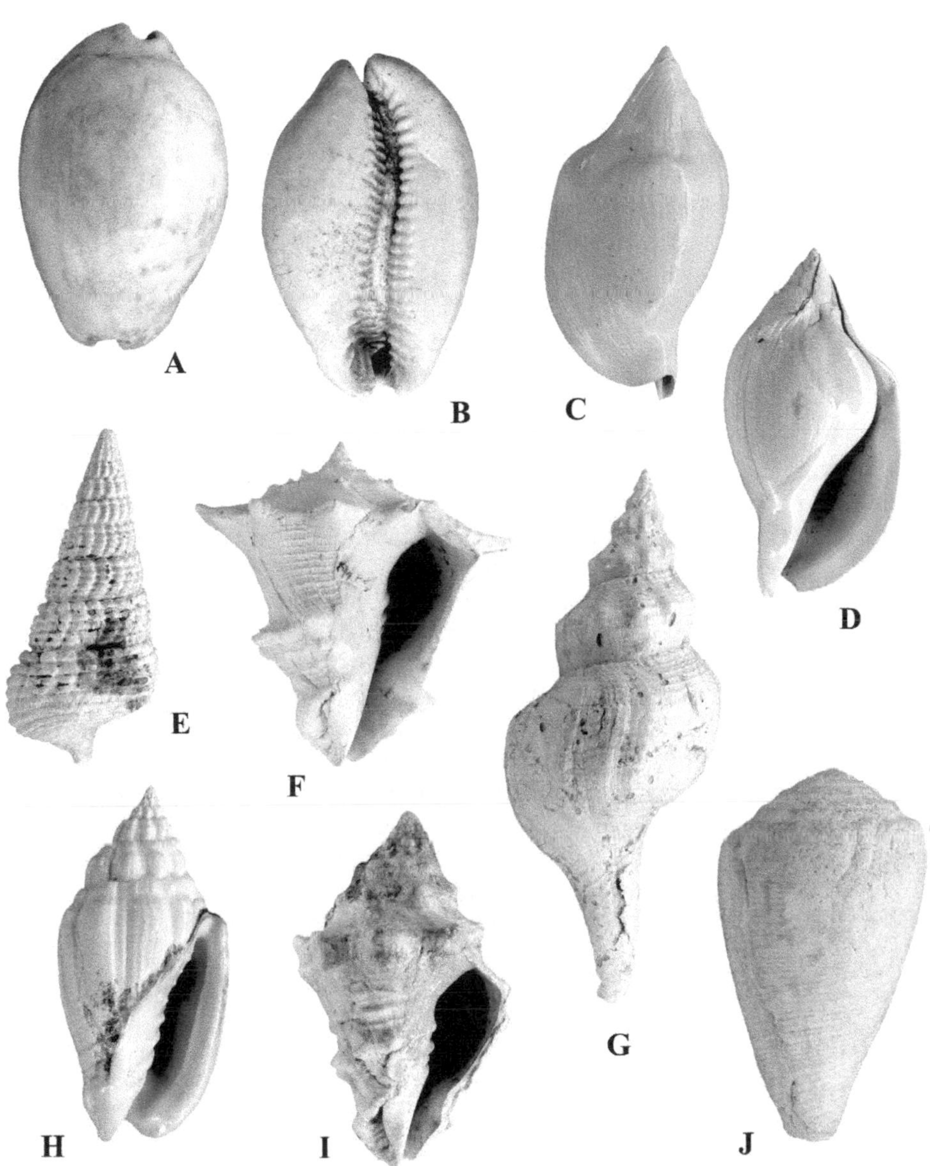

Plate 2. Gastropod index genera and widespread species of the Baitoan Province. A, B= *Trona* Jousseaume, 1884 (*T. fossula* (Ingram, 1947), Cypraeidae), Cantaure Formation, Venezuela; C, D= *Orthaulax* Gabb, 1873 (*O. inornatus* subspecies, Strombidae), Chipola Formation, Florida; E= *Potamides suprasulcatus* (Gabb, 1873) (Potamididae), Cantaure Formation, Venezuela; F= *Vasum haitense* Sowerby, 1849 (Vasinae-Turbinellidae), Chipola Formation, Florida; G= *Triplofusus kindlei* (Maury, 1910) (Fasciolariidae), Chipola Formation, Florida; H= *Falsilyria* Pilsbry and Olsson, 1954 (*F. pycnopleura* (Gardner, 1937), Volutinae-Volutidae), Chipola Formation, Florida; I= *Vasum tuberculatum* Gabb, 1873 (Vasinae-Turbinellidae), Cantaure Formation, Venezuela; J= *Protoconus* daMotta, 1991 (*P. granozonatoides* (Maury, 1912) (Conidae), Cantaure Formation, Venezuela.

California). These subprovinces are described in the following sections and illustrations of some of the index fossils are shown in Chapter 4.

The Onslowian Subprovince

Named for Onslow County, North Carolina, site of the fossil beds of the Belgrade Formation which contain the typical fauna, this northernmost Baitoan subprovince encompassed only the Silverdale Subsea of the Albemarle Sea. Being partially under the influence of the cool water Raritan Current, the Onslowian Subprovince was paratropical in marine climate and lacked well-developed carbonate environments and coral reefs (Petuch, 1997). Of all the Baitoan subprovinces, the Onslowian was the shortest-lived, existing only during the late Chattian Oligocene and early Aquitanian Miocene. After that time, the Raritan Current influence intensified in the Onslowian area and the marine climate cooled substantially. This led to the extinction of many characteristic (and geologically short-lived) endemic groups, such as the gastropod genera *Fenolignum*, *Argyrobessa*, and *Miocenebra* (all Ocenebrinae-Muricidae) and the bivalve genus *Rebeccapecten* (Pectinidae). Although containing several key tropical families and genera, such as *Gradiconus* (*G. postalveatus*, Conidae), "*Fusus*" ("*F.*" *quinquespina*, an undescribed genus, Melongenidae), and *Harpeola* (*H. carolinensis*, Lyriinae-Volutidae), the paratropical nature of the subprovince is demonstrated by the absence of other key tropical groups such as the Cypraeidae, Turbinellidae, Ovulidae, Turbinidae, Modulidae, and Potamididae. The Onslowian Subprovince also shared a large percentage of its molluscan fauna with the oldest part of the Chipolan Subprovince to the south, such as the gastropod genera *Spinifulgur* (Busyconidae), *Scaphella* (Scaphellinae-Volutidae), *Tritonopsis* (Rapaninae-Thaididae), and *Rapanecphora* (Ocenebrinae-Muricidae). Of special interest in the Onslowian gastropod fauna was the oldest-known eastern North American *Cymatophos* (*C. kellumi* n. sp., Buccinidae; see Systematic Appendix) and the oldest-known *Conradconfusus* (*C. hoffmani*, Fasciolariidae; later endemic to the Transmarian Province). By late Burdigalian time, the fauna of the Onslowian area had become completely altered and was replaced by that of the Pungoian Subprovince of the Transmarian Province. Illustrations and discussions of some of the principal Onslowian index species are given in Chapter 3.

The Chipolan Subprovince

Named for the Burdigalian Chipola Formation of the Florida Panhandle, which contains the typical fauna, this subprovince encompassed the Chattahoochee, Chipola, and Walton Subseas of the Choctaw Sea and the Tampa, Arcadia, and Polk Subseas of the Okeechobean Sea (Figure 10). The Chipolan Subprovince was long-lived, ranging from the late Chattian Oligocene to the end of the Langhian Miocene. As a eutropical subprovince, the Chipolan contained all the key molluscan index genera and families and was the only Gulf coast area to have developed extensive zonated coral reef tracts (see Chapters 3 and 4). Being the descendant of the Hernandoan Subprovince of the Antiguan Province, the Chipolan Subprovince also retained many of the older, relictual faunal elements, some of which include the gastropod genera *Omogymna* (Olividae) and *Spinifulgur* (Plate 3, H) (Busyconidae, with a species radiation in the Tampa Subsea and the last two species occurring in the Chipola Subsea) and the coral genus *Antiguastrea* (Montastreidae). A large endemic component had evolved within the subprovincial limits by Chipola time, some of which included the gastropod genera *Busycon* (Plate 3, A) and *Pyruella* (Plate 3, D) (both Busyconidae and with their first appearances in the Chipola Subsea), *Psammostoma* (Peristerniinae, Fasciolariidae), *Floradusta* (Cypraeidae) (new genus, see Systematic Appendix), *Terebraspira* (Plate 3, G) (Fasciolariidae), *Hystrivasum* (Vasinae-Turbinellidae),

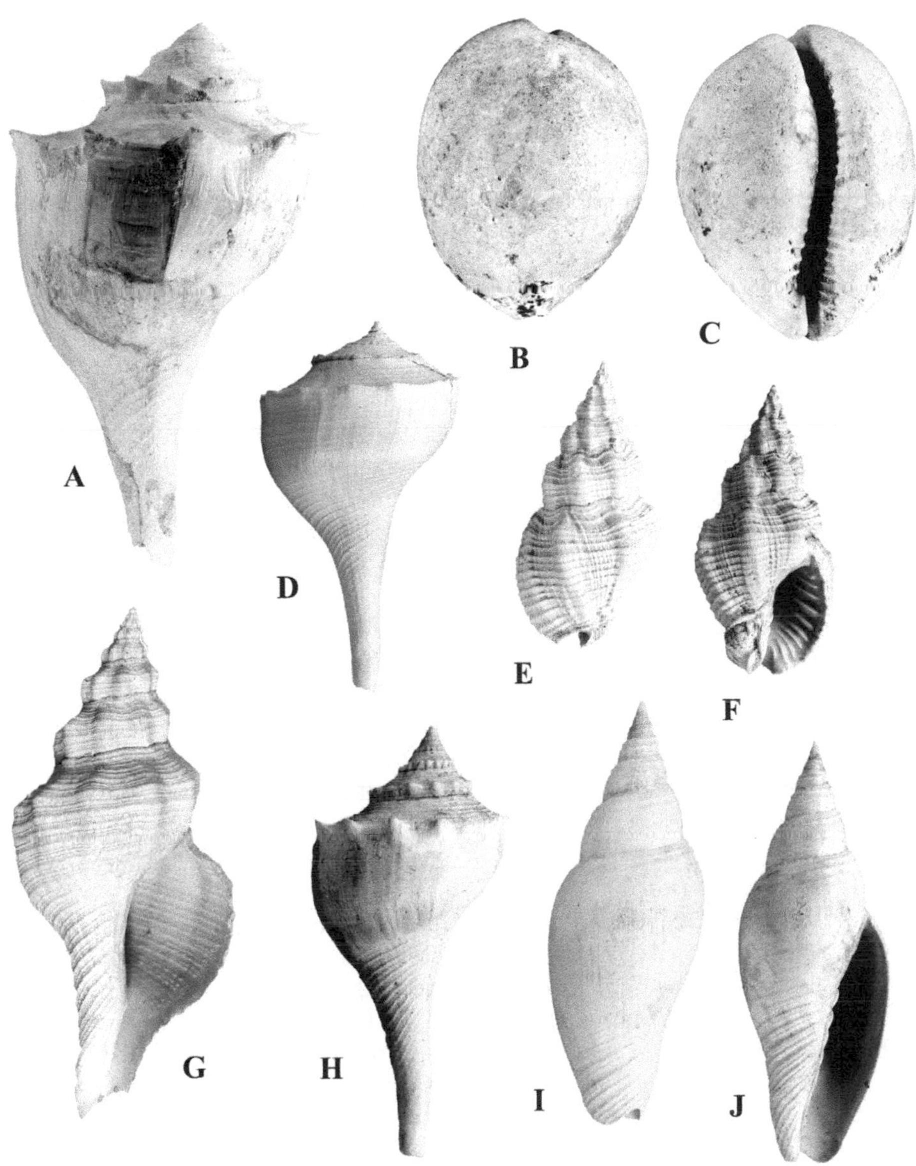

Plate 3. Gastropod index genera of the Chipolan Subprovince of the Baitoan Province. A= *Busycon* Roeding, 1798 (*B. diegelae* Petuch, new species, Busyconidae; see Systematic Appendix); B, C= *Loxacypraea* Petuch, new genus (see Systematic Appendix) (*L. chilona* (Dall, 1900), type of the genus, Cypraeidae); D= *Pyruella* Petuch, 1982 (*P. sicyoides* (Gardner, 1944), Busyconidae); E, F= *Hesperisternia* Gardner, 1944 (*H. chipolana* Gardner, 1944, Buccinidae); G=*Terebraspira* Conrad, 1862 (*T. ramondi* (Maury, 1902), Fasciolariidae); H= *Spinifulgur* Petuch, 1994 (*S. epispiniger* (Gardner, 1944), Busyconidae); I, J= *Pleioptygma* Conrad, 1862 (*P. prodroma* (Gardner, 1937), Pleioptygmatidae). All specimens are from the Chipola Formation of the Florida Panhandle. The genus *Loxacypraea*, although a common and characteristic component of the Chipolan Subprovince, also occurred in the Culebran and Vaquerosian Subprovinces.

Hesperisternia (Plate 3, E and F) (Buccinidae), and *Pleioptygma* (Plate 3, I and J) (Pleioptygmatidae). Several of these survived into the Pliocene and Pleistocene and produced large species radiations in the subsequent Caloosahatchian Province. A strong East Tethyan ("South Pacific-appearing") relictual faunal element was also present on the Chipolan Subprovince reef systems, including the gastropods *Homalocantha* (Muricinae-Muricidae), *Clavocerithium* and *Hemicerithium* (both Cerithiidae), and the cowries *Zoila* and *Lyncina* (Cypraeidae) (Petuch, 1997). The widespread Baitoan strombid genus *Orthaulax* was present within the Chipolan Subprovince only to late Burdigalian time and was not present in the later part of the Chipola Subsea or the following Walton Subsea.

The Agueguexquitean Subprovince

Named for the late Miocene (possibly also Pliocene) Agueguexquite Formation of Veracruz State, Mexico, this subprovince extended from the Texas coast south to Honduras and was centered on the Yucatan and Isthmus of Tehuantepec areas (E. Vokes, 1984) (originally proposed as a Pliocene subprovince of the Gatunian Province; Petuch, 1988) (Figure 10). The descendant of the Alazanian Subprovince of the Antiguan Province, the Agueguexquitean Subprovince was particularly important in the evolution of the subsequent Caloosahatchian Province and the Recent Carolinian and Caribbean Provinces. Here, the first representatives of prominent Gulf of Mexico gastropod genera make their first appearances, including *Scaphella (Clenchina)* (Scaphellinae-Volutidae, *S. americana*, ancestor of the large *Clenchina* species radiation of the Recent Gulf), *Lindafulgur* (new genus, see Systematic Appendix) (Busyconidae, *L. alencasterae*, ancestor of the Recent Gulf of Mexico-restricted *L. candelabra* and *L. lyonsi*) (Perrilliat, 1963, 1973; Petuch, 1987), and *Turbinella hermani* (Turbinellidae, ancestor of the Recent *T. angulata* and of the knobbed chank shells of the Pleistocene). Other subprovincial gastropod endemics included the bizarre, sand-accreting modulid *Psammodulus* (Collins, 1934), *Harpa isthmica* (Harpidae), and *Microrhytis pecki* (Ocenebinae-Muricidae).

The Anguillan Subprovince

Named for the Miocene fossil beds on Anguilla Island, Lesser Antilles (Anguilla Formation), which contain the typical fauna, this eutropical subprovince encompassed the faunas of all the islands of the Greater Antilles and was the descendant of the Guanican Subprovince of the Antiguan Province (Figure 10). The faunas of the classic and well-studied fossil beds of the Baitoa (namesake of the province) and Cercado Formations (Dominican Republic), the Las Cahobas Formation (Haiti), and La Cruz and lower Matanzas Formations (Cuba) are also included within the subprovince. Besides containing the classic provincial index taxa, this area housed a large number of endemic species and species complexes including the *Muracypraea henekeni* complex (Plate 4, C and D) (Cypraeidae), a large complex centered around the tulip shell *Fasciolaria intermedia* (with *F. semistriata* and *F. leura*) (Fasciolariinae-Fasciolariidae), the *Pyruconus haytensis* complex (with *P. domingensis* and *P. politispira*) (Conidae), the *Fusinus henekeni* complex (Fusininae-Fasciolariidae), and the *Strombus proximus* complex (Strombidae). Of particular interest was the presence of the oldest-known *Macrostrombus* species (*M. haitensis*, Strombidae) and *Scaphella* (*S. striata*, Scaphellinae-Volutidae). As in the case of the Chipolan Subprovince, the strombid genus *Orthaulax* became extinct in Langhian time, but may have persisted slightly longer within the Caribbean Basin.

The Culebran Subprovince

Named for the early Miocene Culebra Formation of the Canal Zone, Panama, which

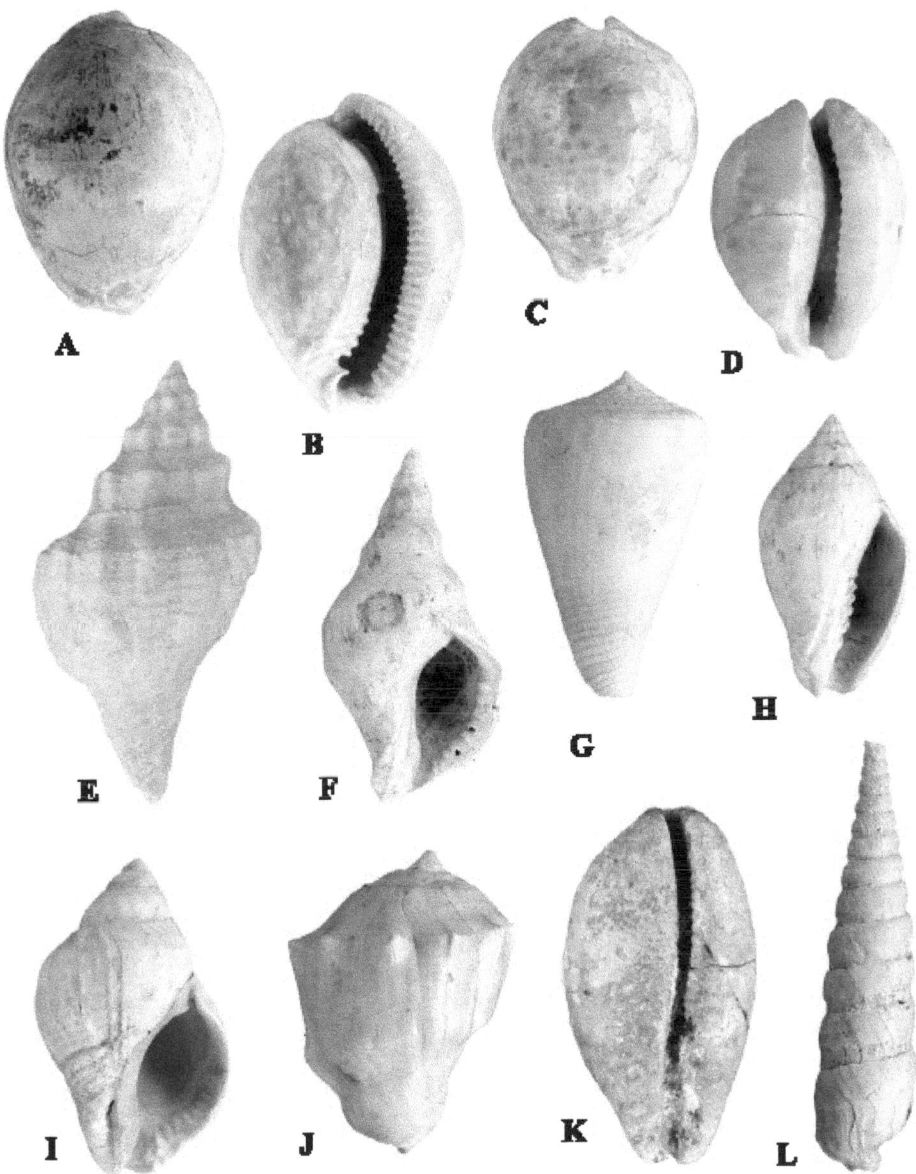

Plate 4. Widespread gastropod index species of the southern subprovinces of the Baitoan Province. A, B= *Marginocypraea wegeneri* (Schilder, 1939) (Ovulidae); C, D= *Muracypraea henekeni* (Sowerby, 1849) (Cypraeidae); E= *Hemifusus antillarum* (Gabb, 1873) (Melongenidae); F= *Vitularia ecuadoriana* Marks, 1951 (Muricopsinae-Muricidae); G= *Pyruconus recognitus* (Guppy, 1874) (Conidae); H= *Voluta cantaurana* Gibson-Smith, 1973 (Volutinae-Volutidae); I= *Macron constrictus* Gibson-Smith, Gibson-Smith, and Vermeij, 1997 (Pseudolovidae); J= *Torquifer barbascoanus* (Gibson-Smith and Gibson-Smith, 1983) (Melongenidae); K= *Macrocypraea trinitatensis* (Mansfield, 1925) (Cypraeidae); L= *Panaterebra cucurrupiensis* (Oinomikado, 1939) (Terebridae). With the exception of *Macrocyprea trinitatensis,* which is from the Grand Bay Formation, Carriacou Island, all specimens are from the Cantaure Formation of Venezuela. These species occurred in the Carriacouan, Cantaurean, Culebran, and Subibajan Subprovinces.

contains the typical fauna, this eutropical subprovince was the descendant of the Bohioan Subprovince of the Antiguan Province and, like its transisthmian predecessor, spanned the Bolivar Strait (Figure 10). Besides the full compliment of Baitoan index taxa, the Culebran Subprovince contained a high percentage of endemic gastropod genera, species, and species complexes, some of which include the genera *Nicema* (Buccinidae) and *Strombinella* (Columbellidae), the *Cymatophos veatchi* complex (Buccinidae), and the oldest-known *Pyruclia* radiation (the *P. cibarcola* complex, Cancellariidae), and species such as *Nanarius acolus* (Nassariidae) and *Enaeta encomia* (Lyriinae-Volutidae) (Woodring, 1964, 1970). Of special interest within the subprovince was a large radiation of cypraeoidean gastropods, including *Loxacypraea* (*L. chilona* complex), *Muracypraea* (with at least four species in the *M. henekeni* complex, including *M. andersoni* and *M. tuberae*) (all Cypraeidae), and *Apiocypraea* (*A. keenae* complex) (Ovulidae).

The Subibajan Subprovince

Named for the early Miocene Subibaja Formation of northern Peru, this paratropical subprovince also encompassed the faunas of the Daule, Zorritos, Tumbez, and Cucurrupi Formations of northern Peru and southern Ecuador (Marks, 1951) (Figure 10). As is typical of paratropical areas, tropically-derived gastropods, such as *Gradiconus* (*G. sophus* complex, Conidae) and *Muracypraea* (*M. angustirima*, Cypraeidae), coexisted with cold water taxa such as "*Chorus*" ("*C.*" *cruziana* complex, Ocenebrinae-Muricidae) and *Aforia*, *Megasurcula* (*M. guayasensis* complex) and *Clinura* (all Turridae). The paratropical nature of Subibajan area was further demonstrated by the absence of the key tropical indicator gastropod families Modulidae, Vasinae (Turbinellidae), Lyriinae (Volutidae), and Ovulidae. Provinciatonal endemics were also present (Marks, 1951), with the bizarre stromboidean genus *Strombiconus* (*S. ecuadorensis*, Strombidae) being the most unusual. The Subibajan Subprovince also shared several species with the Culebran and Cantaurean Subprovinces, principally *Vitularia ecuadoriana* (Plate 4, F) (Muricopsinae-Muricidae) and *Panaterebra cucurrupiensis* (Plate 4, L) (Terebridae).

The Cantaurean Subprovince

Named for the Burdigalian Cantaure Formation of the Paraguana Peninsula, Venezuela, which contains the typical fauna, this subprovince also encompassed the faunas of the Trinidadian Miocene (the Brasso, Couva, and Springvale Formations) (Jung, 1965, 1969). Besides the full compliment of Baitoan provincial index species, the Cantaurean Subprovince contained a large number of highly distinctive endemic genera and species, including *Macron* (Plate 4, I) (*M. constrictus*, the only-known representative ever found in the western Atlantic, Pseudolividae), *Torquifer* (Plate 4, J) (*T. barbascoana*, Melongenidae), the bizarre shell and rubble-accreting turritellid *Springvaleia* (Turritellidae), the oldest-known *Voluta* (Plate 4,H) (*V. cantaurana*, Volutinae-Volutidae), and *Vasum quirosense*. Of special interest within the Cantaurean area was a large radiation of cypraeoidean gastropods, including *Muracypraea* (with the Venezuelan *M. quagga* and *M. hyaena* and the Trinidadian *M. lachrymula* and *M. caroniensis*, all Cypraeidae) and *Marginocypraea* (Plate 4, A and B) (*M. wegeneri*, Ovulidae). Interestingly, members of the gastropod family Strombidae, especially *Orthaulax* and *Strombus*, are rare or completely absent in most of the Cantaurean areas.

The Carriacouan Subprovince

Named for the Burdigalian fossil beds on Carriacou Island, Grenadian Grenadines (Carriacou, Belmont, and Grand Bay Formations) (Jung, 1971), which typify the fauna, this eutropical subprovince spanned the Lesser Antilles. At that time, the islands were in the process of growing and were highly active volcanically. For this reason, the fossil shells and corals on Carriacou are often embedded in volcanic tuffs, indicating that they were buried (and preserved) in pyroclastic flows...essentially the "Pompei" of Caribbean fossils. Although containing widespread Baitoan species such as *Protoconus consobinus* (Conidae), *Malea camura* (Tonnidae), *Nerita (Theliostyla) exuvioides* (Neritidae), and *Hindsiclava consors* (Turridae), the Carriacouan Subprovince contained a highly endemic fauna, some of which include *Ficus carriacouensis* (Ficidae), the *Cymatophos glareosus* species complex (at least three species, Buccinidae), *Cancellomorum coxi* (Moruminae-Harpidae), *Chicoreus jungi* (Muricidae), and *Enaeta trechmanni* (Lyriinae-Volutidae). The Trinidadian Cantaurean cypraeid, *Macrocypraea trinitatensis* (Plate 4, K), the oldest-known member of its genus, was also present on the coral reefs of Carriacou Island. Of special interest was a large endemic radiation of the family Turridae, including species swarms of the genera *Mitratoma, Paraborsonia,* and *Polystira,* and the endemic genus *Caritoma* (*C. antillarum*).

The Piraban Subprovince

Named for the Miocene Pirabas Formation of Ponta Pirabas, Para State, Brazil, which contains the typical fauna, this southernmost Baitoan subprovince encompassed the Amazon Sea and adjacent coasts (Maury, 1924, 1934; Magalhaes and Mezzalira, 1953). Bathed in the warm water of the North Equatorial Current, the Piraban Subprovince contained a eutropical fauna and most of the Baitoan provincial gastropod index taxa. Some of these included *Potamides suprasulcatus* (Potamididae; named "*Cerithium pachecoi*" by Maury), *Orthaulax* (*O. brasiliensis*, Strombidae), *Falsilyria* (*F. musicinoides* and *F. calligona*, Volutinae-Volutidae), *Vasum haitense* (Vasinae-Turbinellidae), and *Melongena consors* (Melongenidae). A large eutropical endemic component also evolved within the subprovince, with some of the more important being *Muracypraea pennae* (Cypraeidae), *Gradiconus whitei* (Conidae), *Cancellomorum harrisi* (Moruminae-Harpide), *Sconsia felix* (Cassidae), and *Cymatium (Septa) williamsi* (Ranellidae). The Amazon Sea embayment, which extended at least half way up the present Amazon River, also contained an endemic gastropod fauna. Largely undescribed, this brackish water fauna contained an impressive cerithiid radiation, with many new genera and with species such as "*Clava*" *williamsi* and "*Cerithium*" *calcivelatum*. A large potamidid radiation was also present and was typified by species such as "*Cerithium*" *pirabicum* (an ornate, *Tympanotonus*-type potamidid).

The Vaquerosian Subprovince

Named for the Oligocene-early Miocene Vaqueros Formation of southern California, which contains the typical fauna, this paratropical subprovince represented the northernmost edge of the Pacific component of the Baitoan Province and extended from Baja California northward to central California (Arnold, 1907; also the Temblor Formation, Wiedey, 1928). The paratropical nature of this subprovince is demonstrated by the lack of many of the key tropical gastropod families, such as the Strombidae, Potamididae, Turbinellidae, and Modulidae, and also by the presence of tropical groups such as the Cerithiidae (*Clavocerithium* species), the Cypraeidae (*Loxacypraea fresnoensis*), the Conidae (*Gradiconus juanensis* and *G. owenianus*), and by the tropical turritellid genus *Torcula* (*T. boesei* complex). Unique to the subprovince was a species complex of an undescribed ocenebrine muricid genus (referred to as "*Rapana*," with examples being "*R.*" *vaquerosensis* and "*R.*" *serrai*; probably close to *Forreria*).

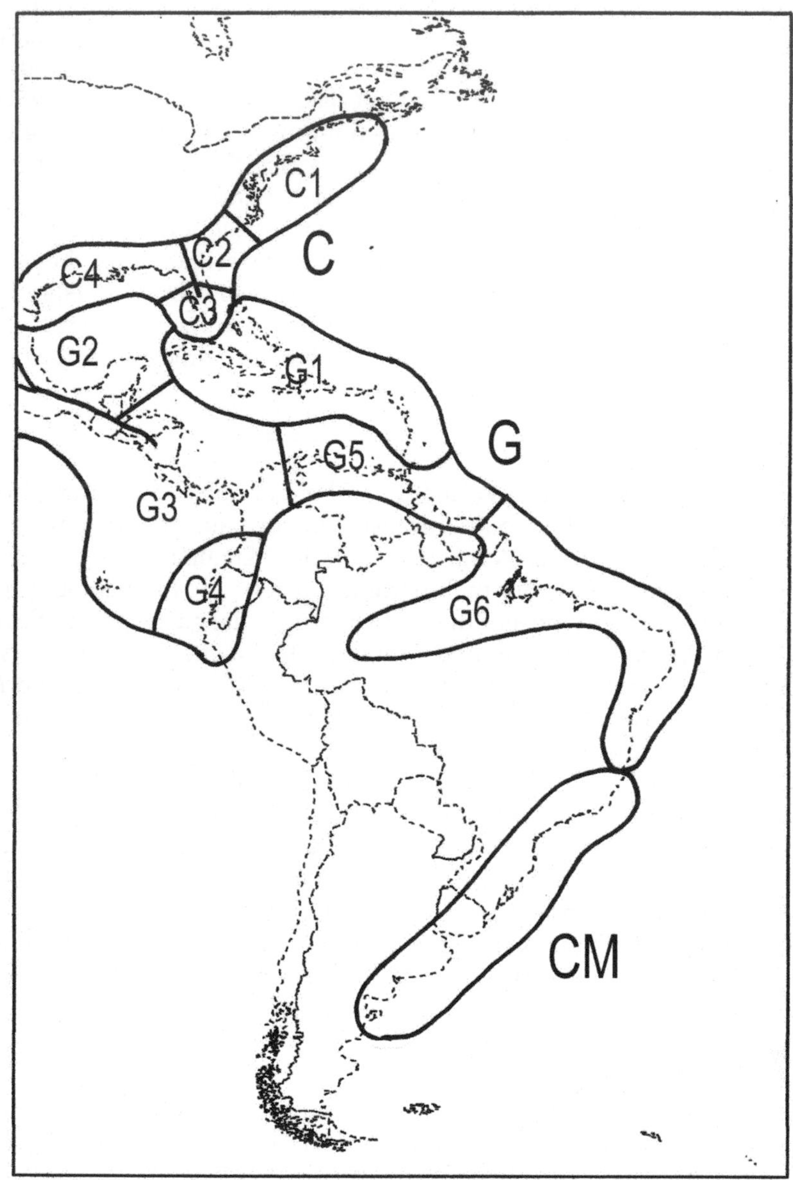

Figure 11. Approximate configuration of the marine molluscan provinces and subprovinces of the western Atlantic and Eastern Pacific during late Tortonian Miocene to late Calabrian Pleistocene time, superimposed upon the outline of the Recent Americas. Biogeographical units include: C= Caloosahatchian Province, C1= Yorktownian Subprovince, C2= Duplinian Subprovince, C3= Buckinghamian Subprovince, C4= Jacksonbluffian Subprovince, G= Gatunian Province, G1= Guraban Subprovince, G2= Veracruzan Subprovince, G3= Limonian Subprovince, G4= Esmeraldan Subprovince, G5= Puntagavilanian Subprovince, G6= Juruaian Subprovince, CM= Camachoan Province. The Imperialian Subprovince of the Gatunian Province, of southeastern California and northern Gulf of California areas, is not shown.

The Platensian Province

Being represented by only a few fossil beds along the Mar de Plata coast of Uruguay, little is known of this Miocene southern paratropical province (Frenguelli, 1946; Figueiras and Broggi, 1976). Named for the Mar de Plata, the range of the province is mostly conjectural, probably extending from Bahia State, Brazil south to the Mar de Plata coasts of Uruguay and Argentina and to the northeasternmost Argentinian Atlantic coast (Figure 10). Since the province is so poorly known, no subprovincial boundaries can be established at this time. The little data that does exist shows that the Platensian Province had a highly endemic fauna with a strong tropical influence and a large number of unusual species complexes. Principal among these were several seen in the gastropod family Strombidae, including an undescribed genus that closely resembles the living genus *Conomurex* (centered around the *C. luhuanus*-appearing "*C.*" *semicoronatus*) (Frenguelli, 1946) and another in an undescribed genus ("*Strombus*" *dallianus* complex). The tropical family Conidae was represented by the genus *Lamniconus* (*L. platensis* complex), which still occurs in these far southern Atlantic waters. A large radiation of the tropical turritellid genus *Torcula* also occurred within the Platensian Province, some of which included *T. indeterminata, T. americana,* and *T. pyramidesia*. Cold water influences included *Neverita* (*N. entreriana* complex, Naticidae) and several species of *Trophon* (Trophoninae-Muricidae).

Marine Provinces of late Tortonian Miocene to Calabrian Pleistocene Time

The late Tortonian Miocene saw another severe worldwide cooling event and the accompanying drops in sea level, mass extinctions, and disruptions of provincial structures (see Chapter 10). Hardest hit was the Transmarian Province, which essentially ceased to exist (Phase 2 of the "Transmarian Extinction" (Petuch, 1993); see Chapter 10). Several characteristic Transmarian genera, however, did manage to survive on into the Pliocene in provinces farther south. At this time, the western Atlantic and Eastern Pacific provinces reconfigured themselves in response to changing water temperatures. A second major cooling event took place, geologically soon after, during the late Messinian Miocene and early Zanclean Pliocene. This solidified the shifting provincial boundaries by forcing the rapid evolution of entire new faunas. During the time of the Tortonian and Messinian climatic degenerations, three new faunal provinces formed and these included the northern *Caloosahatchian Province,* the central *Gatunian Province,* and the southern *Camachoan Province* (Figure 11).

Roughly halfway through the time of the existence of these provinces, during the early Piacenzian Pliocene (Kiegwin, 1978; Vermeij and Petuch, 1986), the Bolivar Strait closed and the newly-formed Isthmus of Panama separated the Eastern Pacific from the western Atlantic. Prior to this time, the current structure was essentially the same as that of the Oligocene and Miocene, with warm water being conveyed as far north as the Albemarle Sea by the northward-flowing Ciboney Current (the Pliocene analogue of the Arawak Current; Petuch, 1997). After the Piacenzian, however, the modern Gulf Stream formed and the coastal current structure became more complex, with more and much stronger southward-flowing cold water countercurrents. Before the closing of the Isthmus of Panama, the North Equatorial Current conveyed warm water all the way to the Imperial Valley of southern California and as far south as Ecuador. The Gatunian and Camachoan Provinces fall outside the scope of this book and will not be discussed in the later chapters. Since all the eastern North American paleoseas of this time period are included within the Caloosahatchian Province, its key index organisms and communities are the only ones discussed and illustrated in Chapters 6, 7, 8, and 9.

The Caloosahatchian Province

Named for the latest Pliocene-earliest Pleistocene Caloosahatchee Formation of southern Florida (Petuch, 1982), which contains the typical fauna, this province extended from Nova Scotia, around the Floridian Peninsula, and into the Gulf of Mexico as far as Texas and encompassed the Albemarle, Charleston, Okeechobean, and Choctaw Seas (Figure 11). The Caloosahatchian Province was the descendant of the combined Agueguexquitean, Chipolan, and Onslowian Subprovinces of the Baitoan Province and of the Transmarian Province. Based upon the ranges and percentages of restricted endemic taxa, the Caloosahatchian Province can be further subdivided into four subprovinces; the northern *Yorktownian Subprovince* (extending from Nova Scotia to Cape Fear, North Carolina), the *Duplinian Subprovince* (extending from Cape Fear to northeastern Florida), the provinciatonal *Buckinghamian Subprovince* (extending from northeastern Florida, across and around the Everglades area, and north to near Tampa, Florida), and the *Jacksonbluffian Subprovince* (the Florida Panhandle west to Texas) (Petuch, 1988; 1997). The boundaries of the Caloosahatchian Province were sharply demarcated by the ranges of a large number of widespread endemic gastropod genera, some of which included *Busycon, Busycotypus, Pyruella* (Plate 5, C), *Fulguropsis, Sinistrofulgur,* and *Brachysycon* (all Busyconidae), *Akleistostoma* (Plate 5, J and K) (Cypraeidae), *Globecphora* (not present in the Jacksonbluffian Subprovince), *Ecphora, Planecphora, Latecphora* (Duplinian and Buckinghamian Subprovinces only), *Trossulasalpinx, Urosalpinx,* and *Vokesinotus* (all Ocenebrinae-Muricidae), *Contraconus* (Plate 5, B) (Conidae), *Heilprinia* (Plate 5, E), *Terebraspira* (Plate 5, I), and *Cinctura* (all Fasciolariidae), *Globinassa, Scalanassa, Paranassa,* and *Ilyanassa* (all Nassariidae), and *Volutifusus* (Scaphellinae-Volutidae). Widespread Caloosahatchian endemic bivalves included *Chesapecten* (only until the mid-Piacenzian) and *Carolinapecten* (Pectinidae), *Dinocardium* (Cardiidae), *Mercenaria* (Veneridae), *Dallarca* and *Cunearca* (both Arcidae), and *Marvacrassatella* (Crassatellidae). Examples of most of these are shown in Chapters 6, 7, 8, and 9. The appearance of the endemic scallop genus *Carolinapecten* (with the oldest-known species, *C. urbannaensis* occurring in both the Charlotte and Rappahannock Subseas) can be used as a "short cut" for determining the beginning of the Caloosahatchian Province.

The Yorktownian Subprovince

Named for the Yorktown Formation of Virginia and northern North Carolina, which contains the typical fauna, this subprovince encompassed the faunas of the Rappahannock, Williamsburg, Yorktown, and Croatan Subseas of the Albemarle Sea and the coastal areas north to Nova Scotia (Figure 11). Under the influence of the cold water, southward-flowing Narragansett Current (the Pliocene analogue of the Miocene Raritan Current) (Petuch, 1997), the Yorktownian Subprovince was faunistically paratropical, lacking key tropical groups such as the Strombidae, Ovulidae, Modulidae, Vasinae-Turbinellidae, and Potamididae. A few tropical influences did exist and some of these included *Akleistostoma* (*A. carolinensis*, Cypraeidae), *Contraconus* (*C. adversarius*) and *Ximeniconus* (*X. marylandicus*) (both Conidae), *Bullata* (*B. antiqua*, Marginellidae) and *Myurella* (*M. unilineata*, Terebridae). Of special interest within the Yorktown Subprovince was the presence of several Transmarian relictual taxa, survivors of the Transmarian Extinction. These were restricted to the Yorktownian area and included the gastropods *Scalaspira* (*S. strumosa*), *Lirosoma* (*L. sulcosa*), and *Chesatrophon* (*C. tetricus*) (all Trophoninae-Muricidae; although some workers place *Scalaspira* and *Lirosoma* in the Buccinidae, I here consider them muricids, based upon their similarity to some living Arctic trophonines and because of their long siphonal canals), *Scaphellopsis* (*S. ricei*, Scaphellinae-Volutidae), and *Mariafusus* (*M. propeparilis*, only in Zanclean time, Fasciolariidae), and the bivalve *Chesacardium* (*C. acutilaqueatum*, Cardiidae). The provinciatonal nature of the Yorktownian fauna was demonstrated by the coexistence of the before-mentioned tropical groups Cypraeidae and Conidae with cold water gastropod groups such as *Bulbus* and *Euspira* (both Naticidae), *Trophonopsis* (Trophoninae-Muricidae), *Solariella* (Trochidae), and *Atractodon* ("*Tortifusus*"; Buccinidae), and the cold water bivalve genera *Placopecten* (Pectinidae), *Mytilus* (*M. edulis* complex, Mytilidae), *Mya* (Myidae), and *Hiatella (Saxicava)* (Hiatellidae) (Campbell, 1993). Provinciatonal endemics included the giant volute genus *Megaptygma* (*M. sinuosa*, Scaphellinae-Volutidae), *Juliamitrella* (Columbellidae), the *Fusinus burnsi* species complex (Fusininae-Fasciolariidae), and a large species complex of the genus *Calliostoma* (centered around *C. mitchelli*, Trochidae).

The Duplinian Subprovince

Named for the Pliocene Duplin Formation of North Carolina, which contains the typical fauna, this subprovince encompassed the Santee, Duplin, and Waccamaw Subseas of the Charleston Sea (Petuch, 1997) (Figure 11). Being within the influence of the Ciboney Current, the Duplinian Subprovince had a warmer marine climate than the Yorktownian Subprovince and shared many genera and species with Florida. Some of these include the gastropods *Akleistostoma pilsbryi* (Cypraeidae), *Busycon filosum* and *B. tritonis* (both Busyconidae), and *Latecphora violetae* and *Ecphora roxaneae* (both Ocenebrinae-Muricidae), and the bivalves *Carolinapecten darlingtonensis*, *C. walkerensis*, and *Chesapecten palmyrensis* (all Pectinidae) (some illustrated and discussed in Chapters 6 and 7). A large percentage of endemic species was also present, including the gastropods *Calliostoma cyclum* (Trochidae), *Terebraspira elegans* (Fasciolariidae), *Uzita smithiana* (Nassariidae), and *Mangelia emissaria* (Turridae), and the bivalve *Chesapecten carolinensis* (Pectinidae). The gastropod families Strombidae, Modulidae, Ovulidae, Melongenidae, and Vasinae-Turbinellidae are all absent from the Duplinian Subprovince. Of particular interest was a rich, highly endemic estuarine fauna that was restricted to the extensive Satilla Lagoon System of the southern part of the Charleston Sea (Dall, 1913) (see Chapter 6). Some of these included *Potamides satillensis* and *Pachycheilus suavis* (Potamididae), *Torcula satilla* (Turritellidae), *Pyrgulopsis satilla* (Rissoidae), and *Syrnola thelma* (Pyramidellidae).

The Buckinghamian Subprovince

Named for the Buckingham Member of the Tamiami Formation, Okeechobee Group (originally considered a separate formation: Mansfield, 1939; Petuch, 1988, 1997) (Figure 11), which contains the typical fauna, this eutropical subprovince was a classic example of a provinciatone and encompassed the Charlotte, Murdock, Tamiami, and Caloosahatchee Subseas of the Okeechobean Sea and the Floridian coastal lagoons as far north as Tampa in the west and Cape Canaveral in the east. Of particular interest within the Buckinghamian Subprovince were the Everglades Pseudoatoll, the largest Pliocene American coral reef tract, and the Myakka Lagoon System, the best-developed North American tropical estuary (see Chapter 7). Because of all the various habitats and biotopes associated with these high-tropical features, the Buckinghamian Subprovince housed the richest and most highly-endemic molluscan fauna found anywhere in the Americas. Some of the genera that are unique to the Buckinghamian Subprovince, and are considered provinciatonal endemics, include *Echinofulgur* and *Tropochasca* (both Echinofulgurinae-Melongenidae), *Pyrazisinus* (Potamididae) (Plate 5, H), *Hystrivasum* (with one Jacksonbluffian exception, Vasinae-Turbinellidae), *Acantholabia* (Muricinae-Muricidae), *Zulloia* (Charlotte Subsea only, Ocenebrinae-Muricidae), *Liochlamys* (Caloosahatchee

Plate 5. Gastropod index genera of the Caloosahatchian Province. A= *Latecphora* Petuch, 1989 (*L. bradleyae hertwecki* (Petuch, 1989), Ocenebrinae-Muricidae); B= *Contraconus* Olsson and Harbison, 1953 (*C. schmidti* Petuch, 1994, Conidae); C= *Pyruella* Petuch, 1982 (*P. rugosicostata* Petuch, 1982, Busyconidae); D= *Hystrivasum* Olsson and Petit, 1964 (*H. violetae* Petuch, 1994, Vasinae-Turbinellidae); E= *Heilprinia* Grabau, 1904 (*H. carolinensis* (Dall, 1892), Fasciolariidae); F, G= *Calusacypraea* Petuch, 1996 (*C. globulina* Petuch, new species, see Systematic Appendix, Cypraeidae); H= *Pyrazisinus* Heilprin, 1886 (*P. miamiensis* Petuch, 1994, Potamididae); I= *Terebraspira* Conrad, 1862 (*T. sparrowi* (Emmons, 1858), Fasciolariidae); J, K= *Akleistostoma* Gardner, 1948 (*A. crocodila* (Petuch, 1994), Cypraeidae). With the exception of *Pyrazisinus miamiensis*, which is from the Bermont Formation of southern Florida, all specimens are from the Tamiami Formation of the Everglades region.

Subsea only, Fasciolariidae), and *Seminoleconus* and *Calusaconus* (new genera, both Conidae; see Systematic Appendix). Of special importance concerning the Buckinghamian Subprovince was the discovery that it contained the single largest radiation of cowries (Cypraeidae) known from any one area on Earth (Petuch, 1997). At present, almost ninety species of cypraeids are known to have existed during the time span of the subprovince. These included large species radiations of the genera *Siphocypraea*, *S.* (*Pahayokea* n. subgen.), and *S.* (*Okeechobea* n. subgen.) (living in *Thalassia* beds), the genus *Akleistostoma* (living in *Thalassia* beds and soft-bottom lagoons), the genera *Calusacypraea* (Plate 5, F and G) and *C.* (*Myakkacypraea* n. subgen.) (confined to estuaries), and the genus *Pseudadusta* n. gen. (coral reefs, reef flats, and estuaries) (for descriptions of new taxa, see Systematic Appendix). Examples of all of these are shown in Chapters 7 and 8.

The provinciatonal nature of the Buckinghamian Subprovince is readily demonstrated by the previously-mentioned provinciatonal endemics and Duplinian influences and also by the presence of a large West Indian-Caribbean faunal component (Gatunian Province, discussed later in this chapter). Some of these, which are found nowhere else in the Caloosahatchian Province, include the gastropod genera *Jenneria* (Ovulidae), *Macrostrombus* (Strombidae), *Parametaria* (Columbellidae), "*Clypeomorus*" (probably an undescribed genus; Cerithiidae), *Malea* (Tonnidae), *Vitularia* (Muricidae), *Scabricola* (Mitridae), and *Bursa* (Bursidae) (see Petuch, 1994) and the coral genera and species (all of which are also found in the Dominican Republic) *Stylophora* (*S. affinis*) and *Pocillopora* (*P. crassoramosa*) (both Seriatoporidae), *Goniopora* (*G. jacobiana*) (Poritidae), *Astrocoenia* (*A. meinzeri*) (Astrocoeniidae), *Thysanus* (*T. excentricus*) (Faviidae), and the fungiids *Antillia* (*A. bilobata*) and *Placocyathus* (*P. variabilis*) (and many other Gatunian coral species; see Chapter 7).

The Jacksonbluffian Subprovince

Named for the Jackson Bluff Formation, Alum Bluff Group of northern Florida (Petuch, 1988, 1997), which contains the typical fauna, this subprovince was the shortest-lived of the Caloosahatchian subdivisions, encompassing only the Alaqua and Jackson Subseas. By late Piacenzian time, the rapidly-growing deltas of the proto-Apalachicola, Choctawhatchee, and Ochlockonee Rivers had completely infilled the entire Choctaw Sea basin, and the subprovincial index fauna became extinct. During the time of its existence, the Jacksonbluffian Subprovince extended from Florida, across the northern Gulf of Mexico, to the Texas coast. Deep well cores from Texas and Louisiana have been found to contain a brackish water fauna that resembles an impoverished version of the faunas of the Myakka and Satilla Lagoon Systems (Dall, 1913), demonstrating that the western part of the subprovince encompassed a giant estuary. The Choctaw Sea basin of the eastern part of the subprovince was under the influence of a current-driven upwelling system and had a cooler marine climate than did the Everglades Pseudoatoll to the south and the estuaries to the west. This cool, nutrient-rich water was incorporated into a branch of the Gulf Loop Current and formed a separate gyre within the Jackson Subsea (the Apalachee Gyre; Petuch, 1997). Based upon the presence of a Yorktownian-type, cool water fauna in the lower part of the Jackson Bluff Formation (the "*Ecphora* Zone"), the Apalachee Gyre was best-developed during the early Piacenzian (equivalent to the Buckingham Formation of the Okeechobee Group farther south). The lack of key tropical families in the "*Ecphora* Zone," such as the Strombidae, Cypraeidae, Melongenidae, Potamididae, and Vasinae-Turbinellidae, readily demonstrates the paratropical nature of the subprovince during that time. Later, during the late Piacenzian (in the "*Cancellaria* Zone" time), the Jackson Subsea warmed considerably and most of the key tropical families reinvaded the Choctaw

Sea Basin. This rewarming points to a change in the current structure of the Gulf of Mexico, with the cessation of permanent upwelling systems and the disappearance of the Apalachee Gyre.

Although sharing many molluscan genera and species with the Buckinghamian area, the Jacksonblufian Subprovince contained a large endemic component. Some of these taxa included a large radiation of the bivalve genera *Lirophora* (with *L. xesta* and at least five species in the *L. ulocyma* complex, Veneridae) and *Dallarca* (the *D. alumensis* complex, Arcidae) and the gastropods *Calliostoma aluminium* (Trochidae), *Eupleura miocenica* (Ocenebinae-Muricidae), and *Ptychosalpinx duerri* (Buccinidae). A colder water, Yorktownian influence is seen in the presence of the Williamsburg and Yorktown Subseas species *Sinum chesapeakensis* (Naticidae), *Volutifusus emmonsi* (Scaphellinae-Volutidae), and *Planecphora mansfieldi* (Ocenebrinae-Muricidae; see Systematic Appendix). The area was only marginally tropical even in the warmest times, as demonstrated by the presence of only one species, each, of the families Strombidae (*Strombus floridanus*), Vasinae-Turbinellidae (*Hystrivasum jacksonense*), and Cypraeidae (*Akleistostoma mansfieldi*) in the "*Cancellaria* Zone" (Mansfield, 1930, 1932).

The Gatunian Province

Named for the early Miocene-late Pliocene Gatun Formation of the Canal Zone, Panama (Petuch, 1982), which contains the typical fauna, this province extended from eastern Mexico and the West Indian Arc south to northern South America as far as central Brazil, and across the Bolivar Strait to southern California in the north and northern Peru in the south (Figure 11). Based upon the ranges of restricted endemic taxa, the Gatunian Province can be further subdivided into seven subprovinces; the *Guraban Subprovince* (extending from Cuba, the Bahamas, and Jamaica south to the Grenadines, Lesser Antilles), the *Veracruzan Subprovince* (eastern Mexico south to Honduras), the *Limonian Subprovince* (Honduras, Costa Rica, Panama, and northern Colombia, across the Isthmus of Panama and northward along western Mexico to the Gulf of California, and the then-emerging Galapagos Islands), the *Esmeraldan Subprovince* (southwestern Colombia, Ecuador, and northern Peru), the *Putagavilanian Subprovince* (Venezuela, Trinidad, to Suriname), the *Juruaian Subprovince* (the Brazilian coast from Para State to Cabo Frio, and the entire Amazonian Sea), and the *Imperialian Subprovince* (northern Gulf of California and the Imperial Valley area of California). The boundaries of the Gatunian province were sharply demarcated by the ranges of a large number of distinctive gastropod index genera, some of which include the cypraeid *Muracypraea* (Plate 6, A and B), the buccinids *Antillophos* (Plate 6, F), *Metaphos* (Plate 6, H), *Fusinosteira* (Plate 6, I), and *Pallacera*, the cancellariids *Euclia* (Plate 6, D) and *Pyruclia* (Plate 6, C), the conids *Protoconus* and *Pyruconus*, the volutid *Voluta* (s.s.) (Plate 6, G and J) (Limonian and Puntagavianian Subprovinces only), and the borsonine turrid *Paraborsonia*.

The Guraban Subprovince

Named for the Pliocene Gurabo Formation of the Dominican Republic, which contains the typical fauna, this eutropical subprovince was the descendant of the Anguillan Subprovince of the Baitoan Province and encompassed the diverse coral reef and carbonate environments of Cuba, Jamaica (Bowden Formation), and Hispaniola (including the well-known Cercado Formation) (Petuch, 1988) (Figure 11). While containing all the key tropical index taxa (Pilsbry, 1922; Woodring, 1928), the Guraban Subprovince also housed an interesting endemic fauna, some of which included a radiation of *Muracypraea* (including *M. porteronis* and *M. noueli*, both Cypraeidae), the genus *Globidrillia* (Turridae), a radiation of an undescribed reef-associated cerithiid genus similar to the Indo-Pacific genus *Clypeomorus* (including *"Cerithium" aduncus* and *"C." harrisi*), *Harpa americana* (Harpidae), the *Turbinella textilis* complex (Turbinellidae), and a large radiation of *Protoconus* (centered around *P. ultimus*, Conidae). Of special interest was the presence of the only-known American representatives of the Indo-Pacific genera *Labiostrombus* (*L. leurus*, Strombidae), *Puncticulis* (*P. pulicarius* complex), *Strioconus* (both Conidae), and *Naquetia* (*N. compactus*, Muricidae) (Petuch, 1988; E. Vokes, 1990). The Guraban Subprovince also shared numerous gastropod species with the Everglades Pseudoatoll reefs of the Buckinghamian Subprovince of the Caloosahatchian Province, including *Pachycrommium guppyi* (Ampullospiridae), *Eustrombus dominator* (ancestor of the Recent *E. gigas*, Strombidae), *Dermomurex olssoni* (Muricidae), *Xenophora dilecta* and *Tugurium imperforatum* (both Xenophoridae), *Modulus basileus* (Modulidae), *Astraea (Lithopoma) aora* (Turbinidae), and *Pseudocyphoma immunita* (Ovulidae) (Petuch, 1994). Some of these are illustrated in Chapter 7.

The Veracruzan Subprovince

Named for the Pliocene fossil beds (Medias Aguas and Conception Inferior Formations) of Veracruz State, Mexico (Figure 11), the fauna of this subprovince is still poorly studied and not well understood. The descendant of, and deriving most of its fauna from, the Agueguexquitean Subprovince of the Baitoan Province, the Veracruzan Subprovince contained a large component of endemic species, including *Chicoreus miltos* (Muricidae) (E. Vokes, 1990), the *Mitra (Tiara) almagrensis* complex (Mitridae), the *Conasprella almagrensis* complex (Conidae), and a complex of undescribed *Lindafulgur* species (in the *L. lindajoyceae* species group, descendants of *L. alencasterae*, Busyconidae). The Veracruzan Subprovince also shared a large number of taxa with the Guraban Subprovince, some of which include *Siratus dominguensis* (Muricidae) (E.Vokes, 1990), the turrids *Ilhycylhara ischna*, *Kurtziella stenotella*, and *Nannodiella amicta*, and the terebrids *Strioterebrum eleutheria* and *Oreoterebra subsulcifera* (Perrilliat, 1973).

The Limonian Subprovince

Named for the Pliocene Limon Formation of Caribbean Costa Rica, which contains the typical fauna, this eutropical subprovince was the descendant of the Culebran Subprovince of the Baitoan Province and straddled the Bolivar Strait (Petuch, 1988, 1997) (Figure 11). Also encompassing the faunas of the Tubera Group of northern Colombia, the upper Gatun Formation of Panama, and the Charco Azul, Armuelles, and La Vaca Formations of Pacific Costa Rica, the Limonian Subprovince exhibited a high level of endemism. Some of the more important endemic gastropods included a species complex of *Muracypraea* (examples being *M. isthmica* and *S. almirantensis*, Cypraeidae), a radiation of the genus *Sconsia* (including *S. bocasensis*, *S. gabbi*, and *S. cocleana*, Cassidae), a radiation of the genus *Metula* (including *M. harrisi*, *M. limonensis*, and *M. cancellata*, Buccinidae), the *Voluta alfaroi-V. eurytera* complex (Volutinae-Volutidae), *Scaphella (Clenchina) costaricana* (Scaphellinae-Volutidae), the *Fasciolaria* (un-named genus?) *gorgasiana* species complex (Fasciolariidae), the *Protoconus toroensis* complex (Conidae), and a radiation of large *Turbinella* species, including *T. scopulus* (the first of the heavily-knobbed *T. angulatus* complex) and *T. magdalenensis* (Turbinellidae). Of special interest within the Limonian Subprovince was the only known American representative of the eastern Atlantic-Mediterranean volutid genus *Ampulla* (*A. americana*, Ampullinae-Volutidae). The fauna of the early Pleistocene Moin Formation of Caribbean Costa Rica, having evolved after the closing of the Isthmus of Panama, contained numerous Limonian relictual taxa.

Plate 6. Gastropod index genera of the Gatunian Province. A, B= *Muracypraea* Woodring, 1959 (*M. isthmica* (Schilder, 1925), Cypraeidae); C= *Pyruclia* Olsson, 1932 (*P. diadela* Woodring, 1970, Cancellariidae); D= *Euclia* H. and A. Adams, 1854 (*E. codazzii* (Anderson, 1929), Cancellariidae); E= *Strombina lessepsiana* complex (*S. gatunensis* Toula, 1911, Columbellidae); F= *Antillophos* Woodring, 1928 (*A. gatunensis* (Toula, 1909), Buccinidae); G, J= *Voluta* Linnaeus, 1758 (*V. alfaroi* Dall, 1900, Volutinae-Volutidae); *H*= *Metaphos* Olsson, 1964 (*M.* un-named species, Buccinidae); I= *Fusinosteira* Olsson, 1932 (*F. protera* (Woodring, 1964), Buccinidae). All specimens are from the Gatun Formation of Panama.

The Esmeraldan Subprovince

Named for the well-known Pliocene Esmeraldas beds of Ecuador (now the Onzole Formation), which contains the typical fauna (Petuch, 1988, 1997), this subprovince was the descendant of the Subibajan Subprovince of the Baitoan Province and represented a provinciatone between the Gatunian and Proto-Peruvian Provinces. While containing a large component of Gatunian gastropod genera such as *Muracypraea* (*M. cayapa*, Cypraeidae), *Pyruconus* (*P. riosantiagensis*, Conidae), the buccinids *Metaphos* (*M. calathus*), *Antillophos* (*A. abundans*), and species radiations of the cancellariids *Pyruclia* (including *P. lacondamini*, *P. pycta*, and *P telembi*) and *Euclia* (including *E. maldonadoi* and *E. fragosa*), the Esmeraldan Subprovince also contained a large provinciatonal endemic component. Some of these, which are unique to the subprovince, include the elongated cancellariid genus *Marksella* (*M. jumala*), the turrid genus *Dallora* (*D. ecuadoria*), the fulgorariine volutid genus *Mysterostropha* (*M. gyrina*), and the elongated cassid genus *Neosconsia* (*N. ecuadoriana*). The paratropical nature of the Esmeraldan fauna is seen in the absence of the tropical indicator families Strombidae, Modulidae, Turbinidae, and Vasinae-Turbinellidae. The Esmeraldan Subprovince also housed a small component of relictual taxa from Subibajan time, including *Perunassa* (*P. bombax*) and *Gordanops* (*G. esmeraldensis*) (both Buccinidae; referred to *Calophos* by some authors, but here considered separate genera). Of special interest within the Esmeraldan subprovince was an endemic cold and deep water Japonic-North Pacific component composed of the turrid genera *Thatcheria* (*T. ecuadoriana*) and *Aforia* (*A. ecuadoriana*), the muricid genera *Genkaimurex* (*G. americanus*) and *Muregina* (*M. ecuadoria*), the trochid genus *Bathybembix* (*B. dalliana*), the buccinid *Kelletia* (*K. ecuadorium*), and the only known American fossil representative of the calliotectinine volutid genus *Teremachia* (*T. fischeri*) (Olsson, 1964; E. Vokes, 1988, 1989).

The Puntagavilanian Subprovince

Named for the Punta Gavilan Formation, Falcon State, Venezuela (Petuch, 1988), which contains the typical fauna, this eutropical province was the descendant of the Cantaurean Subprovince of the Baitoan Province. Also including the faunas of the Mare, Playa Grande, and Cubagua Formations of Venezuela (Weisbord, 1962; Gibson-Smith, 1973) and the Matura, Coubaril, and Talparo Formations of Trinidad (Jung, 1969), the Puntagavilanian Subprovince contained a large component of endemic gastropods. Some of these include a species complex of *Muracypraea* (with *M. grahami*, *M. rugosa*, and several undescribed species, Cypraeidae) (Ingram, 1947), *Chelyconus federalis* and *Pyruconus planitectum* (both Conidae), *Malea mareana* (Tonnidae), a species complex of *Pleuroploca* (sensu stricto, in the *P. salmo* group, with *P. crassinoda* and *P. turamensis*, Fasciolariidae), *Voluta cubaguensis* (ancestor of the Recent Brazilian *V. ebraea*, Volutinae-Volutidae), the last-living species of the agglutinating turritellid *Springvaleia* (*S. secunda*), the only American member of the Indo-Pacific genus *Haustellum* (*H. mimiwilsoni*, Muricinae-Muricidae), and a radiation of *Parametaria* (with *P. prototypus*, *P. schilderi*, *P. rutschi*, and several undescribed species, Columbellidae). Of special interest within the Puntagavilanian Subprovince was the only known American pre-Pleistocene *Charonia* species (*C. lampas weisbordi*, Ranellidae).

The Juruaian Subprovince

Named for the fossil beds along the Jurua River, Amazonas State, Brazil, which contain the typical fauna, this subprovince was composed of two main components; a coastal Brazilian component (of which little is known due to the lack of fossil beds) and the

Amazon Sea component (Figure 11). From the mid-Miocene to the early Pliocene, the Amazon Valley was alternately filled with fresh water lakes or flooded with seawater, and brackish estuarine conditions often existed as far westward as the Peruvian-Brazilian border (Petuch, 1988; Vonhoe, Wesselingh, and Ganssen, 1998; Vermeij and Wesselingh, 2002). This immense embayment, the Amazon Sea, contained its own distinctive endemic fauna, unlike any other known from the Gatunian Province. Collections made along the Yavari and Jurua Rivers and Canama, Brazil at the headwaters of the Amazon (Etheridge, 1924) and in the Pebas Formation of Colombia and Peru (Vermeij and Wesselingh, 2002) revealed an unusual estuarine molluscan fauna, unlike any known from elsewhere in the western Atlantic. Some of the Juruaian endemics included a large radiation of the genus *Neritina* (with *N. ortoni, N. puncta, N. ziczac,* and many undescribed species, Neritidae), a radiation of an undescribed *Tympanotonus*-type potamidid (*"Cerithium" coronatum* complex), a large radiation of the genus *Hemisinus* (with *H. sulcatus, H. septecinctus,* and several undescribed species, Thiaridae), the ornate *Melongena woodwardi* (Melongenidae, originally described as a *Thais*), a radiation of the endemic hydrobiid genus *Dyris*, several undescribed species of *Odostomia* (Pyramidellidae), the thin and smooth-shelled *Nassarius reductus* (Nassariidae), several undescribed species of *Finella* (Cerithiidae), a radiation of the genus *Melanopsis* (with *M. browni, M. bicarinata, M. tricarinata, M. scarioides,* and many undescribed species, Melaniidae), and pachydontine corbiculid bivalves in the genera *Panamicorbula* and *Anticorbula*. By the early Pliocene, the Amazon Sea was almost completely infilled with sediments from the eroding Andes Mountains and the Juruaian fauna ceased to exist.

The Imperialian Subprovince

Named for the Imperial Formation (with its Latrania and Coyote Mountain Members) of Imperial County, California (Hanna, 1926; Petuch, 1988), which contains the typical fauna, this northernmost Gatunian subprovince occupied an immense, tropical embayment (the Imperial Sea) that was separated from the Pacific Ocean by the Baja Californian Peninsula and the coastal mountain ranges of southern California. Within the Imperial Sea, which extended northward from the present Gulf of California into the Californian Central Valley and Colorado Desert area, eutropical conditions existed and this allowed for the development of extensive reef complexes of endemic, Caribbean-type corals. Some of these included *Dichocoenia merriami* and *D. crassisepta*, and *Eusmilia carrizensis* (all Eusmiliidae), *Solenastrea fairbanki, S. minor, S. columnaris,* and *S. normalis* (Montastreidae), *Siderastrea californica, S. mendenhalli,* and *S. minor* (Agariciidae), *Porites carrizensis* (Poritidae), and *Meandrina bowersi* (Faviidae) (Hanna, 1926). The Caribbean appearance of the Imperial Sea was further enhanced by the presence of the echinoid *Clypeaster* and the gastropods *Cassis* (*C. subtuberosa*, Cassidae, which also has been found in Esmeraldan Subprovince), *Muracypraea* (*M. amandusi*, Cypraeidae), and *Spuriconus durhami* (closely related to the Buckinghamian-Guraban *S. yaquensis*) and *S. brankampi* (both Conidae). Other endemic gastropods, some of which are ancestral to species in the Recent Panamic Province, included *Taenioturbo mounti* (Turbinidae), *Strombus obliteratus* (Strombidae), *Vasum pufferi* (Vasinae-Turbinellidae), and *Triplofusus* un-named species (Fasciolariidae).

At the same time that the eutropical reef systems were flourishing within the Imperial Sea, the outer Pacific coast of California had a much colder, temperate marine climate. This is readily demonstrated by the faunas of the Pliocene Etchegoin Group of the San Joaquin Valley embayment of California (Nomland, 1917), which lack all the tropical index taxa and which contain cold water, North Pacific genera such as the gastropods *Neptunea* (Neptuneidae), *Boreotrophon* (Trophoninae-Muricidae), *Forreria, Ocenebra,* and *Pteropurpura* (all Ocenebrinae-Muricidae), *Boreoscala* (Epitoniidae), and *Megasurcula* (Turridae), and the bivalve *Swiftopecten* (Pectinidae). The Imperial Sea must have been open only in the south, in the tropical Limonian region along Mexico, and was closed in the north, with no connections to the Pacific Ocean. This allowed a warm-water offshoot of the North Equatorial Current to produce a gyre within the Imperial Sea, producing eutropical conditions until the late Pliocene.

The Camachoan Province

Named for the late Miocene-Pliocene Camacho Formation of Uruguay (Figueras and Broggi, 1985), which contains the typical fauna, this southern province was originally described as a subprovince of the Gatunian Province (Petuch, 1988) and extended from Cabo Frio, Brazil south to the Mar de Plata area of Uruguay and Argentina (Figure 11). Although little is known about this South Atlantic paratropical province, enough data now exists (mostly from well cores in Parana and Rio Grande do Sul States, Brazil and from Buenas Aires, Argentina) to show that the Camachoan Province contained a high enough level of endemism to warrant full provincial status. The Camachoan fauna was a mixture of both tropical groups and those with cold water Magellanic and subantarctic affinities. Some of the tropically-derived endemics included *Lamniconus patagonicus* and *L. iheringi* (both Conidae), *Strioterebrum calcaterrai* (Terebridae), *Turricula rebuffoi* (Turridae) and the *Torcula iheringiana* species complex (Turritellidae). Cold water, subantarctic groups included *Pleurotomella parodizi* (Turridae), the *Trophon*-like "*Paziella*" *eliseoduartei* complex (Muricidae), and the *Adelomelon patagonicus* complex (Volutidae). Because of the lack of data and well-preserved fossil beds, subprovincial subdivisions of the Camachoan Province cannot be determined at this time.

Marine Provinces of early Pleistocene to Recent Time

The final provincial reorganization, and the accompanying extinctions of index taxa, took place during the Aftonian Stage of the Pleistocene (see Chapter 10). Prior to this time during the Calabrian, a series of worldwide cooling events was initiated and the resultant glaciations and eustatic drops became progressively more severe and sequentially closer together. This culminated in a prolonged major glaciation during the Nebraskan Stage. At that time, the Caloosahatchian, Gatunian, and Camachoan Provinces disappeared and four new provinces took their places. These included the western Atlantic *Carolinian Province*, the *Caribbean Province*, and the *Brazilian Province* (Figure 12) (Petuch, 1988, 1997), and the Eastern Pacific *Panamic Province*. After the late Pliocene closing of the Isthmus of Panama, the Gatunian faunas that were isolated along western Central America and northwestern South America formed the Panamic Province. Although this Caribbean-appearing province contains many Limonian, Guraban, and Puntagavilanian relict taxa (Woodring, 1965; Petuch, 1982; Vermeij and Petuch, 1986), it falls outside the scope of this book and will not be discussed further. The modern eastern North American oceanic current structure was established by Aftonian time, with a northward-flowing warm Gulf Stream and a southward-flowing cool countercurrent converging along the coast of the Mid-Atlantic States. The only major variable during the Pleistocene was the convergence point of these two currents, which varied from the Delmarva Peninsula area during some warm interglacial times to Cape Fear during some cold glacial times. For most of the Pleistocene stages (and in the Recent), however, the two currents converged at Cape Hatteras. These provinces, which have been discussed at length in the literature of the Recent marine molluscan faunas (Valentine, 1973; Petuch, 1988, 1997; Briggs, 1995), are only briefly outlined here for comparison with the older, ancestral provincial boundaries.

The Carolinian Province

Named for the Carolinas, which contain the typical fauna, this province extends from Cape Cod, Massachusetts, southward around Florida, and into the Gulf of Mexico as far as Cabo Catoche, Yucatan, Mexico (Figure 12) (Petuch, 1997). Based upon the ranges and percentages of restricted endemic taxa, the Recent Carolinian Province can be further subdivided into six subprovinces; the *Virginian Subprovince* (extending from Cape Cod to Cape Hatteras, and including a relict fauna on Sable Island, Nova Scotia), the *Georgian Subprovince* (extending from Cape Hatteras to Palm Beach, Florida), the *Floridian Subprovince* (extending from Palm Beach to Marco Island, Florida), the *Suwannean Subprovince* (extending from Naples, Florida to the Mississippi Delta), the *Texan Subprovince* (extending from the Mississippi Delta to Tampico, Mexico), and the *Yucatanean Subprovince* (extending from Tampico to Cabo Catoche, Yucatan, Mexico). The boundaries of the Carolinian Province are demarcated by the ranges of a large number of endemic gastropod taxa, some of which include the entire family Busyconidae (genera *Busycon, Lindafulgur, Busycoarctum, Sinistrofulgur, Fulguropsis,* and *Busycotypus*), the genus *Rexmela* (Melongenidae), the genera *Scaphella* (s.s.) and *Aurinia* (Scaphellinae-Volutidae) (Caribbean "*Scaphella*" species belong to the subgenus *Clenchina*), the genera *Urosalpinx* and *Vokesinotus* (Ocenebrinae-Muricidae), *Hexaplex fulvescens* (Muricinae-Muricidae), and *Macrocypraea cervus* (Cypraeidae), and the bivalve genus *Mercenaria* (Veneridae). The coral reef-dominated Floridian Subprovince is a provinciatone, sharing many eutropical taxa with the Caribbean Province and containing a large component of provinciatonal endemics (Petuch, 1987). From Aftonian to Sangamonian Pleistocene time, the eastern United States coast contained three transitional Carolinian subprovinces; the *Canepatchian Subprovince* (ancestor of the Virginian and Georgian Subprovinces), the *Bermontian Subprovince* (the ancestor of the Floridian Subprovince), and the *Manatean Subprovince* (ancestor of the Suwannean Subprovince) (Petuch, 1997). All three of these had, essentially, a modern fauna but contained some now-extinct relict Caloosahatchian taxa.

The Caribbean Province

Named for the Caribbean Sea basin, which contains the typical fauna, this province extends from Cabo Catoche, Yucatan, Mexico, to Cuba and the Bahamas, the Antilles Arc, eastern Central America, northern South America as far as the Amazon River mouth, and to the isolated Bermuda Islands (Figure 12) (Petuch, 1988, 1997). Based upon the ranges and percentages of restricted endemic taxa, the Caribbean Province can be further subdivided into seven subprovinces; the *Bahamian Subprovince* (Bahamas Banks, Turks and Caicos Banks, and northern Cuba; named for the Bahamas), the *Bermudan Subprovince* (Bermuda), the *Antillean Subprovince* (eastern coast of the Yucatan Peninsula and the Belizean Barrier Reef, southern Cuba, Jamaica, the Greater Antilles, and the Virgin Islands; named for the Antilles Arc), the *Nicaraguan Subprovince* (Bay Islands of Honduras to the Gulf of Uruba, Colombia-Panama border, and the central Caribbean banks; named for Nicaragua), the *Venezuelan Subprovince* (Gulf of Uruba to Guyana, and including Aruba, Curacao, and Bonaire, the Venezuelan coastal islands, and Trinidad; named for Colombia, the *Grenadian Subprovince* (Anguilla to Grenada, and including Tobago and Barbados; named for Grenada and the Grenadines), and the *Surinamian Subprovince* (Guyana, Suriname, and Cayenne, to Cabo Norte, Brazil; named for Suriname). The boundaries of the province are demarcated by the ranges of gastropod index taxa, some of which include *Eustrombus gigas* (Strombidae), *Cassis flammea* (Cassidae), *Charonia variegata* (Ranellidae), the genera *Cenchritis* and *Tectininus* (both Littorinidae), *Puperita* (Neritidae), the *Oliva (Strephona) reticularis* species complex and the genera *Cariboliva* and *Jaspidella* (all Olividae), *Vasum murica-*

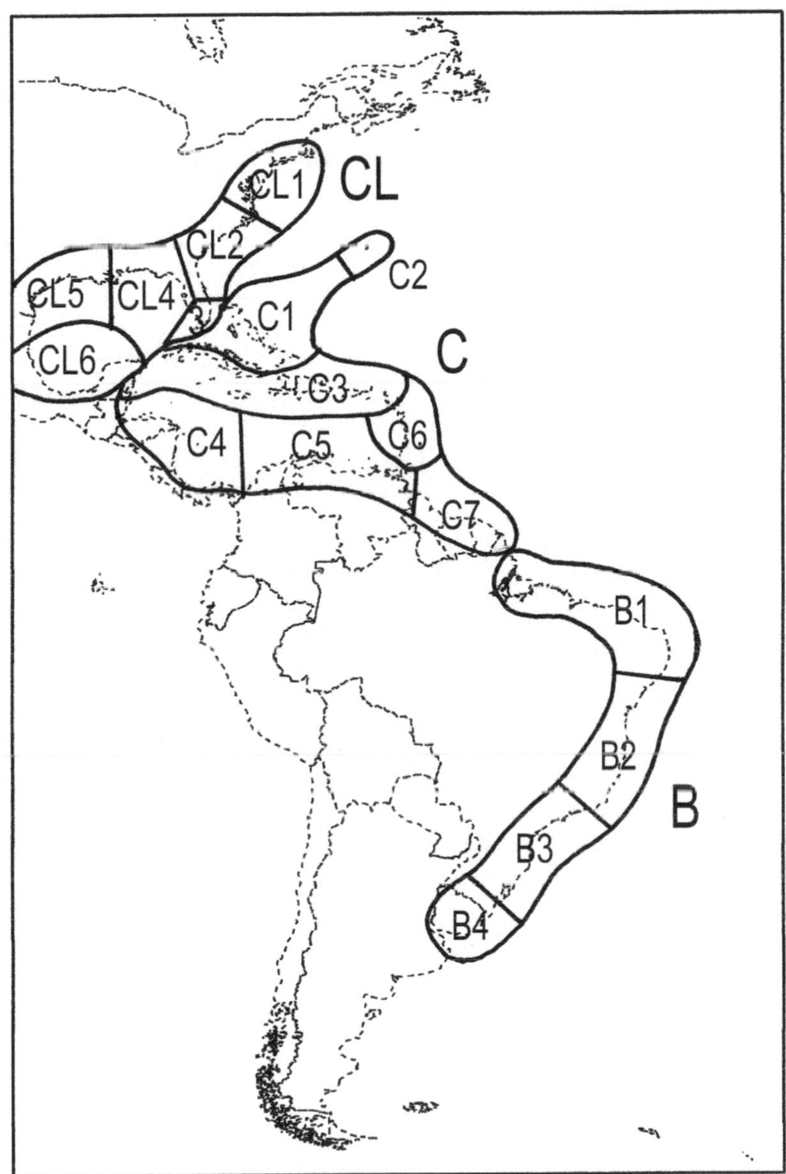

Figure 12. Configuration of the late Pleistocene and Recent marine molluscan provinces and subprovinces of the western Atlantic and Eastern Pacific, superimposed upon the outline of the Americas. Biogeographical units include: CL= Carolinian Province, CL1= Virginian Subprovince, CL2= Georgian Subprovince, (CL)3= Floridian Subprovince, CL4= Suwannean Subprovince, CL5= Texan Subprovince, CL6= Yucatanean Subprovince, C= Caribbean Province, C1= Bahamian Subprovince, C2= Bermudan Subprovince, C3= Antillean Subprovince, C4= Nicaraguan Subprovince, C5= Venezuelan Subprovince, C6= Grenadian Subprovince, C7= Surinamian Subprovince, B= Brazilian Province, B1= Cearaian Subprovince, B2= Bahian Subprovince, B3= Paulinian Subprovince, B4= Uruguayan Subprovince.

tum and *V. capitellum* (both Vasinae-Turbinellidae), *Melongena melongena* (Melongenidae), *Jaspidiconus jaspideus*, *Stephanoconus regius*, *Hermes granulatus*, and *Chelyconus ermineus* (Conidae) and the conid genera *Purpuriconus* and *Protoconus*, and the bivalve genus *Lindapecten* (Pectinidae). The southern Caribbean subprovinces contain many relictual Gatunian influences and these are discussed in Chapter 10.

The Brazilian Province

Named for Brazil (Figure 12), this area was only recently recognized as having full provincial status (Petuch, 1988). Previously, it had been considered to be simply a subregion of the Caribbean Province (Warmke and Abbott, 1962), but more detailed studies have demonstrated the presence of a large endemic component (Petuch, 1987, 1988; Rios, 1994). Based upon the ranges and percentages of restricted endemic taxa, the Brazilian Province can be further subdivided into four subprovinces: the *Cearaian Subprovince* (Cabo Norte to northern Bahia State; named for Ceara State, Brazil), the *Bahian Subprovince* (northern Bahia State to Cabo Frio, and the offshore Abrolhos Islands; named for Bahia State), the *Paulinian Subprovince* (Cabo Frio to southern Rio Grande do Sul State; named for Sao Paulo State), and the *Uruguayan Subprovince* (southern Rio Grande do Sul State, Brazil to Uruguay and the Mar de Plata; named for Uruguay). The Brazilian provincial boundaries are sharply demarcated by the ranges of a large number of endemic taxa, some of which include the gastropods *Vasum cassiforme* and *Turbinella laevigata* (both Turbinellidae), the genus *Lamniconus* (Conidae), the genus *Bullata* (Marginellidae), the genus *Northia* (the only Atlantic species, Buccinidae), the *Oliva (Strephona) circinata* species complex (Olividae), *Pleuroploca aurantiaca* (Fasciolariidae; the only true living *Pleuroploca* s.s. in the Atlantic), the odd volutid subgenus *Plochelaea (Plicoliva)* (*P. zelindae*, Plicolivinae) and genus *Odontocymbiola* (Odontocymbiolinae), and *Strombus worki* and *Titanostrombus goliath* (Strombidae). Although having a small Caribbean influence in the northern subprovinces, the Brazilian Province fauna more closely resembles that of the Recent Panamic Province and contains many Gatunian relicts. The Brazilian offshore island complexes of Fernando de Noronha and Atol das Rocas, with *Malea (Quimalea)* (Tonnidae) and endemic members of the *Nerita ascensionis* species complex (Neritidae), may constitute a separate subprovince.

Chapter 3. Oligocene and Earliest Miocene Seas

Of the nine known eastern North American subseas of Oligocene and earliest Miocene age (Rupelian-early Aquitanian), only three contain fossiliferous formations that are well-preserved enough to allow for paleoceanographic analyses. These include the Bainbridge Subsea of the Choctaw Sea (Rupelian-early Chattian age), represented by the Suwannee Formation, the Silverdale Subsea of the Charleston Sea (late Chattian-early Aquitanian), represented by the Belgrade Formation, and the Tampa Subsea of the Okeechobean Sea (late Chattian-early Aquitanian), represented by the Tampa Formation. The River Bend Subsea of the Albemarle Sea (Rupelian-early Chattian) and the Old Church Subsea of the Salisbury Sea (late Chattian-early Aquitanian) also contain fossiliferous formations, but these are characteristically leached, poorly preserved, or moldic, making any detailed ecological studies virtually impossible.

Communities and Environments of the Bainbridge Subsea

The oldest paleosea to have well-preserved fossil beds, the Bainbridge Subsea, was the most complex of all the eastern North American pericontinental seas. The dominant geomorphological features of the Bainbridge Subsea were Orange Island on the Florida Platform (Vaughan, 1919: 267), the wide Suwannee Strait that separated Orange Island from southern Georgia, and the Flint River Embayment (Figure 13). All three areas were dominated by high-tropical coralline environments, including two main zonated coral reef tracts, the Flint River Reef System along the Flint River Embayment (see Dall, 1916 and Vaughan, 1919) and the Orange Island Reef System along the western coast of Orange Island (see Vaughan, 1919 and Mansfield, 1937) (Figure 13). The Suwannee Formation, deposited in the areas of the Suwannee Strait and the western and southern coasts of Orange Island, contains the best-preserved and richest fossil faunas from Rupelian time (Petuch, 1997). This formation will be the main source of the ecological data reported here. Most of the fossils in the Flint River and Suwannee Formations are siliceous pseudomorphs, having been completely replaced by agate.

Six distinct shallow water tropical marine communities have been preserved within the Suwannee Formation. These include (from the shoreline, across the reef lagoon, to the coral reef tract): the *Telescopium hernandoense* Community (mangrove coastline mud flats), the *Lucina perovata* Community (*Thalassia* (Turtle Grass) beds within the reef lagoon), the *Orthaulax hernandoensis* Community (shallow carbonate sand bottom areas within the reef lagoon), the *Apicula bowenae* Community (turritellid beds in deeper water carbonate sand bottom areas), the *Stylophora minutissima* Community (sheltered back reef and lagoonal bioherms), and the *Goniopora decaturensis* Community (reef platform). Based upon mixed assemblages encountered in the Suwannee Formation (Petuch, 1997), the *Lucina perovata* (Turtle Grass), *Orthaulax hernandoensis* (open sand), and *Stylophora minutissima* (coral bioherm) Communities, and their accompanying biotopes and lithofacies, interfingered within the Orange Island lagoons.

The Telescopium hernandoense Community

The western edge of Orange Island was fringed with extensive forests of mangroves and intertidal mud flats and housed the *Telescopium hernandoense* Community. This

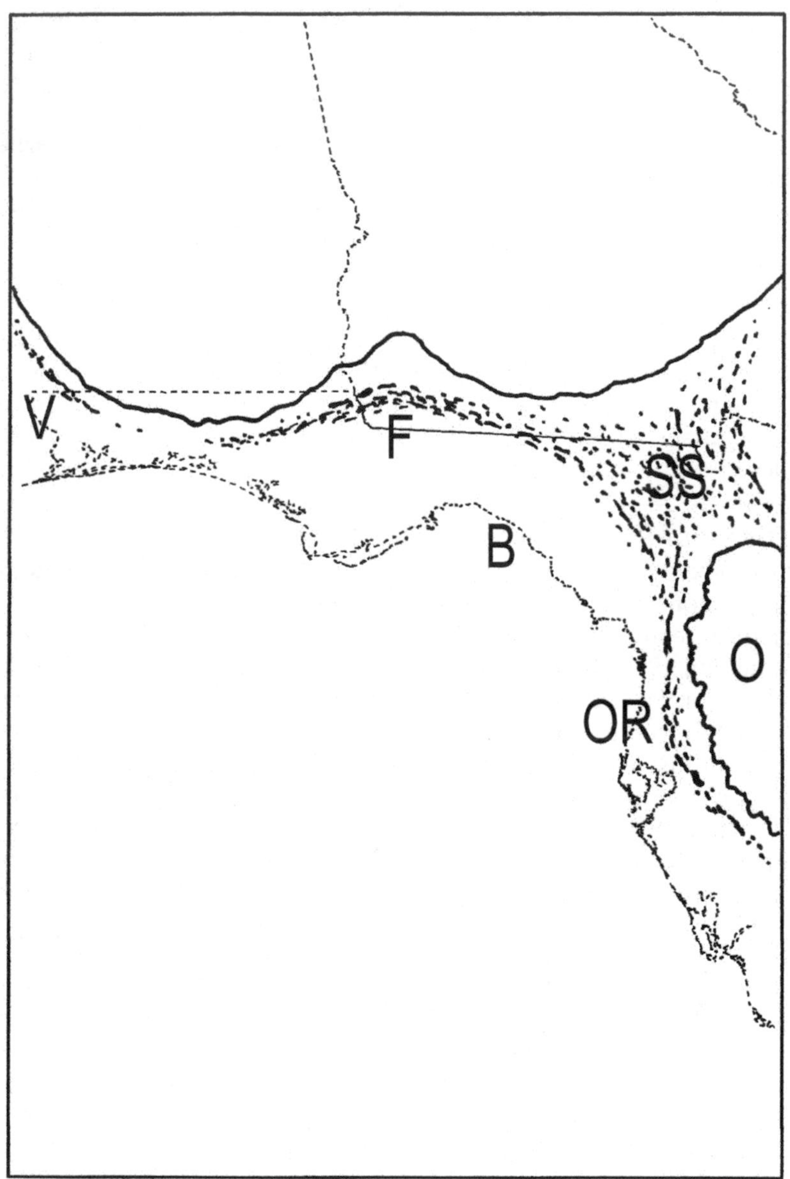

Figure 13. Approximate configuration of the coastline of the northeastern Gulf of Mexico during Rupelian to earliest Chattian Oligocene time, superimposed upon the outline of the Recent Gulf coast. Prominent features included: B= the Bainbridge Subsea of the Choctaw Sea, SS= the Suwannee Strait (also referred to as the Gulf Trough), F= Flint River Embayment and Reef Tract, O= Orange Island, OR= the Orange Island Reef tract, V= Vicksburg Sea (out of the scope of this book). At this time, the Suwannee Strait was very shallow and filled with coralline complexes composed of both bioherms and zonated reef tracts (speckling on map). The western coast of Orange Island, protected by the offshore reefs, was fringed with dense mangrove forests and mud flats. A wide, shallow lagoon separated the Orange Island Reef Tract from the coastal mangrove swamps. The coral reef complexes extended across the Suwannee Strait and Flint River Embayment to the Alabama coast of the Vicksburg Sea (Salt Mountain Formation).

muddy habitat, which closely resembled that of Recent coastal northwestern Australia (Petuch, 1997), supported immense beds of several potamidid gastropods, including *Telescopium hernandoense* (Mansfield, 1937) (Plate 7, E), *Telescopium blackwaterense* (Mansfield, 1937) (Plate 7, D), *Potamides* (?) *cookei* (Dall, 1916), and the oldest-known *Pyrazisinus, P. kendrewi* Petuch, 1997 and the bizarre, *Alaba*-like cerithiid gastropod *Prismacerithium prisma* Petuch, 1997 (Plate 7, I). Like their living relatives, these amphibious gastropods fed on algal and bacterial films that grew on the tidally-exposed mud and mangrove roots. The Orange Island endemic bubble shell *Suwannescapha lindae* Petuch, 1997 (Plate 7, H) also was an abundant animal in this muddy habitat. Scattered along the mud flats, in close proximity to the mangrove forests, were large numbers of the tubiform teredinid bivalve *Kuphus* species, large clumps of the mytilid mussel *Modiolus* cf. *silicatus* Dall, 1898, and oyster bars composed of *Ostrea paroxis* Lesueur in Dockery, 1984 (Plate 7, A) and *Ostrea vicksburgensis* Conrad, 1848 (Plate 7, J). Feeding on the oysters, as in Recent Australia, were several large carnivorous gastropods, including the melongenid *Myristica crassicornuta* (Conrad, 1848) (Plate 7, B and C) and the muricid *Pterynotus propeposti* (Mansfield, 1937) (Plate 7, F and G). Similar mangrove and mud flat-associated assemblages have been found in the lagoons behind the Flint River Reef System (Flint River Formation; see Dall, 1916). Here, two other *Telescopium* species, *T. halense* (Dall, 1916) and *T. vaughani* (Dall, 1916), occured together on mud flats near oyster bars.

The *Lucina perovata* Community

Farther out in the lagoon, in sheltered, shallow water areas (1-5 m) with normal salinity, were the Turtle Grass (*Thalassia*) beds, typified by the *Lucina perovata* Community. Like similar Turtle Grass communities found in the Recent Caribbean Sea, especially those on Roatan Island, Honduras (Petuch, 1997), several characteristic small gastropods lived on the individual grass blades. Feeding on epiphytic algae and diatoms were the modulid *Modulus liveoakensis* Mansfield, 1937 (Plate 8, I), the turbinid *Astraea (Astralium) polkensis* Petuch, 1997, and a host of small cerithiids, including *Cestumcerithium pascoensis* (Mansfield, 1937) (Plate 8, E), *C. liveoakensis* (Mansfield, 1937) (Plate 8, F), *C. brooksvillensis* (Mansfield, 1937), and *Cerithium (Thericium) insulatum* Dall, 1916. Feeding on epibionts such as ectoprocts, hydrozoans, and sponges were small carnivorous gastropods, especially the marginellids *Persicula suwanneensis* Petuch, 1997 (Plate 8, D), *Persicula macneili* Petuch, 1997, *Dentimargo dalli* Petuch, 1997, and *Hyalina silicifluvia* (Dall, 1916). Living buried within the sediments of the *Thalassia* rhizome root mass were numerous characteristic grass bed bivalves, including *Lucina perovata* (Dall, 1916) (the community namesake) (Plate 8, G), *Chione bainbridgensis* Dall, 1916 (Plate 8, J), *Trachycardium brooksvillense* (Mansfield, 1937) (Plate 8, A), and *Semele* cf. *silicata* Dall, 1900 (Plate 8, H). Feeding on tubiculous polychaetes and enteropneustes was the largest gastropod of the *Thalassia* beds, *Turbinella suwannensis* (Mansfield, 1937) (Plate 8, B). Also resident in the sea grass beds were large sirenians such as manatees and dugongs (Plate 8, C), whose ribs are commonly encountered in *Thalassia* facies of the Suwannee Formation.

The *Orthaulax hernandoensis* Community

As in Recent tropical western Atlantic and Indo-Pacific reef lagoons, large areas of open carbonate sand bottom were adjacent to the Turtle Grass beds, particularly in tidal channels. This open sea floor biotope supported a third rich and diverse ecosytem, with the strombid gastropod *Orthaulax hernandoensis* Mansfield, 1937 (the community namesake, Plate 9, B and C) being the most abundant and conspicuous organism. As in the genus *Strombus* today, *O. hernandoensis* occurred in large aggregations on the sand surface,

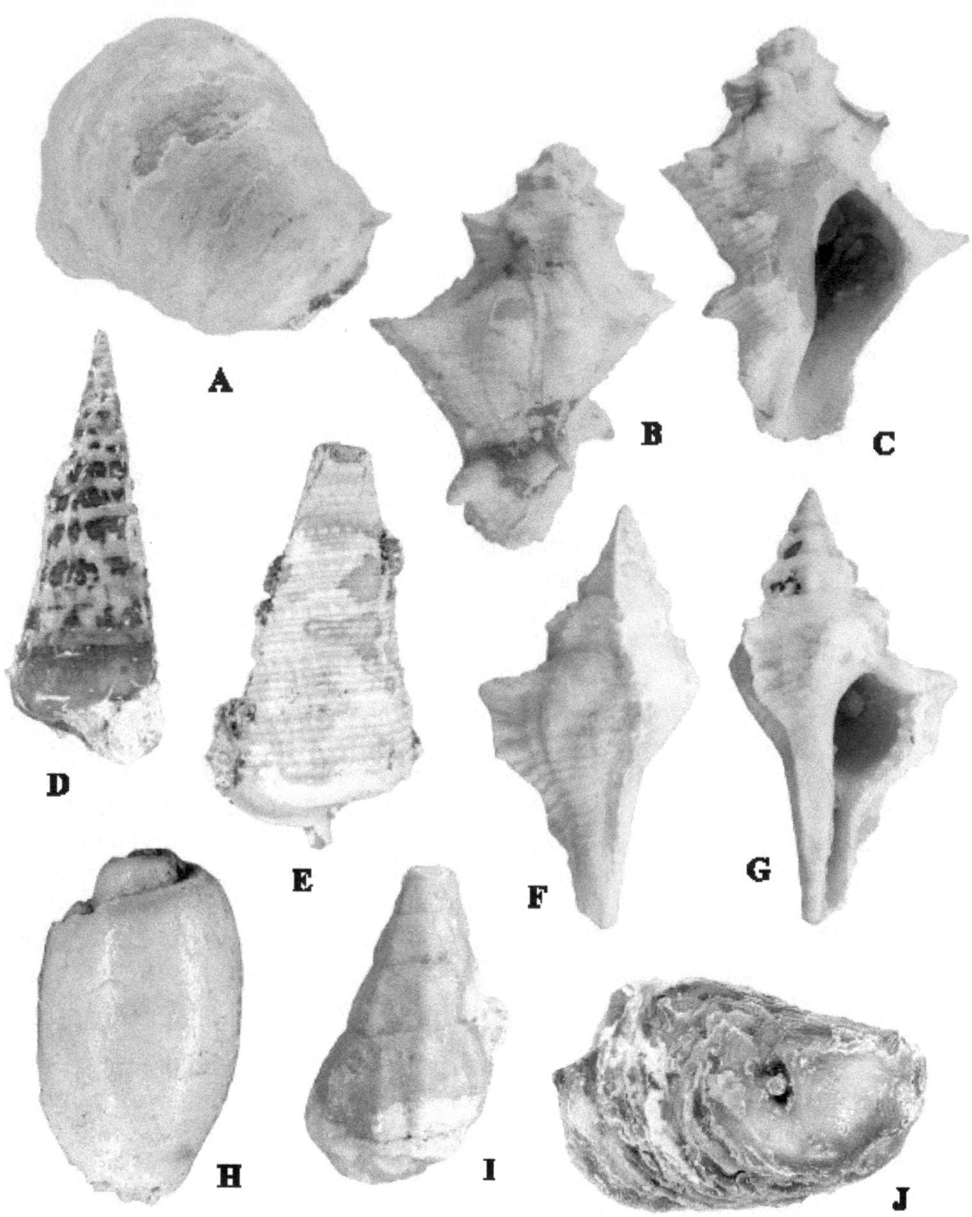

Plate 7. Index fossils and characteristic organisms of the *Telescopium hernandoense* Community (mud flat and mangrove forest environment) of the Bainbridge Subsea. Included are: A= *Ostrea paroxis* Lesueur in Dockery, 1984, length 73 mm; B,C= *Myristica crassicornuta* (Conrad, 1848), length 49 mm; D= *Telescopium hernandoense* (Mansfield, 1937), length 49 mm; E= *Telescopium blackwaterense* (Mansfield, 1939), length 43 mm; F,G= *Pterynotus propeposti* (Mansfield, 1937), length 35 mm; H= *Suwanneescapha lindae* Petuch, 1997, 18 mm; I= *Prismacerithium prisma* Petuch, 1997, length 19 mm; J= *Ostrea vicksburgensis* (Conrad, 1848), length 68 mm. All specimens are replaced by agate and were collected at the Terramar Pit in rural Polk County, near Zephyrhills, Florida.

Plate 8. Index fossils and characteristic organisms of the *Lucina perovata* Community (Turtle Grass bed environment) of the Bainbridge Subsea. Included are: A= *Trachycardium brooksvillense* (Mansfield, 1937), length 28 mm; B= *Turbinella suwannensis* (Mansfield, 1937), length 107 mm; C= Sirenian rib fragment, length 130 mm; D= *Persicula suwanneensis* Petuch, 1997, length 10 mm; E= *Cestumcerithium pascoensis* (Mansfield, 1937), length 37 mm; F= *Cestumcerithium liveoakensis* (Mansfield, 1937), length 24 mm; G= *Lucina perovata* (Dall, 1916), length 32 mm; H= *Semele* cf. *silicata* Dall, 1900, length 41 mm; I= *Modulus liveoakensis* Mansfield, 1937, length 15 mm; J= *Chione bainbridgensis* Dall, 1916. All specimens are replaced by agate and were collected at the Terramar Pit in rural Polk County, near Zephyrhills, Florida.

feeding on surficial algal films. Co-existing with the *Orthaulax*, and also feeding on surficial algal films, were the seraphsid stromboidean *Terebellum hernandoensis* Mansfield, 1937 and the large cerithiid gastropod *Cerithioclava eutextile* (Dall, 1916) (Plate 9, J). The glycymerid bivalve *Glycymeris suwanensis* Mansfield, 1937) (Plate 9, F) was the major infaunal organism, forming immense, shallowly buried beds. This small bivalve underwent heavy predation by a host of carnivorous gastropods, including the naticids *Naticarius caseyi* (MacNeil, 1984) and *Amauropsis mansfieldi* (Petuch, 1997) (Plate 9, G) and the busyconids *Spinifulgur gemmulatum* Petuch, 1997 and *Spinifulgur* (?) *proterum* (Gardner, 1944). The irregular echinoids *Rhyncholampas gouldii* (Bouve, 1846) (Plate 9, A) and *R. ericsoni* (Fischer, 1951) were especially common on the shallow sand bottoms and were the main food source for the echinoderm-feeding cassid gastropods *Cassis flintensis* Mansfield, 1940 (Plate 9, E) and *Phalium caelatura* (Conrad, 1848) (Plate 9, I). The other carnivorous gastropod group that was most abundant on the open bottoms was the sand-burrowing Olividae, including both *Omnogymna brooksvillensis* (Mansfield, 1937) (Plate 9, D) and *Olivella liveoakensis* Mansfield, 1937). Similarly today, shallow carbonate sand areas in reef lagoons on the Bahama Banks, the Philippines, and Indo-Pacific islands house large swarms of olivids of many genera. The ampullospirid gastropod *Pachycrommium dalli* Petuch, 1997 (Plate 9, H) also occurred on these carbonate banks, probably feeding on algal growths. This same community also occurred in the Flint River lagoon system (Dall, 1916).

The Apicula bowenae Community

In the deepest areas of the reef lagoon (10-30 m), on open carbonate sand bottoms, was a fourth community centered around the suspension-feeding turritellid gastropods *Apicula bowenae* (Mansfield, 1937) (the community namesake, Plate 10, I), *Torcula mississippiensis* (Conrad, 1848) (Plate 10, J), and *Torcula caseyi* (MacNeil, 1984). Like the lagoons along the Recent Gulf of Venezuela (Petuch, 1976), this biotope housed immense turritellid beds that covered large areas of sea floor. These virtually sessile gastropods served as prey items for a number of large, active, carnivorous gastropods including the ficid *Ficus mississippiensis* Conrad, 1848 (Plate 10, F) and the muricid *Poirieria (Dallimurex) rufirupicolus* (Dall, 1916) (see Petuch, 1997). Abundant polychaete annelid worms and nemerteans occurred within the turritellid beds, as evidenced by the presence of a large number of vermivorous gastropods, including the conoideans *Gradiconus cookei* (Dall, 1916) (Plate 10, A), *G. alveatus* (Conrad, 1865), *Conorbis porcellanus* (Conrad, 1848) (Plate 10, E), *Terebrellina divisura* (Conrad, 1848) (Plate 10, H), and *Pleurofusia brooksvillensis* (Mansfield, 1937), and the volutoideans *Fusimitra conquisita* (Conrad, 1848) (Plate 10, D), *Clavolithes vicksburgensis* (Conrad, 1848) (Plate 10, C), and *Conomitra staminea* (Conrad, 1848). Scattered among the turritellid beds were small shoals of the mobile, gregarious pectinid bivalves *Dimarzipecten brooksvillensis* (Mansfield, 1937) (Plate 10, B) and *Leptopecten flintensis* (Mansfield, 1937). The large ampullospirid gastropod, *Ampullinopsis flintensis* (Mansfield, 1937) (Plate 10, G) also occurred on sand patches between the turritellid beds, feeding on algae.

The Stylophora minutissima Community

Farther from shore, the water became shallower along the quiet back-reef areas of the Flint River, Orange Island, and Suwannee Straits Reef Tracts, providing the perfect habitat for the development of forests of delicate branching corals, the fifth environmental type. Here, the seriatoporid *Stylophora minutissima* Vaughan, 1900 (namesake of the community, Plate 11, J) and the acroporids *Acropora panamensis* Vaughan, 1919 (Plate 11, D) and *Acropora (Rhabdocyathus) saludensis* Vaughan, 1919 (Plate 11, I) formed dense coral thickets that housed a molluscan fauna reminiscent of those found in similar habitats in the Recent

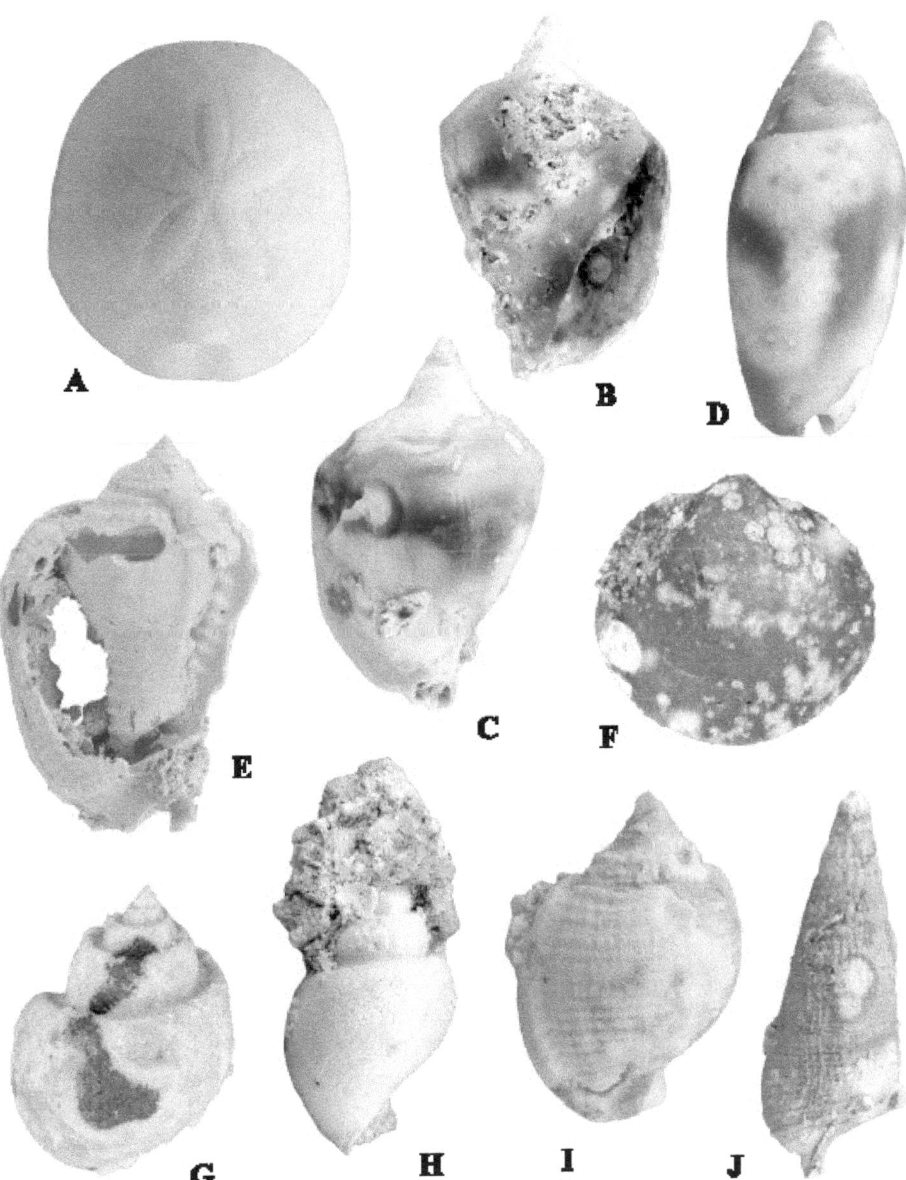

Plate 9. Index fossils and characteristic organisms of the *Orthaulax hernandoensis* Community (shallow, carbonate sand bottom environment) of the Bainbridge Subsea. Included are: A= *Rhyncholampas gouldii* (Bouve, 1846), length 42 mm; B,C= *Orthaulax hernandoensis* Mansfield, 1937, length 70 mm; D= *Omogymna brooksvillensis* (Mansfield, 1937), length 20 mm; E= *Cassis flintensis* Mansfield, 1940 (broken specimen), length 58 mm; F= *Glycymeris suwannensis* Mansfield, 1937, length 15 mm; G= *Amauropsis mansfieldi* (Petuch, 1997), length 28 mm; H= *Pachycrommium dalli* Petuch, 1997, length 26 mm; I= *Phalium caelatura* (Conrad, 1848), length 29 mm; J= *Cerithioclava eutextile* (Dall, 1916), length 57 mm. All specimens are replaced by agate and were collected at the Terramar Pit in rural Polk County, near Zephyrhills, Florida.

Plate 10. Index fossils and characteristic organisms of the *Apicula bowenae* Community (deep lagoon environment) of the Bainbridge Subsea. Included are: A= *Gradiconus cookei* (Dall, 1916). Length 38 mm; B= *Dimarzipecten brooksvillensis* (Mansfield, 1937), length 22 mm; C= *Clavolithes vicksburgensis* (Conrad, 1848), length 89 mm; D= *Fusimitra conquisita* (Conrad, 1848), length 35 mm; E= *Conorbis porcellanus* (Conrad, 1848), length 36 mm; F= *Ficus mississippiensis* Conrad, 1848, length 23 mm; G= *Ampullinopsis flintensis* (Mansfield, 1937), length 32 mm; H= *Terebrellina divisura* (Conrad, 1848), length 54 mm; I= *Apicula bowenae* (Mansfield, 1937), length 39 mm; J= *Torcula mississippiensis* (Conrad, 1848), length 51 mm. All specimens are replaced by agate and were collected at the Terramar Pit in rural Polk County, near Zephyrhills, Florida.

Caribbean Sea. Especially noteworthy was the presence of two species of the Antiguan Province endemic volute genus *Falsilyria*, *F. mansfieldi* (Dall, 1916) (Plate 11, A and B) and *F. kendrewi* Petuch, 1997 (Plate 11, G and H). These molluscivorous gastropods preyed upon a wide variety of small interstitial gastropods, including *Calliostoma silicatum* Mansfield, (1937) (Plate 11, F), *Neritopsis persculpturata* (Dall, 1916), and *Arene halensis* (Dall, 1916), and the coral-nesting bivalve *Lima halensis* Dall, 1916. Several muricid gastropods, such as *Chicoreus stetopus* (deGregorio, 1890) (Plate 11, E) and *Dermomurex (Takia) portelli* E. Vokes, 1992 (Plate 11, C) also occurred within the coral thickets, drilling into and feeding upon sessile bivalves such as *Chama* species (Mansfield, 1937: 239) and *Spondylus filiaris* Dall, 1916.

The Goniopora decaturensis Community

The shallow, higher-energy reef platforms of the Flint River, Orange Island, and Suwannee Straits Reef Tracts contained over thirty different species of massive hermatypic corals (Vaughan, 1919). Growth forms varied between ramose types with thick, short branches, such as *Goniopora decaturensis* Vaughan, 1919 (namesake of the community, Plate 12, J), *Actinactis alabamiensis* Vaughan, 1900 (Plate 12, I), and *Stylophora ponderosa* Vaughan, 1900 (Plate 12, G), and low, rounded types such as *Montastrea bainbridgensis* Vaughan, 1919, *Antiguastrea silicensis* Vaughan, 1919, and *Astrocoenia decaturensis* Vaughan, 1919 (Plate 12, E). Living among these corals was a large variety of gastropod mollusks, including the cypraeid *Cypraeorbis kendrewi* Petuch, 1997 (Plate 12, A and B), the small buccinid *"Solenosteira" suwanneensis* Petuch, 1997 (probably belonging to a new genus), and the large cerithiid *Cerithium corallicolum* Dall, 1916. Judging from the number of vermivorous gastropods present within this facies of the Suwannee Formation, interstitial reef polychaetes and nemerteans must have been abundant. The most commonly encountered of these were the conids *Hermes kendrewi* (Petuch, 1997) (Plate 12, C) and *Protoconus vaughani* (Dall, 1916) (Plate 12, F) and the turbinellid *Vasum suwanneensis* Petuch, 1997 (Plate 12, D). Ophiuroids, asteroids, and small echinoids must also have been abundant on the reef platforms and these were the main prey items of molluscivorous and echinoderm-feeding gastropods such as the bursid *Bursa victrix* Dall, 1916 and the ranellid *Cymatium cecilianum* Dall, 1916. The coral-nesting limid bivalve *Lima halensis* Dall, 1916 commonly occurred in coral rubble areas on the reef platform. Of special interest was the presence of the reef-dwelling tridacnid bivalve *Hippopus gunteri* Mansfield, 1937, the only-known record of the Indo-Pacific genus *Hippopus* in the Americas.

Communities and Environments of the Silverdale Subsea

The Silverdale Subsea contained subtropical to warm-temperate environments and represented the northernmost edge of the Baitoan Province and its tropical faunal influences. Embodying the incipiency of the Albemarle Sea, the Silverdale Subsea was geomorphologically simple and spatially small, covering only a fraction of the area of the later subseas. Besides the main Silverdale Basin, the only other major feature was the Trent Lagoon System (named for the Trent Marls) at the southern end (Figure 14). Here, the fossiliferous and well-preserved Haywood Landing Member of the Belgrade Formation was deposited in sheltered areas behind an archipelago of sand islands. Lithologically, the Trent Lagoon contained a mixed bag of sediments, having both siliciclastics and fine particulate carbonates. No coral reef complexes were present and the sea may have been too cool in the winter for any extensive coral growth. Although the Silverdale fauna has its closest affinity to the fauna of the contemporaneous Tampa Subsea, and shares many species in common, it also contained a high level of endemism. Faunal surveys of the

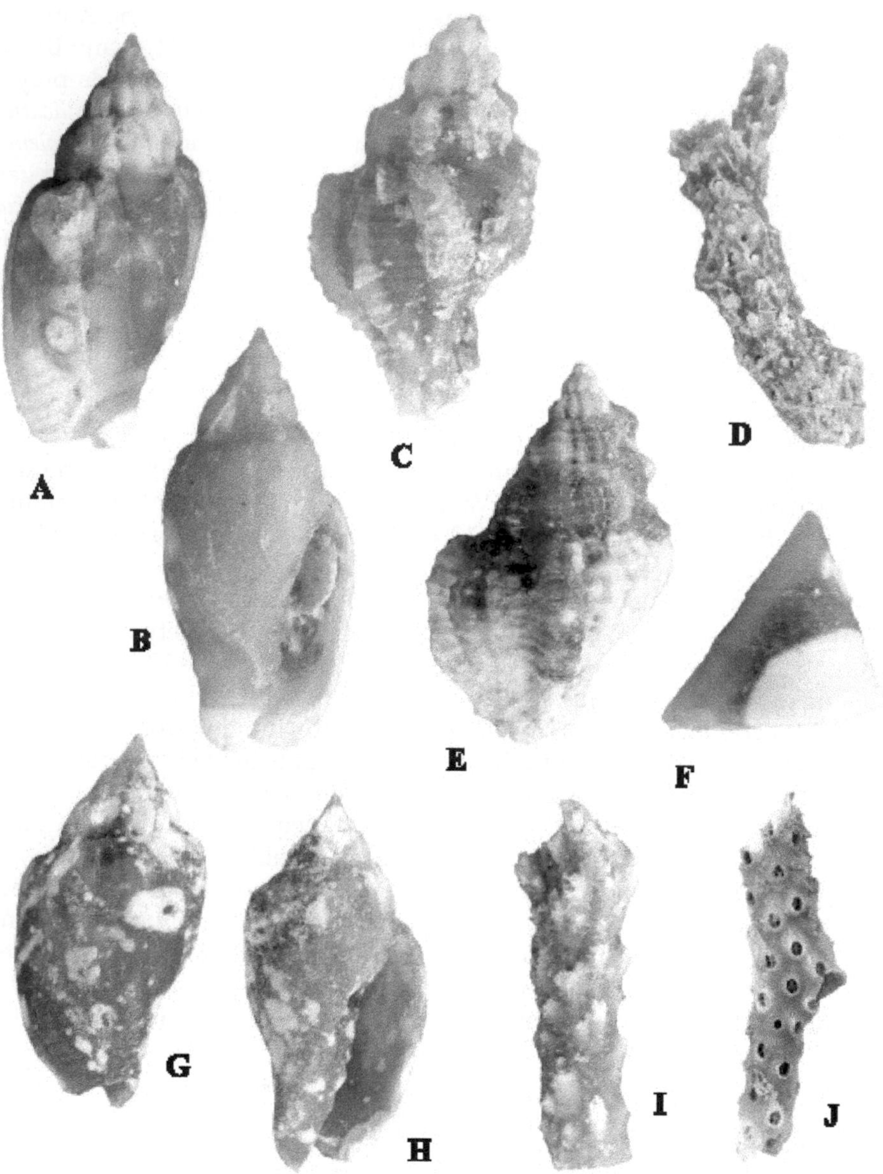

Plate 11. Index fossils and characteristic organisms of the *Stylophora minutissima* Community (back-reef environment) of the Bainbridge Subsea. Included are: A,B= *Falsilyria mansfieldi* (Dall, 1916), length 32 mm; C= *Dermomurex (Takia) portelli* E. Vokes, 1992, length 27 mm; D= *Acropora panamensis* Vaughan, 1919, length 41 mm; E= *Chicoreus stetopus* (de Gregorio, 1890), length 32 mm; F= *Calliostoma silicatum* Mansfield, 1937, length length 14 mm; G,H= *Falsilyria kendrewi* Petuch, 1997, length 36 mm; I= *Acropora saludensis* Vaughan, 1919, length 22 mm; J= *Stylophora minutissima* Vaughan, 1900. All specimens are replaced by agate and were collected at the Terramar Pit in rural Polk County, near Zephyrhills, Florida.

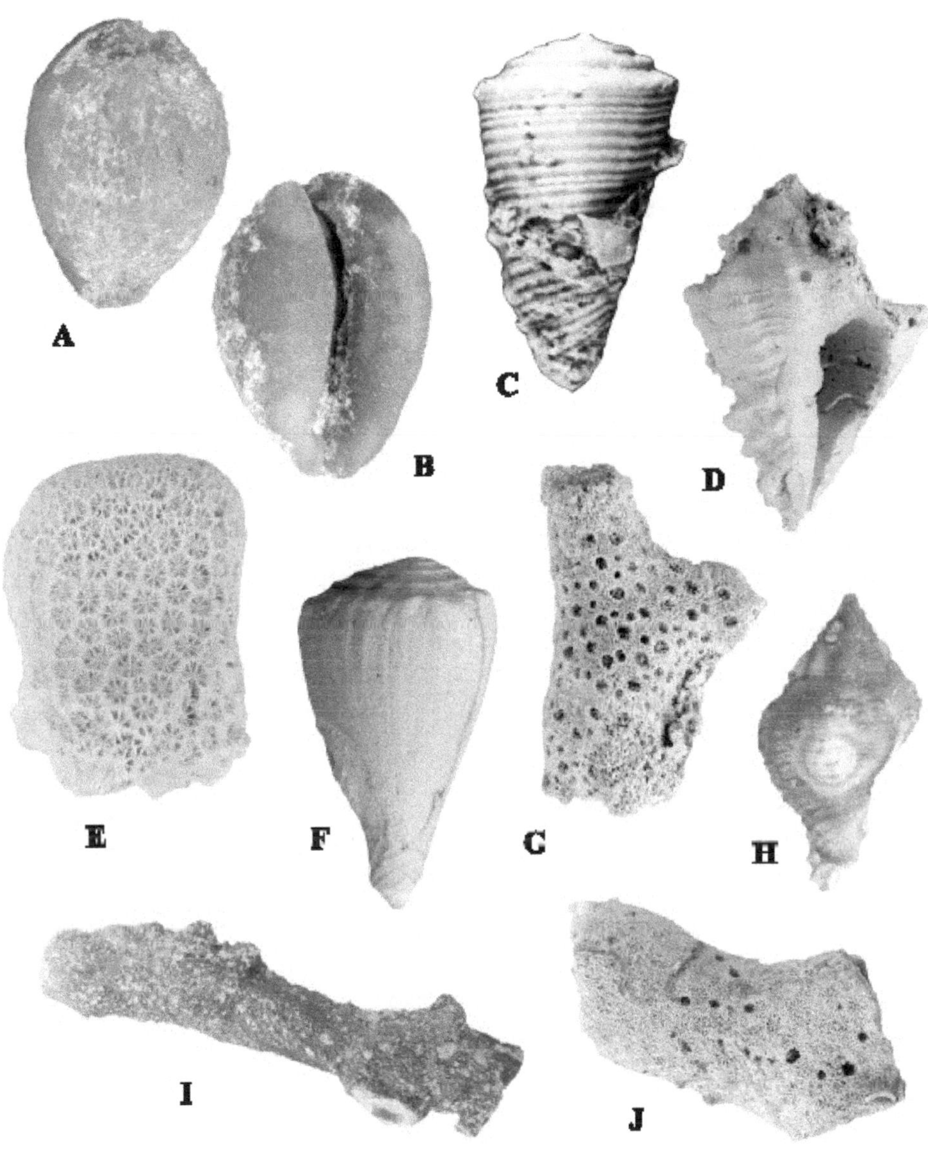

Plate 12. Index fossils and characteristic organisms of the *Goniopora decaturensis* Community (reef platform environment) of the Bainbridge Subsea. Included are: A,B= *Cypraeorbis kendrewi* Petuch, 1997, length 26 mm; C= *Hermes kendrewi* (Petuch, 1997), length 35 mm; D= *Vasum suwanneensis* Petuch, 1997, length 45 mm; E= *Astrocoenia decaturensis* Vaughan, 1919, length 23 mm; F= *Protoconus vaughani* (Dall, 1916), length 29 mm; G= *Stylophora ponderosa* Vaughan, 1900, length 33 mm; H= *"Solenosteira " suwanneensis* Petuch, 1997, length 27 mm; I= *Actinactis alabamensis* Vaughan, 1900, length 73 mm; J= *Goniopora decaturensis* Vaughan, 1919, length 34 mm. All specimens are replaced by agate and were collected at the Terramar Pit in rural Polk County, near Zephyrhills, Florida.

Haywood Landing Member (Kellum, 1926; Ward, 1992; Vermeij and Vokes, 1997) are used here for paleoecological reconstructions. Only three main marine community types have been recognized within the Silverdale Subsea area. These include (from deeper water areas, across the shallow lagoons, to shoreline): the *Calvertitella fuerta* Community (deep lagoon (10-30 m) and turritellid beds), the *Gradiconus postalveatus* Community (shallow lagoon (5-10 m), open sand bottom), and the *Mercenaria capax* Community (intertidal sand flats and shallow (0-5 m) sand areas).

The Calvertitella fuerta Community

As is typical of most of the American paleoseas, the deeper water, open sand bottom areas of lagoons were covered with immense beds of turritellid gastropods. In the case of the Silverdale Subsea, the large species *Calvertitella fuerta* (Kellum, 1926) (the community namesake, Plate 13, G) dominated the sea floor and probably served as the main prey item for a number of molluscivorous gastropods. These included a host of endemic species such as the volutid *Scaphella stromboidella* Kellum, 1926 (Plate 13, A), the muricids *Miocenebra silverdalensis* (E.Vokes, 1963), *Odontopurpura festivoidea* (E. Vokes, 1963), *Fenolignum umbilicatum* Vermeij and E. Vokes, 1997, and *Phyllonotus davisi* (Richards, 1943) and the fasciolariids *Conradconfusus hoffmani* (Ward, 1992) and *Fusinus* species (Plate 13, E). Shared with the Tampa Subsea fauna were other molluscivorous gastropods such as the naticid *Sinum imperforatum* Dall, 1915 (Plate 13, F; possibly not a driller) and the ocenebrine muricid *Ecphorosycon tampaensis* (Dall, 1892). Nestled within the turritellid beds were shallow-burrowing bivalves such as the crassatellid *Marvacrassatella silverdalensis* (Kellum, 1926), the glycymerid *Glycymeris anteparilis* Kellum, 1926, and the carditid *Cyclocardia trentensis* Ward, 1992. These also served as prey for the molluscivorous gastropods.

The Gradiconus postalveatus Community

Closer to shore, the shallow water sand bottom areas housed a rich faunal assemblage that also contained a large number of endemic mollusks. Principal among these were the vermivorous conid gastropod *Gradiconus postalveatus* (Kellum, 1926) (namesake of the community, Plate 13, B), generalist molluscivores/scavengers such as the volutid *Harpeola carolinensis* (Kellum, 1926) and the buccinid *Cymatophos kellumi* Petuch, n.sp. (Plate 13, H) (see Systematic Appendix), and bivalves such as the arcid *Cunearca silverdalensis* (Kellum, 1926), the astartids *Astarte onslowensis* Kellum, 1926 and *A. claytonrayi* Ward, 1992, and the lucinid *Pseudomiltha nocariensis* (Kellum, 1926). The endemic busyconid gastropod *Spinifulgur onslowensis* (Kellum, 1926) (Plate 13, C) would have been a top predator on these shallowly infaunal bivalves. Also occurring in shallow water areas, probably in or near beds of sea grass (*Zostera*?), were small shoals of the widespread western Atlantic pectinid, *Dimarzipecten crocus* (Cooke, 1919). This species may have also been a prey item of *Harpeola* and *Spinifulgur*. The area also contained the only large biohermal structures found in the Trent Lagoon System; combined "reefs" of the chamid bivalve *Chama tampaensis* Dall, 1903, giant oysters of the *Crassostrea gigantissima* species complex, and giant barnacles (*Chesaconcavus* species). These probably were the main prey of the rapanine thaidid gastropods *Tritonopsis gilletti* (Richards, 1943) (Plate 13, J) and *T. biconica* (Dall, 1915) (Plate 13, I, and shared with the Tampa fauna), and the muricid *Argyrobessa kellumi* (Richards, 1943).

The Mercenaria capax Community

The third community type of the Trent Lagoon System encompassed the intertidal sand flat biotopes along the sheltered lagoon side of the barrier islands (Figure 14). This low species diversity community was centered around the abundant venerid bivalve

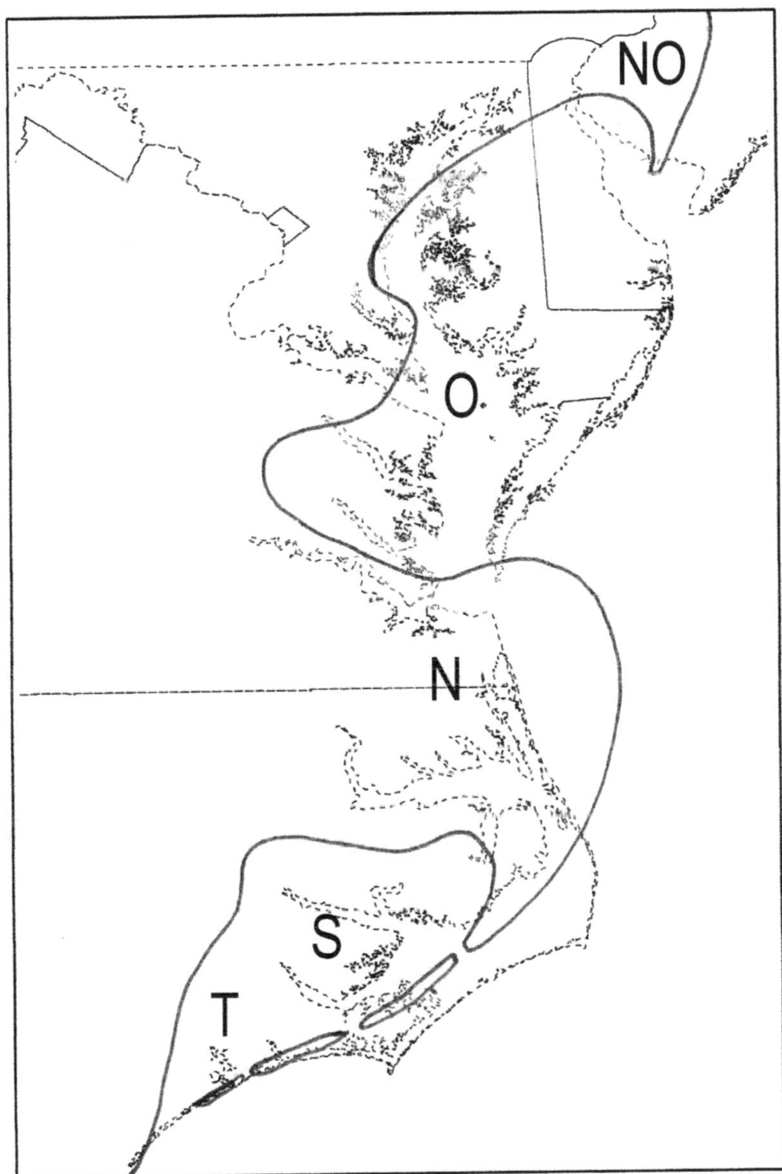

Figure 14. Approximate configuration of the coastline of central eastern North America during late Chattian Oligocene and early Aquitanian Miocene time, superimposed upon the outline of Recent Virginia, Maryland, Delaware, and southern New Jersey. Prominent features included: O= Old Church Subsea of the Salisbury Sea, S= Silverdale Subsea of the Albemarle Sea, T= Trent Lagoon System and Trent Barrier Islands, NO= Normandy Arch and Peninsula, N= Norfolk Arch and Peninsula.

Mercenaria capax (Conrad, 1843), and was composed primarily of shallow-burrowing bivalves. According to Ward (1992), *M. capax* was also present in the Old Church Subsea and represented a northern influence in the otherwise subtropical Silverdale Subsea. Some of the other bivalves living with *M. capax* included the cardiid *Dinocardium taphrium* (Dall, 1900) and the small venerids *Anomalocardia floridana* (Conrad, 1846) and *Chione spada* Dall, 1903, all of which were also present in the Tampa Subsea. These occurred along with the Silverdale endemic venerid *Macrocallista tia* Kellum, 1926 and the endemic myid *Mya wilsoni* Ward, 1992. Feeding on these bivalves was the main gastropod predator of the intertidal flats, the melongenid *"Fusus" quinquespina* (Dall, 1890; an undescribed genus) (Plate 13, D). Like its *Dinocardium* and *Anomalocardia* prey, this bizarre, elongated melongenid was also found in the Tampa Subsea. Evidence for high energy surf zones near the *Mercenaria capax* Community is given by the presence of the beach-dwelling donacid bivalve *Donax idoneus* Conrad, 1872 in the Haywood Landing Member (Ward, 1992). These would have washed over and around the barrier islands during storms and later were deposited in the lagoon.

Communities and Environments of the Tampa Subsea

Contrasting with the stark, low diversity siliciclastic-rich environments of the Silverdale Subsea were the high-tropical carbonate and coral reef environments of the Tampa Subsea. Like the Bainbridge Subsea, this southernmost Gulf paleosea was geomorphologically complex and contained large coral reef tracts, archipelagos of low coral keys, wide carbonate lagoons, and extensive coastal mangrove jungles. Some of the primary structural features included the offshore Tampa Reef Tract (named for Tampa, Florida), the Tampa Archipelago, which formed on the reef tract, and the Hillsborough Lagoon System (named for Hillsborough County, Florida), which formed behind the reef tract and archipelago (Figure 15). Farther south, only two small and relatively undeveloped reef systems, the Immokalee Reef Tract (named for Immokalee, Collier County, Florida) and the Dade Reef Tract (named for Dade County, Florida) (Figure 15), grew around a deep lagoon. These two coralline complexes represented the incipiency of the Everglades Pseudoatoll (see Chapter 7). The Tampa Formation, deposited within the area of the Tampa Reef Tract and Tampa Archipelago, contains the richest late Chattian-early Aquitanian mollusk and coral faunas found anywhere along eastern North America (Dall, 1915; Vaughan, 1919; Mansfield, 1937). All fossils are siliceous pseudomorphs and the agatized Tampa coral has been designated as the Florida State Rock.

Seven distinct communities have been preserved within the Tampa Formation. These include (from island shoreline, across the reef lagoon, to the coral reef tract): the *Cerion anodonta* Community (shoreline grasses and hypersaline evaporation pools), the *Pyrazisinus campanulatus* Community (mangrove coastline mud flats), the *Periglypta tarquinia* Community (Turtle Grass), the *Venericardia serricosta* Community (shallow carbonate sand areas within the reef lagoon), the *Calvertitella tampae* Community (deeper water carbonate sand bottom and turritellid beds), the *Porites anguillensis* Community (sheltered back reef and lagoonal bioherms), and the *Montastrea tampaensis* Community (reef platform). Similar to the Orange Island reef lagoon, the biotopes and lithofacies of the Turtle Grass, open sand bottom, and coral bioherm environments interfingered within the Hillsborough Lagoon System. For details on the Tampa corals, mollusks, and foraminiferans, see Dall, 1915, Vaughan, 1919, and Mansfield, 1937.

Plate 13. Index fossils and characteristic organisms of the *Calvertitella fuerta* Community (sand bottom deep lagoon environment), the *Gradiconus postalveatus* Community (sand bottom shallow lagoon environment), and the *Mercenaria capax* Community (intertidal sand flat environment) of the Silverdale Subsea. Included are: A= *Scaphella stromboidella* (Kellum, 1926), length 62 mm; B= *Gradiconus postalveatus* (Kellum, 1926), length 39 mm; C= *Spinifulgur onslowensis* (Kellum, 1926), length 42 mm; D= "*Fusus*" *quinquespina* (Dall, 1890), length 47 mm; E= *Fusinus* species, length 37 mm; F= *Sinum imperforatum* Dall, 1915, length 12 mm; G= *Calvertitella fuerta* (Kellum, 1926), length 64 mm; H= *Cymatophos kellumi* Petuch, new species (see Systematic Appendix), length 46 mm; I= *Tritonopsis biconica* (Dall, 1915), length 19 mm; J= *Tritonopsis gilletti* (Richards, 1943), length 25 mm. All specimens were collected in the Silverdale Marl Company quarry, Silverdale, Onslow County, North Carolina.

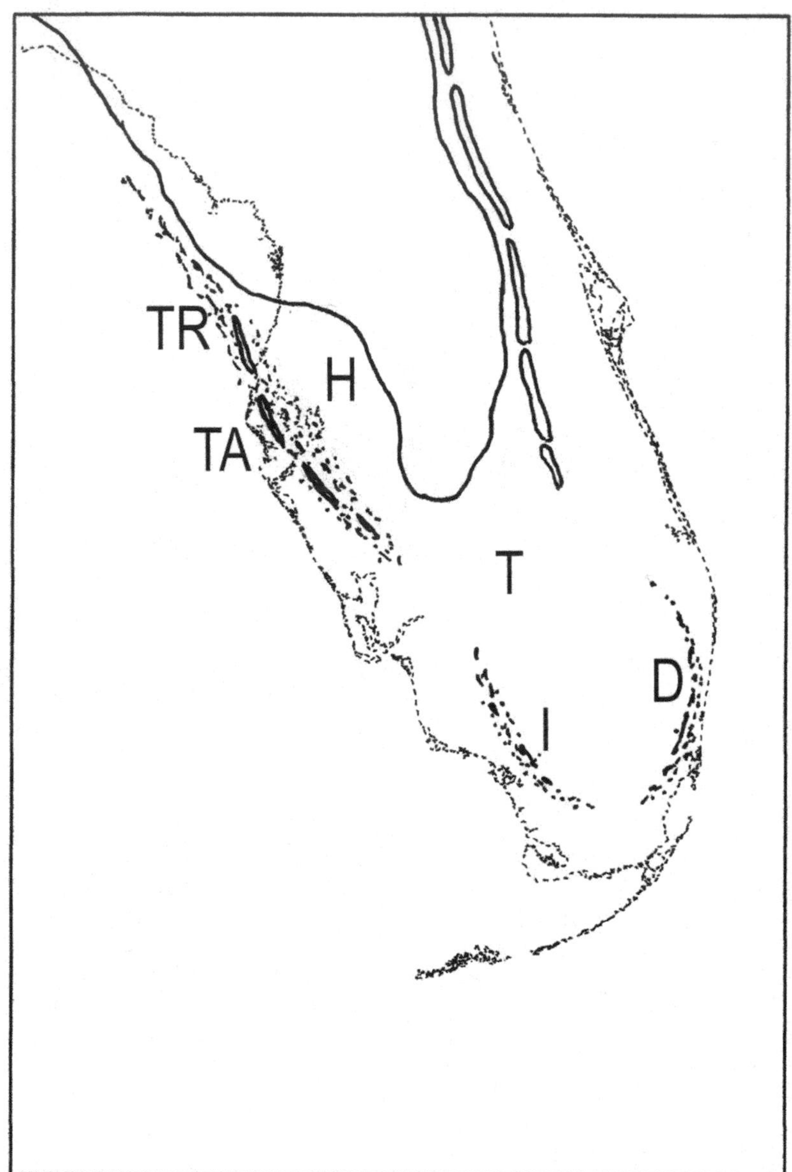

Figure 15. Approximate configuration of the Floridian Peninsula during late Chattian Oligocene to early Aquitanian Miocene time, superimposed upon the outline of Recent Florida. Prominent features included: T= Tampa Subsea of the Okeechobean Sea, TA= Tampa Archipelago, TR= Tampa Reef Tract, H= Hillsborough Lagoon System, I= Immokalee Reef Tract, D= Dade Reef Tract. At this time, the Tampa, Immokalee, and Dade Reef Tracts contained both bioherms and zonated true reefs (speckling on map).

The Cerion anodonta Community

As in the Recent Florida Keys and Florida Bay, large salt ponds and hypersaline evaporation pools existed along both the lagoon sides of the Tampa Archipelago islands and the coast of the Hillsborough Lagoon. Here halophilic grasses and saltworts (*Salicornia*) offered habitats for a species complex of the salt-tolerant cerionid land snails, including *Cerion anodonta* Dall, 1890 (namesake of the community, Plate 14, G and H), *C. floridanum* Dall, 1890, and *Microcerion floridanum* Dall, 1915. Another endemic complex of salt-tolerant land snails occurred in the helicid genus *Cepolis*, with at least three species living along the Tampa Archipelago. In the adjacent hypersaline pools, large numbers of small gastropods occurred, including the hydrobioids *Assiminea aldra* Dall, 1915 and *"Amnicola" adesta* Dall, 1915, the cerithiids *Bittium priscum* Dall, 1890 and *B. sora* Dall, 1915, and the potamidids *Lampanella transecta* Dall, 1890 and *"Potamides" hillsboroensis* (Heilprin, 1886). Salt marsh and estuarine bivalves, including the corbiculid *Polymesoda floridana* (Dall, 1900) and the mactrid *Rangia pompholyx* (Dall, 1900), formed extensive monoculture beds in less saline areas of the Hillsborough Lagoon.

The Pyrazisinus campanulatus Community

Immediately adjacent to the salt marshes and hypersaline pools were the mangrove jungles and accompanying organic-rich mud flats. This area supported immense beds of the potamidid gastropods *Pyrazisinus campanulatus* (Heilprin, 1886) (namesake of the community, Plate 14, B and C), *P. cornutus* (Heilprin, 1886), and *P. acutus* (Dall, 1892). Like the closely-related Recent potamidid genera *Tympanotonus* of West Africa and *Pyrazus* of Australia, these three *Pyrazisinus* species fed on the algal and bacterial films that formed on the mud surface and on the mangrove roots. The prop roots of the mangroves, probably in the Red Mangrove (*Rhizophora*) complex, were heavily encrusted with the ornate mangrove oyster, *Crassostrea rugifera* (Dall, 1898) (Plate 14, D and F). As in the living Caribbean mangrove oyster, *Crassostrea rhizophorae* (Guilding, 1828), the lower valve of *C. rugifera* often bears a concave depression (Plate 14, D), corresponding to the shape of the mangrove root to which it was attached. Scattered across the mud flats, between the beds of *Pyrazisinus*, were large "reefs" of the flattened oyster *Crassostrea vaughani* (Dall, 1915) (Plate 14, E). Both living and feeding on the mangrove oysters and the mud flat oyster bars were the carnivorous drilling muricid gastropods *Vokesinotus magna* (Dall, 1890), *Pterynotus posti* (Dall, 1900), and *Trachypollia scabrosa* (Dall, 1915). Also living on the manrove roots and mangrove oysters were the lacunid periwinkle *Lacuna precursor* Dall, 1915 and the mangrove nerite *Nerita (Theliostyla) tampaensis* Dall, 1892. The largest species and the primary gastropod predators on the oyster bars were the high-spired melongenid gastropods *Melongena turricula* Dall, 1890 (Plate 14, A) and *"Fusus" quinquespina* (Dall, 1892). The entire ecological structure of the *Pyrazisinus campanulatus* Community closely resembled that of the *Telescopium hernandoense* Community of the Bainbridge Subsea mud flats and that of the Recent mangrove estuaries of northern Australia and New Guinea (Vermeij, 1973).

The Periglypta tarquinia Community

In deeper water farther out in the Hillsborough Lagoon, on carbonate sand and mud bottoms, were extensive beds of Turtle Grass (*Thalassia*). These housed a rich interstitial bivalve fauna, dominated, as in similar Recent Caribbean environments, by venerids and lucinids. Principal among these was the venerid *Periglypta tarquinia* (Dall, 1900) (namesake of the community, Plate 15, I) and the lucinids *Codakia scurra* Dall, 1915, *Pseudomitha hillsboroensis* (Heilprin, 1886), and *Lucina "domingensis"* (Dall, 1903). Living on the grass

Plate 14. Index fossils and characteristic organisms of the *Cerion anodonta* Community (hypersaline pool, shore grass environment) and *Pyrazisinus campanulatus* Community (mud flat and mangrove forest environment) of the Tampa Subsea. Included are: A= *Melongena turricula* Dall, 1890, spiny adult specimen, length 90 mm; B,C= *Pyrazisinus campanulatus* (Heilprin, 1886), length 41 mm; D= *Crassostrea rugifera* (Dall, 1898), showing convex impression of mangrove root attachment, length 52 mm; E= *Crassostrea vaughani* (Dall, 1915), length 47 mm; F= *Crassostrea rugifera* (Dall, 1898), upper valve, length 42 mm; G,H= *Cerion anodonta* Dall, 1890, length 28 mm. All specimens are replaced by agate and were collected at Ballast Point, Pinellas County, Florida.

blades were numerous classic Turtle Grass environment index species, including the large modulid *Modulus turbinatus* (Heilprin, 1886), the cerithiid *Cerithium praecursor* Heilprin, 1886, the rissoid *Alvania plectrum* (Dall, 1915), and the trochid *Calliostoma tampicum* Dall, 1915. As in Recent Floridian Turtle Grass beds, the small muricid gastropod *Calotrophon* (in this case *C. heilprini* (Cossmann, 1903)) was the major predator on small interstitial bivalves and small gastropods. The eurytopic predatory gastropod, *Melongena turricula*, was also present in the Turtle Grass community and may have fed on the interstitial venerid and lucinid bivalves. As in Recent grass beds and those of Orange Island, the large turbinellid gastropod genus *Turbinella* (here represented by *T. polygonatus* Heilprin, 1886) was the primary predator of interstitial enteropneust and polychaete worms. The giant archaiid foraminiferan, *Archaias floridanus* (Conrad, 1846) was also abundant in this community, living attached to the Turtle Grass blades.

The Venericardia serricosta Community

As in the Recent Florida Keys, a rich molluscan fauna lived on, and buried within, the open sand bottom areas interspersed between the sea grass beds. Especially abundant on these shallow water sand flats were the venerids *Lirophora ballista* (Dall, 1903) (Plate 15, C), *Chione nuciformis* (Heilprin, 1886), and *Chione spada* Dall, 1903, all of which formed dense, shallowly buried beds. Interspersed between the small venerids were numerous other bivalves, principal of which were the carditid *Venericardia serricosta* (Heilprin, 1886) (namesake of the community, Plate 15, J) and the arcid *Anadara latidentata* (Dall, 1898) (Plate 15, G). Feeding on these dense bivalve beds were several types of molluscivorous gastropods, including the muricids *Phyllonotus tritonopsis* (Heilprin, 1886) and *Chicoreus trophoniformis* (Heilprin, 1886), and the naticid *Sinum imperforatum* Dall, 1915) (a molluscivore, but probably not a driller). Other carnivorous gastropods included the scavenger/predator *Oliva (Strephona) posti* Dall, 1915 (Plate 15, F) and the vermivorous conids *Gradiconus planiceps* (Heilprin, 1886) and *G. illiolus* Dall, 1915, and a species radiation of the busyconid genus *Spinifulgur*, including *S. tampaensis* (Dall, 1890), *S. nodulatum* (Conrad, 1849), *S. perizonatum* (Dall, 1890), and *S. stellatum* (Dall, 1890). Of primary interest within this community was the presence of two algivorous stromboidean gastropods; the large, common *Orthaulax pugnax* (Heilprin, 1886), which occurred in large aggregations and the smaller and more uncommon *Doxander liocyclus* (Dall, 1915), which was the only member of its genus to be found in the western Atlantic. The ampullospirid gastropods *Ampullinopsis streptostoma* (Heilprin, 1886) and *Pachycrommium floridana* (Dall, 1892) also occurred here and fed on algal growths.

The Calvertitella tampae Community

In deeper open sand areas of the Hillsborough Lagoon, suspension-feeding turritellid gastropods formed immense beds that carpeted the bottom for kilometers. These beds were made up almost entirely of three species; *Calvertitella tampae* (Heilprin, 1886) (namesake of the community, Plate 15, H), *Eichwaldiella atacta* (Dall, 1915) (Plate 15, D), and *Eichwaldiella medioconstricta* (Dall, 1915). Other, rarer turritellids also occurred within these beds, including *Apicula megabasis* (Dall, 1892), *Torcula litharia* (Dall, 1915), and *T. systoliata* (Dall, 1915). Feeding on the turritellids were carnivorous drilling gastropods such as the muricine muricid *Viator sexangula* (Dall, 1915), the ocenebrine muricid *Ecphorosycon tampaensis* (Dall, 1892) (Plate 15, E), the naticid *Amauropsis solidula* (Dall, 1892), and the rapanine thaidid *Tritonopsis biconica* (Dall, 1915) (Plate 15, A). Interspersed between the turritellids, on open sand patches, were small beds of several species of scallops, including *Dimarzipecten crocus* (Cooke, 1919), *D. marionensis* (Mansfield, 1937), and *Chesapecten bur-*

Plate 15. Index fossils and characteristic organisms of the *Periglypta tarquinia* Community (Turtle Grass bed environment), the *Venericardia serricosta* Community (shallow carbonate sand bottom environment), and the *Calvertitella tampae* Community (deep carbonate lagoon environment) of the Tampa Subsea. Included are: A= *Tritonopsis biconica* Dall, 1915), length 27 mm; B= *Gradiconus planiceps* (Heilprin, 1886), length 49 mm; C= *Lirophora ballista* (Dall, 1903), length 22 mm; D= *Eichwaldiella atacta* (Dall, 1915), length 31 mm; E= *Ecphorosycon tampaensis* (Dall, 1892), length 21 mm; F= *Oliva (Strephona) posti* Dall, 1915, length 23 mm; G= *Anadara latidentata* (Dall, 1898), length 30 mm; H= *Calvertitella tampae* (Heilprin, 1886), length 64 mm; I= *Periglypta tarquinia* (Dall, 1900), length 44 mm; J= *Venericardia serricosta* (Heilprin, 1886), length 28 mm. All specimens are replaced by agate and were collected at Ballast Point, Pinellas County, Florida.

netti (Tucker, 1934). The large vermivorous turrid gastropod, *Polystira tampensis* Mansfield, 1937, also lived among the scallop beds on the open bottom areas in deeper water and was the principal predator on polychaete worms. The large ampullospirid gastropod, *Ampullina amphora* (Heilprin, 1886) also occurred here, feeding on algal growths.

The Porites anguillensis Community

The sheltered back-reef areas of the Tampa Reef Tract, on the Hillsborough Lagoon side, supported dense forests of delicate branching corals. Some of the more common of these included *Porites anguillensis* Vaughan, 1919 (namesake of the community, Plate 16, E), *Stylophora* species, *Acropora* species (*A. saludensis* complex), *Goniopora* cf. *canalis* Vaughan, 1919, and *Alveopora* species. In the sandy areas adjacent to the back-reef community, several solitary cup corals were also present, including *Antillia* species and *Endopachys* species (Vaughan, 1919: 211). This quiet water coralline habitat housed numerous reef-loving gastropods, the most important being the cypraeids *Floradusta ballista* (Dall, 1915) (Plate 16, B and C) and *Loxacypraea tumulus* (Heilprin, 1886) (see Systematic Appendix), the harpid *Cancellomorum tampanum* (Mansfield, 1937), the muricids *Homalacantha crispangula* (Heilprin, 1886) and *Chicoreus larvaecosta* (Heilprin, 1886), the volutids *Falsilyria musicina* (Heilprin, 1886) ((Plate 16, F), *Lyria tampaensis* Mansfield, 1937, *L. silicata* Dall, 1915, *L. heilprini* Dall, 1915, and *Enaeta* species, and a large species radiation of the reef-dwelling fasciolariid genus *Latirus (Polygona)*, including *L. floridanus* (Heilprin, 1886), *L. multilineatus* Dall, 1890, *L. rugatus* Dall, 1890, and *L. callimorphus* Dall, 1890. The large spondylid bivalve *Spondylus tampensis* Mansfield, 1937 also occurred in this back-reef community, often producing long and delicate spines.

The Montastrea tampaensis Community

The higher energy, main reef platform of the Tampa Reef Tract was dominated by large, massive, rounded coral heads of over twelve species, including *Montastrea tampaensis* Vaughan, 1919 (namesake of the community, Plate 16, H and I) and *M. silicensis* Vaughan, 1919, *Antiguastrea cellulosa* (Duncan, 1863) (Plate 16, G), *Siderastrea silicensis* Vaughan, 1919 and *S. hillsboroensis* Vaughan, 1919, *Meandrina* cf. *dumblei* (Vaughan, 1919), *Galaxea* species, at least two species of *Solenastrea*, and *Syzygophyllia* species (see Vaughan, 1919: 211). Living among the coral heads and feeding on interstitial polychaetes were a number of prominent, large vermivorous gastropods such as the turbinellids *Vasum subcapitellum* Heilprin, 1886 and *V. engonatum* Dall, 1890 (ancestor of the widespread Burdigalian *V. haitense*) and the conids *Dauciconus designatus* (Dall, 1915) and *D. wakullensis* (Mansfield, 1937). Several coral-loving algae grazers also lived on the main reef platform, the most conspicuous being the turbinid *Turbo (Marmarostoma) crenorugatus* Heilprin, 1886 (Plate 16, A) and the cerithiid *Hemicerithium adela* (Dall, 1915). Large biohermal masses of the ornately-sculptured chamid bivalves *Chama tampaensis* Dall, 1903, (Plate 16, J), *C. hillsboroughensis* Mansfield, 1937 and *Pseudochama* cf. *draconis* (Dall, 1903) grew among the coral heads and created the habitat for a large number of small, cryptic gastropods such as the liotiids *Arene coronata* (Dall, 1892) and *A. solariella* (Heilprin, 1886), and the trochids *Calliostoma tampicum* Dall, 1915, *C. metrium* Dall, 1892, and *Tegula heliciformis* (Heilprin, 1886).

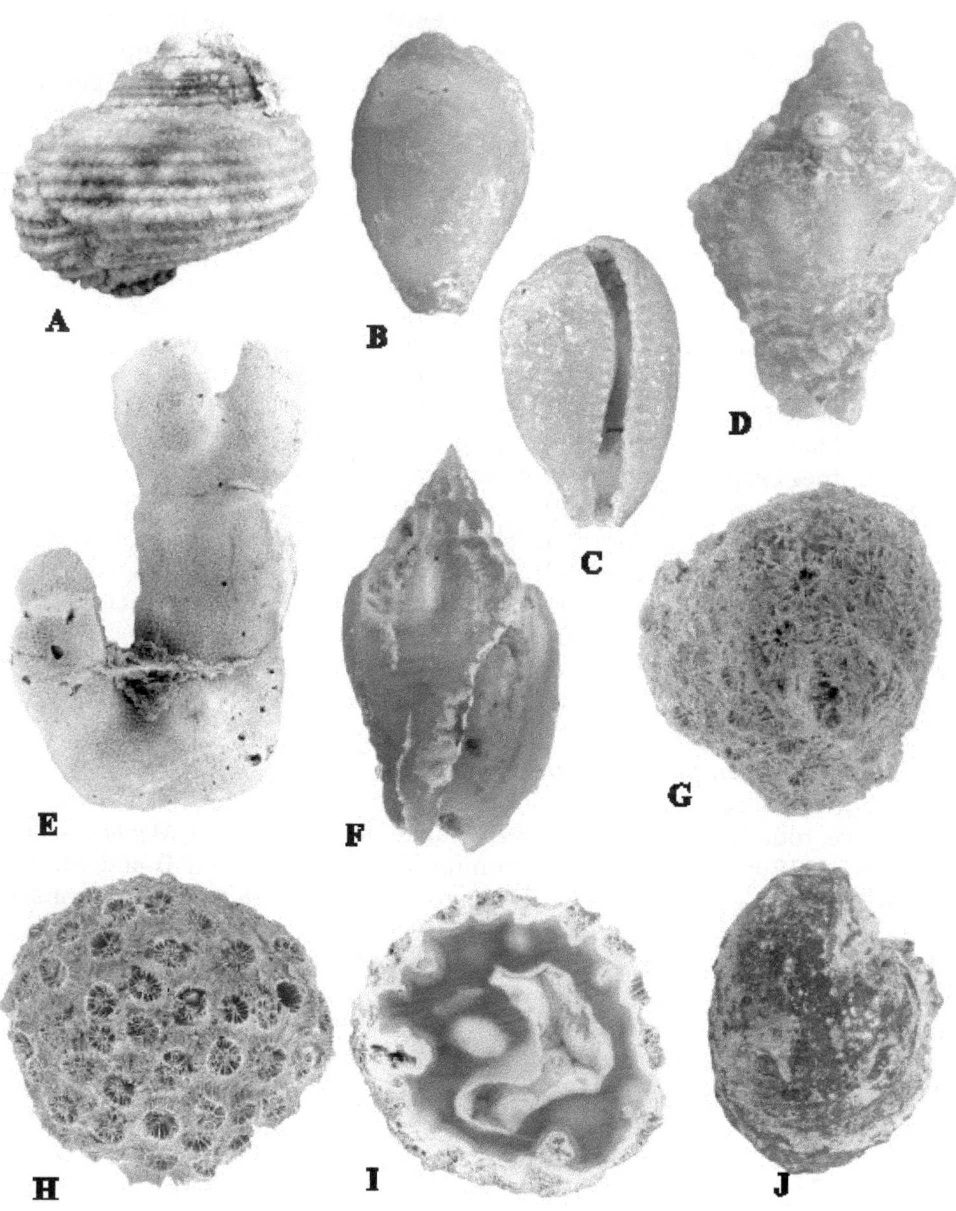

Plate 16. Index fossils and characteristic organisms of the *Porites anguillensis* Community (back-reef environment) and the *Montastrea tampaensis* Community (reef platform environment) of the Tampa Subsea. Included are: A= *Turbo (Marmarostoma) crenorugatus* Heilprin, 1886, length 28 mm; B,C= *Floradusta ballista* (Dall, 1915) (see Systematic Appendix), length 23 mm; D= *Vasum subcapitellum* Heilprin, 1886, length 36 mm; E= *Porites anguillensis* Vaughan, 1919, length 158 mm; F= *Falsilyria musicina* (Heilprin, 1886), length 41 mm; G= *Antiguastrea cellulosa* (Duncan, 1863), length 71 mm; H,I= *Montastrea tampaensis* Vaughan, 1919, length 87 mm, showing geode-like agatized interior; J= *Chama tampaensis* Dall, 1903. All specimens are replaced by agate and were collected at Ballast Point, Pinellas County, Florida.

Chapter 4. Early Miocene Seas

Of the six known eastern North American subseas of early Miocene age (Burdigalian-Langhian), only three contain fossiliferous formations that are well-preserved enough to allow for paleoceanographic analyses. These include the Chipola Subsea of the Choctaw Sea (represented by the Chipola and Oak Grove Formations), the Pamlico Subsea of the Albemarle Sea (represented by the Pungo River Formation), and the Calvert Subsea of the Salisbury Sea (represented by the Calvert Formation). As in the case of the heavily-leached sediments of the River Bend and Old Church Subseas, the formations of the Arcadia Subsea of the Okeechobean Sea, the Coosawhatchee Subsea of the Charleston Sea, and the Raritan Sea are poorly preserved, making any detailed paleoecological studies virtually impossible. These will not be covered in the following sections.

Communities and Environments of the Chipola Subsea

Containing one of the richest and most diverse tropical faunas of the Burdigalian Americas (Petuch, 1997), the Chipola Subsea was also geomorphologically complex (Figure 16). By this time, the Suwannee Strait had infilled and the American mainland finally connected to the Floridian islands (Orange Island complex). Part of this connecting area, particularly along the Gulf of Mexico coast, was still under marine influence as a series of brackish lagoons containing extensive oyster bars (Torreya Formation; Scott, 1988) and mangrove jungles (the Torreya Lagoon System). The dominant geomorphological feature of the Chipola Subsea was the Chipola Reef System, a series of zonated coral reefs that developed on top of the topographic highs laid down by the older Flint River Reef System. Behind the Chipola Reef System, and extending into present-day southwestern Georgia, was a large open lagoon containing all the typical tropical environments of the Baitoan Province.

Based upon the molluscan and coral fossils of the Chipola Formation in Calhoun County, Florida Panhandle, six distinct tropical marine communities are recognized. These include (from shoreline, across the lagoon, to the reef tract): the *Orthaulax gabbi* Community (shallow, intertidal carbonate sand flats and algal beds), the *Modulus willcoxii* Community (*Thalassia* (Turtle Grass) beds), the *Torcula subgrundifera* Community (deep lagoon turritellid beds), the *Hemicerithium akriton* Community (sheltered back reef and lagoonal bioherms), the *Vasum haitense* Community (reef platform), and the *Loxacypraea hertleini* Community (reef lagoon in late Chipola time). Mixed assemblages in the Chipola Formation demonstrate that the *Modulus willcoxii* (Turtle Grass), *Orthaulax gabbi* (open sand), and *Hemicerithium akriton* (coral bioherm) Communities, and their accompanying biotopes and lithofacies, interfingered within the Chipola lagoons.

The *Orthaulax gabbi* Community

Along the shoreline, in quiet water shallow areas (0-5 m) with normal salinity, reef-derived carbonate sediments accumulated and, under the influence of tidal currents, produced extensive sand banks. The banks closest to the shoreline and mangrove forests were rich in organic matter and supported immense beds of the potamidid gastropods *Terebralia dentilabris* (Gabb, 1873) and *Potamides suprasulcatus* (Gabb, 1873). Farther from shore, in cleaner, less organic-rich areas, these shallow banks housed a community that

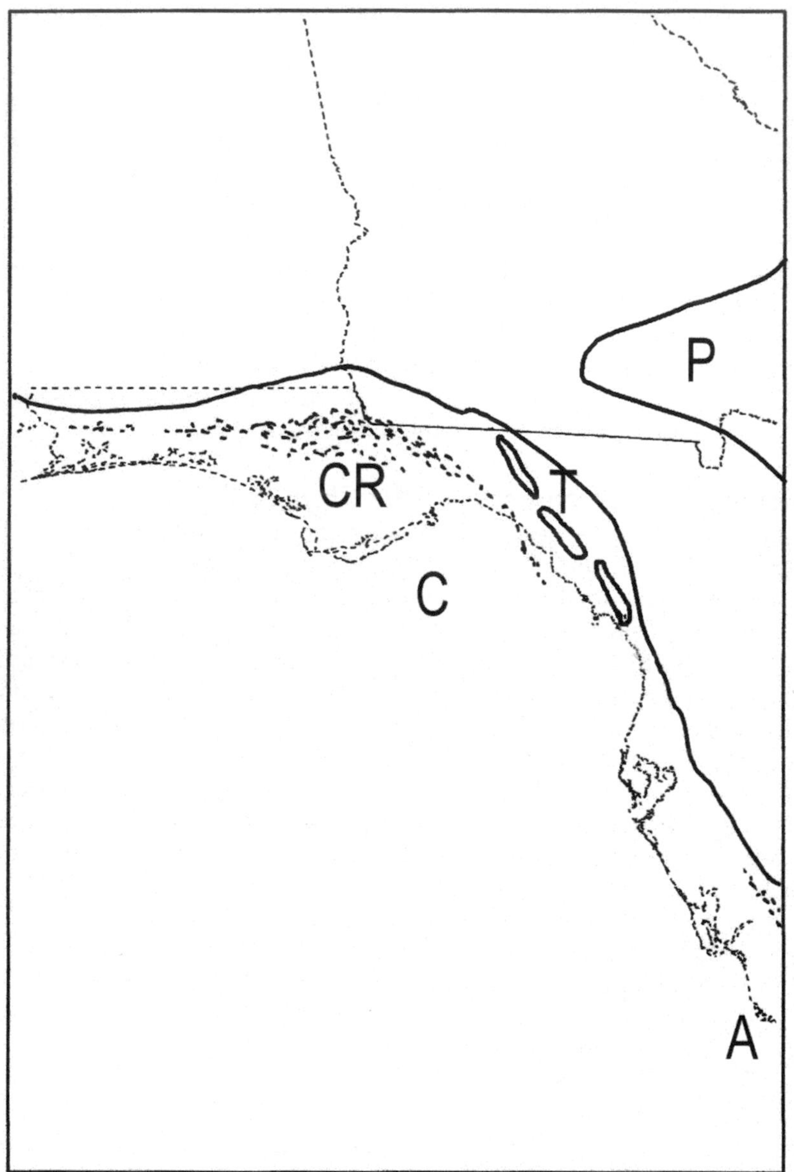

Figure 16. Approximate configuration of the northeastern Gulf of Mexico during Burdigalian Miocene time, superimposed upon the outline of the Recent Gulf coast. Prominent features included: C= the Chipola Subsea of the Choctaw Sea, CR= the Chipola Reef Tract, T= the Torreya Lagoon System, P= Parachucla Embayment, and A= the Arcadia Subsea of the Okeechobean Sea.

was unique within the Baitoan Province; one dominated by stromboidean gastropods. Here, *Orthaulax gabbi* Dall, 1890 (Plate 17, B and C, namesake of the community), *O. inornatus* subspecies (Plate 2 C; Chapter 2), *Strombus chipolanus* Dall, 1890 (Plate 17, E and F), *S. aldrichi* Dall, 1890 (Plate 17, G and H), *S. mardieae* Petuch, new species (see Systematic Appendix; Plate 17, A, I, and J), and an undescribed large *Strombus* species (all Strombidae) formed immense aggregations that grazed upon algal growths, much like the Recent Floridian *Strombus alatus* (Gmelin, 1791) and Caribbean *S. pugilis* Linnaeus, 1758. In some facies of the Chipola Formation, the stromboideans make up almost the entire bioclastic component, attesting to their abundance on these carbonate banks. Occurring with the strombids were a host of other tropical sand-dwelling species such as the vermivore conid *Ximeniconus chipolanus* (Dall, 1896) (*X. tornatus* species group) (Plate 17, K), the carnivore/scavenger olivids *Omogymna calhounensis* (Drez, 1981) (Plate 17, L), *Strephonella martensii* (Dall, 1903), *Oliva (Strephona) liodes* (Dall, 1903), and *Oliva* (un-named subgenus) *vokesorum* Drez, 1981 (Plate 17, L), and the algivore cerithiid *Clavocerithium burnsii* (Dall, 1892) (Plate 17, D). The large busyconid *Busycon diegelae* Petuch, new species (see Systematic Appendix) (Plate 3, A, Chapter 2) was the main predator of the strombids.

The Modulus willcoxii Community

In shallow (2-10 m) subtidal areas within the lagoon, extensive beds of Turtle Grass (*Thalassia*) developed, as evidenced by the presence of at least three species of the sea grass-dwelling modulid genus *Modulus*, including *Modulus willcoxii* Dall, 1892, (namesake of the community, Plate 18, H), *M. biconicus* Gardner, 1944, and *M. compactus* Dall, 1892. Living interstitially within the *Thalassia* rhizome mats were a host of characteristic grass bed bivalves, including the arcid *Dallarca geraetera* Gardner, 1926 (Plate 18, L), the cardiid *Dinocardium chipolanum* (Dall, 1900) (Plate 18, E), the venerid *Clementia grayi* Dall, 1900 (Plate 18, C), the lucinids *Lucina glenni* (Dall, 1903), and *Codakia chipolana* (Dall, 1903), and the tubiform teredinid *Kuphus incrassatus* Gabb, 1873. As in the *Thalassia* beds of the Recent Gulf of Venezuela (Petuch, 1988), grass bed-dwelling cypraeids occurred in large aggregations, feeding on the grass blades and the grass epibionts (see Chapter 10). In the Chipola Subsea, this grass bed cowrie niche was occupied by *Loxacypraea arlettae* (Dolin, 1991) (Plate 18, A and B). Primary predators of the interstitial bivalves included the busyconids *Spinifulgur armiger* Petuch, new species (Plate 18, I) (see Systematic Appendix), *S. epispiniger* (Gardner, 1944) (Plate 3, H; Chapter 2), and *Pyruella sicyoides* (Gardner, 1944) (Plate 18, K), and *Triplofusus kindlei* (Maury, 1910) (Plate 2, G; Chapter 2). Feeding on grass bed-dwelling irregular echinoids, as in the Recent Caribbean region, was the cassid *Cassis delta* Parker, 1948 (Plate 18, F and G).

The Torcula subgrundifera Community

On open sand bottom areas in the deeper parts of the lagoon (10-20 m), immense beds of several species of turritellids accumulated, including *Torcula subgrundifera* (Dall, 1892) (namesake of the community, Plate 18, D), *Apicula mixta* (Dall, 1892), and *A. chipolana* (Dall, 1892). Living in scattered patches of open sand between the turritellid beds, much as in the Recent Gulf of Venezuela (see Chapter 10), were a large number of carnivorous gastropods, including the ficid *Ficus eopapyratia* Gardner, 1947 (Plate 18, J), and the echinoderm-feeding cassids *Phalium aldrichi* (Dall, 1890) and *Sconsia paralaevigata* Gardner, 1947, the olivid *Turrancilla chipolana* (Dall, 1900), the muricids *Siratus chipolanus* (Dall, 1890), *S. dasus* (Gardner, 1947), and *Talityphis linguiferus* (Dall, 1890), and the vermivore conids *Conasprella aneuretos* (S. Hoerle, 1976), *C. rapunculus* (S. Hoerle, 1976), *Gradiconus ambonos* (S. Hoerle, 1976), and *Spuriconus cracens* (S. Hoerle, 1976). In shallower open sand areas

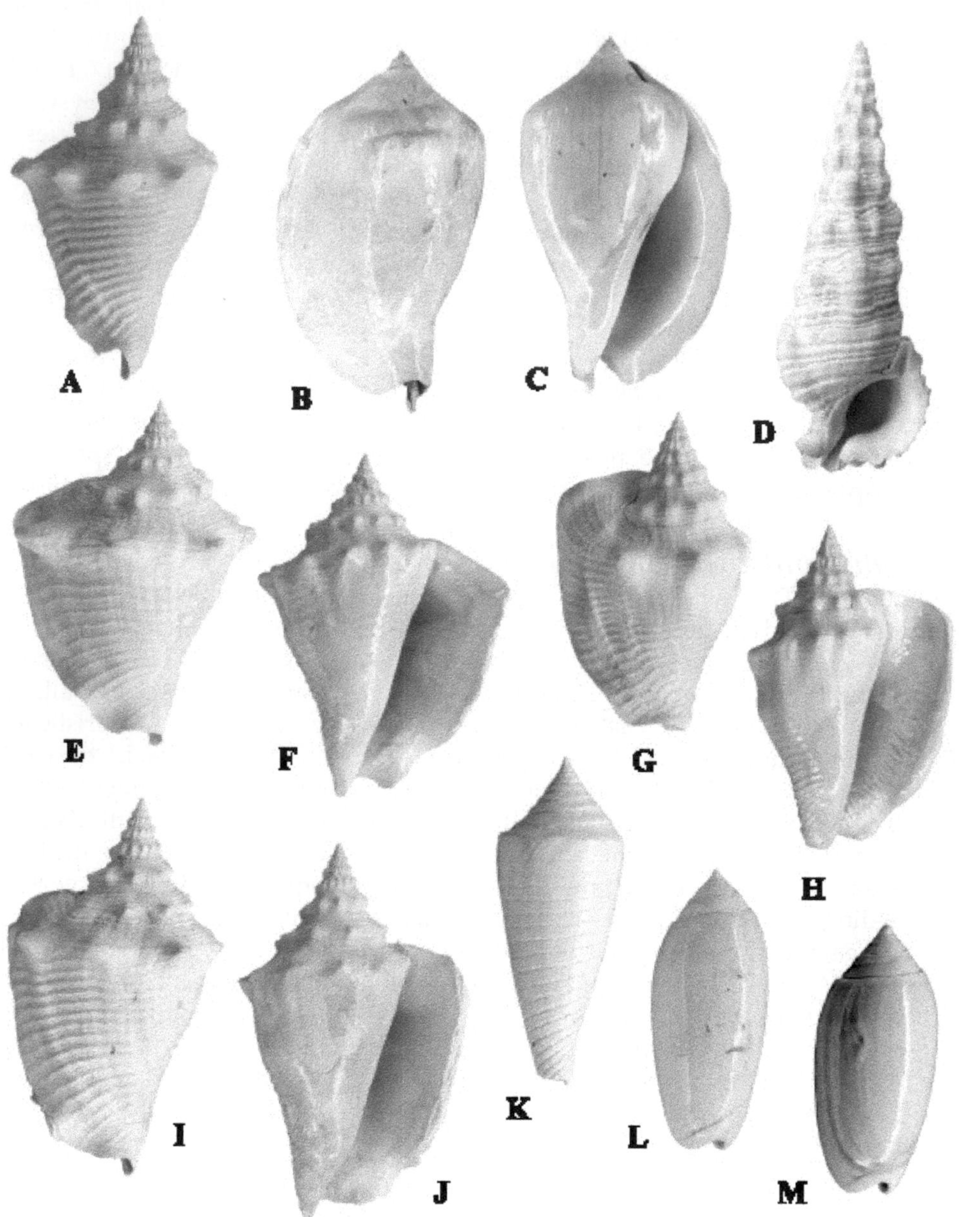

Plate 17. Index fossils and characteristic organisms of the *Orthaulax gabbi* Community (intertidal carbonate sand flats and algal beds) of the Chipola Subsea. Included are: A= *Strombus mardieae* Petuch, new species, length 32 mm; B,C= *Orthaulax gabbi* Dall, 1890, length 72 mm; D= *Clavocerithium burnsii* (Dall, 1892), length 70 mm; E,F= *Strombus chipolanus* Dall, 1890, length 60 mm; G,H= *Strombus aldrichi* Dall, 1890, length 50 mm; I,J= *Strombus mardieae* Petuch, new species, holotype, length 66 mm; K= *Ximeniconus chipolanus* (Dall, 1896), length 35 mm; L= *Oliva vokesorum* Drez, 1981, length 41 mm; M= *Omogymna calhounensis* (Drez, 1981), length 28 mm. All specimens were collected in the Chipola Formation, along the Chipola River, Calhoun County, Florida.

Plate 18. Index fossils and characteristic organisms of the *Modulus willcoxii* Community (Turtle Grass beds) and *Torcula subgrundifera* Community (deep lagoon turritellid beds) of the Chipola Subsea. Included are: A,B= *Loxacypraea arlettae* (Dolin, 1991), length 41 mm; C= *Clementia grayi* Dall, 1900. length 45 mm; D= *Torcula subgrundifera* (Dall, 1892), length 58 mm; E= *Dinocardium chipolanum* (Dall, 1900), length 56 mm; F,G= *Cassis delta* Parker, 1948, length 33 mm; H= *Modulus willcoxii* Dall, 1892, length 15 mm; I= *Spinifulgur armiger* Petuch, new species, holotype, length 42 mm; J= *Ficus eopapyratia* Gardner, 1947, length 72 mm; K= *Pyruella sicyoides* (Gardner, 1944). Length 38 mm; L= *Dallarca geraetera* (Gardner, 1926). All specimens were collected in the Chipola Formation, along the Chipola River, Calhoun County, Florida.

near the back reef environment, large beds of several species of solitary cup corals grew, principally the flabellid *Flabellum chipolanum* Weisbord, 1971 and the fungiid *Antillophyllia chipolana* Weisbord, 1971. Also occurring in this ecotone between the reef and the lagoon were dense growths of a tubular dasycladacean alga (H. Vokes, 1977). Attached to the algal tubes were large aggregations of the lophid oyster *Lopha frondicula* H. Vokes, 1977. These were probably a major food resource for the naticid gastropod *Amauropsis burnsi* (Dall, 1892). The ampullospirid gastropod *Globularia fischeri* (Dall, 1892) also occurred in these open sand bottom areas and fed on algal growths.

The Hemicerithium akriton Community

Because of the wide topographic high produced by the underlying buried Flint River Reef System platform, the most extensive biotope of the Chipola Subsea was the back reef environment. Here, widespread forests of delicate branching corals developed, including such species as *Stylocoenia pumpellyi* (Vaughan, 1900), *Stylophora imperatoris* Vaughan, 1919 and *S. undata* Weisbord, 1971, and *Porites chipolanum* Weisbord, 1971. These coral thickets housed a very large and diverse molluscan fauna, equivalent in species-richness to the back reef faunas of the Recent Indo-Pacific. Principal among these was a radiation of the reef-dwelling cerithiid genus *Hemicerithium* (now confined to the Indo-Pacific), including *H. akriton* S. Hoerle, 1972 (namesake of the community, Plate 19, D), *H. cossmanni* (Dall, 1892), *H. craticulum* S. Hoerle, 1972, and *H. pagodum* S. Hoerle, 1972. Of special importance within this environment was a large radiation of cypraeids, including *Floradusta willcoxi* (Dall, 1890) (see Systematic Appendix for description of the new endemic genus *Floradusta*), *F. heilprini* (Dall, 1890) (Plate 19, B, C), *F. emilyae* (Dolin, 1991) (Plate 19, K and L), *F. praelatior* (Dolin, 1991), *Trona calhounensis* (Dolin, 1991) (Plate 19, F, G), *Zoila tapeina* (Gardner, 1947) (Plate 19, I, J; originally thought to have come from the younger Shoal River Formation; Gardner, 1947), and several undescribed species. A reef-associated volute fauna was also present within the branching coral thickets and coral rubble and included the volutinine volutids *Falsilyria anoptos* S. Hoerle and E. Vokes, 1978 (Plate 19, A) and *F. pycnopleura* Gardner, 1937 (Plate 2, H; Chapter 2) and the lyriinine volutids *Lyria limata* S. Hoerle and E. Vokes, 1978 and *Enaeta isabellae* (Maury, 1910). These were the main predators on small cryptic gastropods such as columbellids, buccinids, and marginellids. Feeding on interstitial small bivalves such as *Dimya*, *Plicatula*, *Chama*, and *Spondylus* and small gastropods was a large muricid fauna, some of which included the ornately-sculptured muricines *Chicoreus dujardinoides* (E. Vokes, 1963) (Plate 20, J), *C. lepidotus* (E. Vokes, 1963), and *C. infrequens* (E. Vokes, 1963) (Plate 20, K) and the ergalitaxine *Lindapterys vokesae* Petuch, 1987. Living on and within the coral rubble was a host of other gastropods, with the most prominent being the algivore turbinid *Astraea (Lithopoma) chipolana* (Dall, 1892) (Plate 19, H), the harpid *Cancellomorum chipolanum* (Gardner, 1947) (Plate 20, A and B), the vermivore conid *Dauciconus demiurgus* (Dall, 1896) (Plate 19, E), the echinoderm-feeding (probably ophiuroids) ranellid *Cymatium ritteri* Schmeltz, 1989, and the fasciolariid barnacle predator *Psammostoma costatum* (Dall, 1890). The nautiloid *Aturia* (*A.* cf. *curvilineata*; photograph at end of chapter) also occurred in this back reef environment and was a predator on reef-associated crustaceans.

The Vasum haitense Community

Farther from shore, in the higher-energy area under the influence of wave surge, a rich fauna of massive corals grew on the main reef platform. Some of these corals included *Montastrea cavernosa* subspecies, *M. costata* (Duncan, 1863), *Antiguastrea silicensis* (Vaughan, 1919), *Goniopora calhounensis* (Weisbord, 1971), the brain corals *Diploria* species, and

Plate 19. Index fossils and characteristic organisms of the *Hemicerithium akriton* Community (back reef and lagoonal bioherms) of the Chipola Subsea. Included are: A= *Falsilyria anoptos* S. Hoerle and E. Vokes, 1978, length 55 mm; B,C= *Floradusta heilprini* (Dall, 1890), length 43mm; D= *Hemicerithium akriton* S. Hoerle, 1972, length 18 mm; E= *Dauciconus demiurgus* (Dall, 1896), length 42 mm; F,G= *Trona calhounensis* (Dolin, 1991) , length 47 mm; H= *Astraea (Lithopoma) chipolana* (Dall, 1892), length 19 mm; I,J= *Zoila tapeina* (Gardner, 1947), length 48 mm (originally thought to have come from the younger Shoal River Formation, but now found to be from the Chipola Formation); K,L= *Floradusta emilyae* (Dolin, 1991), length 42 mm. All specimens were collected in the Chipola Formation, along the Chipola River, Calhoun County, Florida.

Plate 20. Index fossils and characteristic organisms of the *Hemicerithium akriton* Community (back reef and lagoonal bioherms) and *Vasum haitense* Community (reef platform) of the Chipola Subsea. Included are: A,B= *Cancellomorum chipolanum* (Gardner, 1947), length 39 mm; C= *Vasum haitense* Sowerby, 1847, length 72 mm; D,E= *Lyncina theresae* (Dolin, 1991), length 49 mm; F,G= *Floradusta shirleyae* (Dolin, 1991), length 31 mm; H= *Chelyconus tapetus* (S. Hoerle and E. Vokes, 1976), length 47 mm; I= *Artemidiconus isomitratus* (Dall, 1896), length 28 mm; J= *Chicoreus dujardinoides* (E. Vokes, 1963), length 38 mm; K= *Chicoreus infrequens* (E. Vokes, 1963), length 36 mm. All specimens were collected in the Chipola Formation, along the Chipola River, Calhoun County, Florida.

Meandrina species, and the encrusting ahermatypes *Astrangia calhounensis* Weisbord, 1971 and *Symbiangia vaughani* (Weisbord, 1971). Prominent on the open coral bottom of the reef platform was the vasinine turbinellid gastropod *Vasum haitense* Sowerby, 1847 (namesake of the community, Plate 20, C). Like their living Caribbean descendants, these large gastropods occurred in large aggregations, feeding on interstitial polychaete and nemertean worms. Living on and under coral heads, as on Recent reefs, were a large number of gastropods, some of which included a large fauna of vermivorous conids such as *Chelyconus tapetus* (S. Hoerle, 1976) (Plate 20, H), *C. trajectionis* (Maury, 1910), *Dauciconus aquoreus* (S. Hoerle, 1976), *Protoconus praecipuus* (S. Hoerle, 1976), and *Artemidiconus isomitratus* (Dall, 1896) (Plate 20, I), the algivorous neritopsid *Neritopsis vokesorum* R. Hoerle, 1972, the large, drilling, rapanine thaidid *Edithais drezi* (E. Vokes, 1972) (originally described as a species of *Concholepas*), and the cypraeids *Lyncina theresae* (Dolin, 1991) (Plate 20, D and E; the only-known American member of the Indo-Pacific genus *Lyncina*) and *Floradusta shirleyae* (Dolin, 1991) (Plate 20, F and G).

The Loxacypraea hertleini Community

During the mid-Langhian, lowered sea levels and increased shoaling of the carbonate bank systems produced chains of low sand islands within the Chipola Subsea. These covered much of the back reef and reef platform areas (as the uppermost beds of the Chipola Formation, best developed and highest at Alum Bluff) and formed the bases of the Alum Bluff Islands of the later subsequent Jackson Subsea. Extensive beds of *Thalassia* grew within these shallow carbonate lagoons and housed a radiation of sea grass-dwelling cypraeid gastropods, including *Loxacypraea hertleini* (Ingram, 1948) (namesake of the community, Plate 21, M and N), *L. apalachicolae* (Ingram, 1948) (Plate 21, K and L), *L. chilona* (Dall, 1900) (Plate 3, B and C; Chapter 2), and *Floradusta alumensis* (Ingram, 1948) (Plate 21, B and C). Also living in the Turtle Grass beds was a large fauna of molluscivorous gastropods, including the melongenid *Melongena sculpturata* Dall, 1890 (Plate 21, H), the drilling muricid *Ceratostoma virginiae* (Maury, 1910) (Plate 21, G), and the busyconid *Busycon burnsii* Dall, 1890 (Plate 21, F). The large, ornate cerithiid *Clavocerithium louisae* (Schmelz, 1997) (Plate 21, E), which most probably competed with the cypraeids for grass blade epiphytic algae, was one of the main prey items of the busyconid and melongenid gastropods. The sand patches scattered among the Turtle Grass beds supported large aggregations of the algivore strombid *Orthaulax gabbi* subspecies (Plate 21, A; the last of its genus in the United States), the vermivore conid *Gradiconus sulculus* (Dall, 1896) (Plate 21, D), and the carnivore/scavenger pleioptygmatid miter *Pleioptygma apalachicolae* Petuch, new species (Plate 21, I and J) (see Systematic Appendix).

Communities and Environments of the Calvert Subsea

Contrasting greatly with the eutropical, carbonate-rich Chipola Subsea was the paratropical, siliciclastic-rich Calvert Subsea, the farthest-north Miocene paleosea to have well-preserved fossil beds. From the paleoenvironments contained in these beds, it can be seen that the Calvert Subsea was geomorphologically complex, having several distinctive biotopes and a wide range of bathymetries. Of primary importance within the Calvert Subsea was the Delmarva Archipelago (Figure 17), a chain of sand barrier islands that protected the enclosed, lagoonlike sea from heavy oceanic swell and that formed the core of the Recent Delmarva Peninsula. Also of importance was the Millville Delta (Figure 17), which filled the Calvert seafloor with clays and muds and which produced the Delmarva Archipelago by feeding sediments into the southward-flowing longshore current. From the earliest development of the Calvert Subsea, during the deposition of the Fairhaven

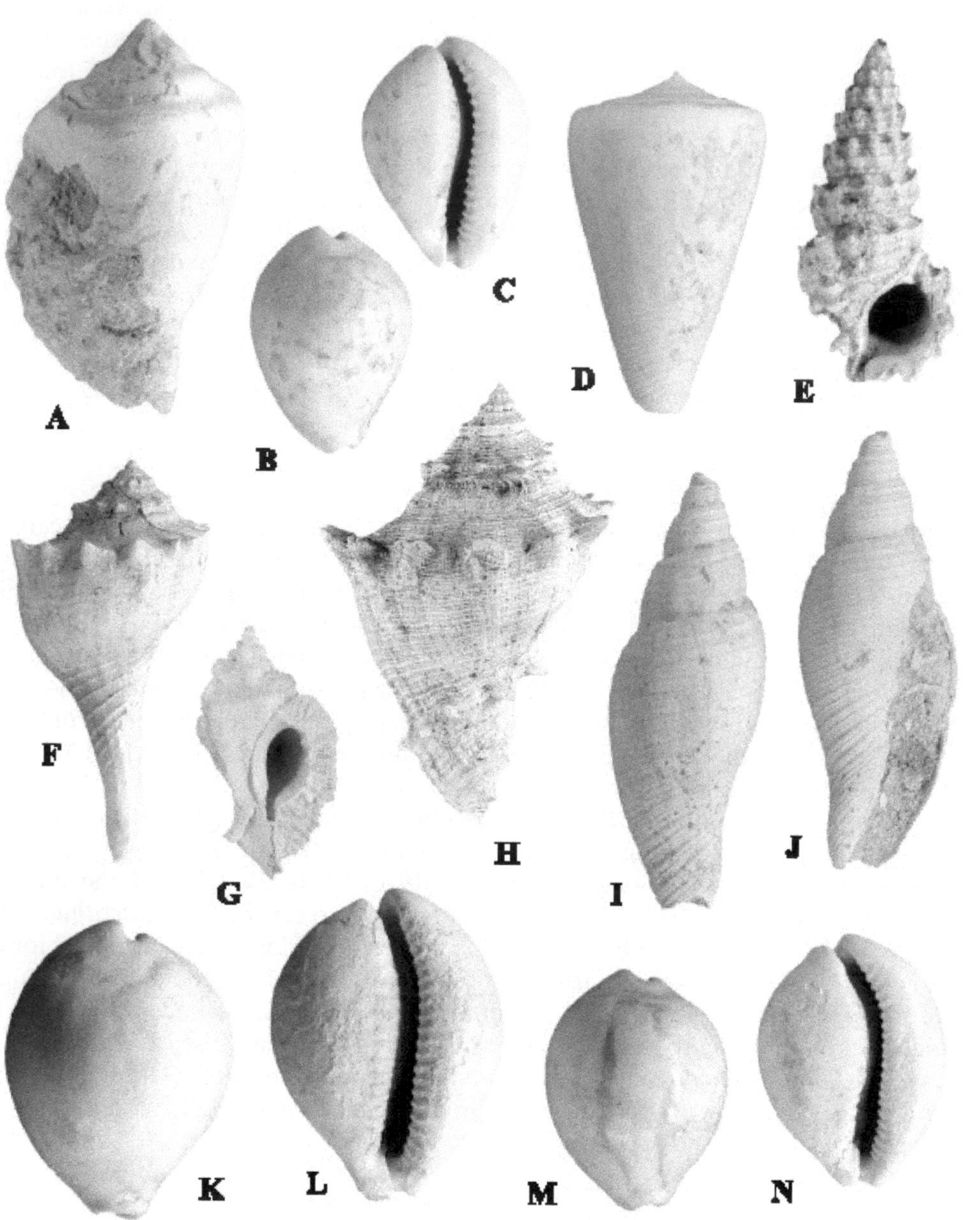

Plate 21. Index fossils and characteristic organisms of the *Loxacypraea hertleini* Community (carbonate sand banks) of the Chipola Subsea. Included are: A= *Orthaulax gabbi* subspecies, length 92 mm; B,C= *Floradusta alumensis* (Ingram, 1948), length 22 mm; D= *Gradiconus sulculus* (Dall, 1896), length 38 mm; E= *Clavocerithium louisae* (Schmelz, 1997), length 57 mm; F= *Busycon burnsi* Dall, 1890, length 59 mm; G= *Ceratostoma virginiae* (Maury, 1910), length 21 mm; H= *Melongena sculpturata* Dall, 1890, length 80 mm; I, J= *Pleioptygma apalachicolae* Petuch, new species, holotype, length 64 mm; K,L= *Loxacypraea apalachicolae* (Ingram, 1948), length 52 mm; M.N= *Loxacypraea hertleini* (Ingram, 1948), length length 40 mm. All specimens were collected in the lowest bed (at water level) at Alum Bluff, Liberty County, Apalachicola River, Florida (upper part of the Chipola Formation).

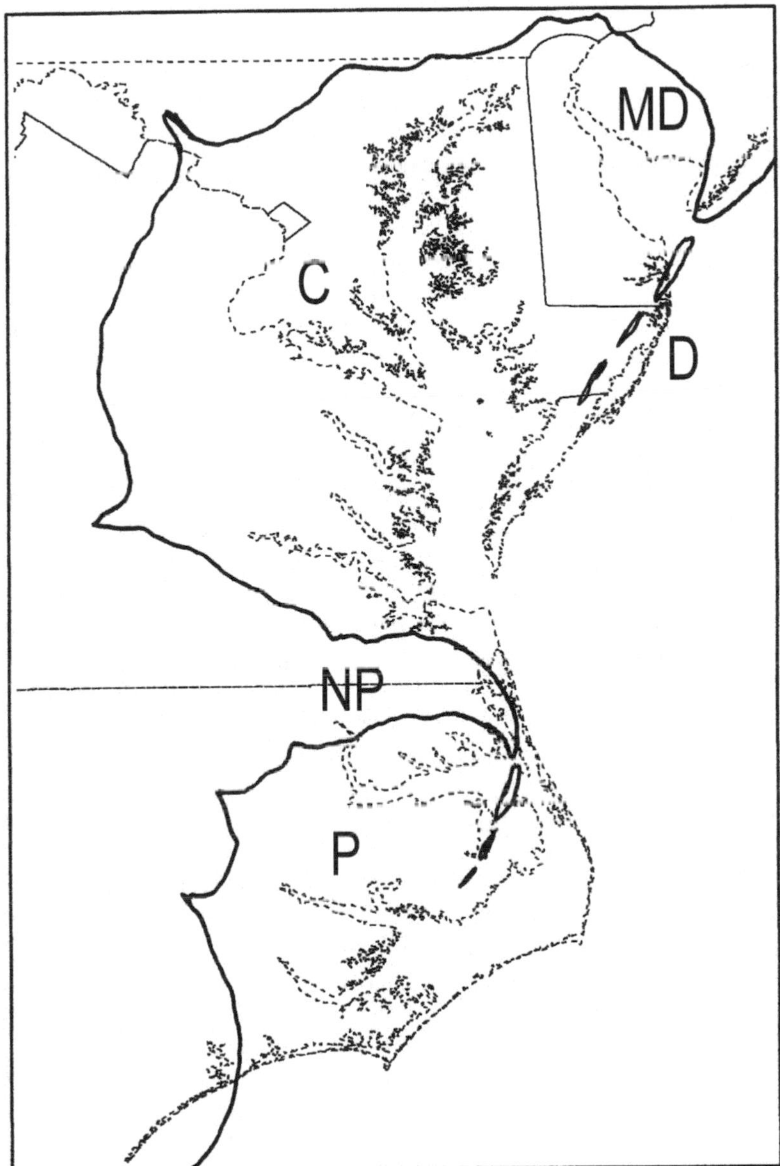

Figure 17. Approximate configuration of the coastline of the mid-Atlantic United States during late Burdigalian and Langhian Miocene time, superimposed upon the outlines of the Recent Delaware Bay, Chesapeake Bay, and Pamlico Sound areas. Prominent features included: C= Calvert Subsea of the Salisbury Sea, P= Pamlico Subsea of the Albemarle Sea, MD= Millville Delta, D= Delmarva Archipelago, NP= Norfolk Arch and Peninsula.

Member of the Calvert Formation (Shattuck Zones 1-9), two main communities are recognized; the *Hyotissa percrassa* Community (intertidal oyster bioherms during Shattuck Zone 4 time) and the *Caryocorbula elevata* Community (shallow muddy lagoons during Shattuck Zones 5 and 6 time). In the fossil beds of the species-rich Plum Point Member (Shattuck Zone 10) of the Calvert Formation (Maryland and Virginia) and the Kirkwood Formation (Delaware and New Jersey), three separate marine communities can be discerned, including the *Bulliopsis variabilis* Community (deltaic intertidal mud flats), the *Torcula exaltata* Community (deep lagoon turritellid beds), and the *Chesapecten coccymelus* Community (deep lagoon scallop beds). In the younger Calvert Beach Member (Shattuck Zones 12 and 14) of the Calvert Formation, two more marine communities are present, including the *Ecphora calvertensis* Community (late Calvert Subsea (Shattuck Zone 12) intertidal mud flats), and the *Ecphora chesapeakensis* Community (latest Calvert Subsea (Shattuck Zone 14) intertidal mud flats). For listings of the Calvert and Kirkwood molluscan faunas, see Whitfield (1894), Martin (1904), Ward (1992, 1998), and Petuch (1989, 1997).

The *Hyotissa percrassa* Community

During late Burdigalian time, the marine climate of the Salisbury Sea basin was cool and only marginally paratropical and sea levels were still low, with much of the Calvert Subsea being emergent. At this time, immense beds of the ostreid oyster *Hyotissa percrassa* (Conrad, 1840) (namesake of the community) formed extensive biohermal structures along the estuarine areas of the western coastline. The drilling ocenebrine muricids *Trisecphora* species (possibly *T. chamnessi* (Petuch, 1989)) and *Ecphora* species (wide-ribbed *E. calvertensis* complex) (both preserved mostly as fragments) were the main gastropod predators of the oysters and, together, occupied the same niche as does *Urosalpinx* (oyster drills) in the Recent. Because of the heavily-leached nature of the Zone 4 sediments, little is known of the ecology of this community, as only calcitic fossils are preserved.

The *Caryocorbula elevata* Community

Later in Fairhaven Member time (latest Burdigalian-earliest Langhian), sea levels rose and the marine climate began to warm. Extensive organic-rich mudflats and shallow muddy lagoons formed along the entire western coast at this time and these supported immense aggregations of the corbulid bivalve *Caryocorbula elevata* (Conrad, 1838) (namesake of the community). These monoculture corbulid beds are especially well developed in Shattuck Zones 5 and 6 (particularly at Chesapeake Beach, Calvert County, Maryland) and served as the principal food resource for several drilling molluscivorous gastropods, including the ocenebrine muricids *Ecphora* species (wide-ribbed *E. calvertensis* complex), *Trisecphora chamnessi* (Petuch, 1989) (also found in the lowest beds of the Pungo River Formation of North Carolina), and *Chesathais* species (*C. ecclesiasticus* complex) and the naticid *Euspira* species (*E. heros* complex). On open areas between the corbulid beds, small beds of the larger astartid bivalve *Astarte cuneiformis* Conrad, 1840 occurred and these were also prey items for the ocenbrine muricids and naticids. As sea levels continued to rise and temperatures warmed at the beginning of the Langhian, this low diversity system gradually became enriched with migrants from the Albemarle Sea to the south.

The *Bulliopsis variabilis* Community

Along the northern coast of the Calvert Subsea, in the Millville Delta area, extensive intertidal mud flats and cool-water estuaries predominated. These contained a rich fauna of mud-dwelling gastropods, including the scavenger buccinid *Bulliopsis variabilis* (Whitfield, 1894) (namesake of the community), the nassariids *Ilyanassa elongata* (Whitfield,

1894), and *I. sopora* (Pilsbry and Harbison, 1933), the columbellid *Mitrella mediocris* (Pilsbry and Harbison, 1933), and the algivore cerithiid *Bittium insulaemaris* (Pilsbry and Harbison, 1933). These deltaic estuaries also housed extensive beds of the oyster *Crassostrea mauricensis* (Gabb, 1860), the mytilid mussel *Perna incurva* (Conrad, 1839) and the brackish water mytilopsid mussel *Mytilopsis erimiocenicus* H. Vokes, 1985 (Ward, 1998). The drilling thaidid gastropod *Cymia woodi* (Gabb, 1860) and the busyconid *Coronafulgur scalaspira* (Conrad, 1863) (see Systematic Appendix) were essentially confined to the Millville Delta area and were the primary predators on the oysters and mussels. The large crocodilian *Thecachampsa contusor* Cope, 1867, the manatee sirenian *Trichechus giganteus* (DeKay, 1842), and the soft-shelled turtle *Trionyx cellulosus* Cope, 1867 also inhabited the deltaic and mud flat areas.

The *Torcula exaltata* Community

By the mid-Langhian, the Calvert Subsea was at its maximum bathymetric development and contained its warmest marine climate. In deeper water (5-20 m), high productivity areas during Plum Point time (Shattuck Zone 10), open muddy bottoms predominated and these supported immense beds of suspension-feeding turritellid gastropods. Some of these include the large, very elongated *Torcula exaltata* (Conrad, 1841) (namesake of the community, Plate 22, H), and others such as *Torcula cumberlandia* (Conrad, 1863), *Calvertitella indenta* (Conrad, 1841) (Plate 22, I), *C. chesapeakensis* (Oleksyshyn, 1959), and *Mariacolpus aequistriata* (Conrad, 1863) (Plate 22, E). Feeding on the turritellids and small mud-loving bivalves such as *Caryocorbula*, *Bicorbula*, *Mactra*, and *Erycina* was a large fauna of drilling ocenebine muricids. These included large species such as *Ecphorosycon pamlico* (Wilson, 1987) (Plate 22, A and B), *Trisecphora tricostata* (Martin, 1904) (Plate 22, J and K), and *Chesathais ecclesiastica* (Dall, 1915) (Plate 23, J and K), which all fed on large bivalves and turritellids, and smaller species such as *Trisecphora eccentrica* (Petuch, 1989) (Plate 22, C and D), *Trisecphora prunicola* (Petuch, 1988) (Plate 22, F and G), and *Ecphora wardi* Petuch, 1989 (Plate 23, C and D), which all fed on smaller, more delicate bivalves and turritellids. Other predators on the turritellids included the drilling naticid gastropods *Neverita* sp. and *Euspira interna* (Say, 1824) and the giant skate *Raja dux* Cope, 1867. Also occurring on open sand patches and small sea grass beds (*Zostera*?) between the turritellid beds were several other small gastropods, including algivores such as the fossarid *Isapis dalli* (Whitfield, 1894) and the xenophorid *Xenophora* species, small echinoderm-feeders such as the ficid *Ficus harrisi* (Martin, 1904) and the personiid *Sassia centrosa* (Conrad, 1868), the colonial urochordate-feeding eratoid *Erato perexigua* (Conrad, 1841), the bivalve-feeding small trophonine muricid *Patuxentrophon* species, and sand-dwelling zoantherian-feeders such as the epitoniids *Amaea prunicola* (Martin, 1904) and *A. reticulata* (Martin, 1904), *Cirsotrema calvertensis* (Martin, 1904), *Epitonium marylandicum* (Martin, 1904), and *Stenorhytis pachypleura* (Conrad, 1841) and the architectonicid *Heliacus trilineatum* (Conrad, 1841).

The *Chesapecten coccymelus* Community

In the deeper areas of the Calvert Subsea (20-100 m) during Plum Point time, large beds of scallops (pectinids) often carpeted the sea floor. Forming the basis of the entire ecosystem, these beds were composed of several species, including *Chesapecten coccymelus* (Dall, 1898) (namesake of the community, Plate 23, E), *C. sayanus* (Dall, 1898), *Pecten humphreysii* (Conrad, 1842), *P. woolmani* (Heilprin, 1888), and *Eburneopecten cerinus* (Conrad, 1869). Feeding on the scallops were a number of large molluscivorous gastropods, including the fasciolariids *Conradconfusus devexus* (Conrad, 1843) (Plate 23, F) and

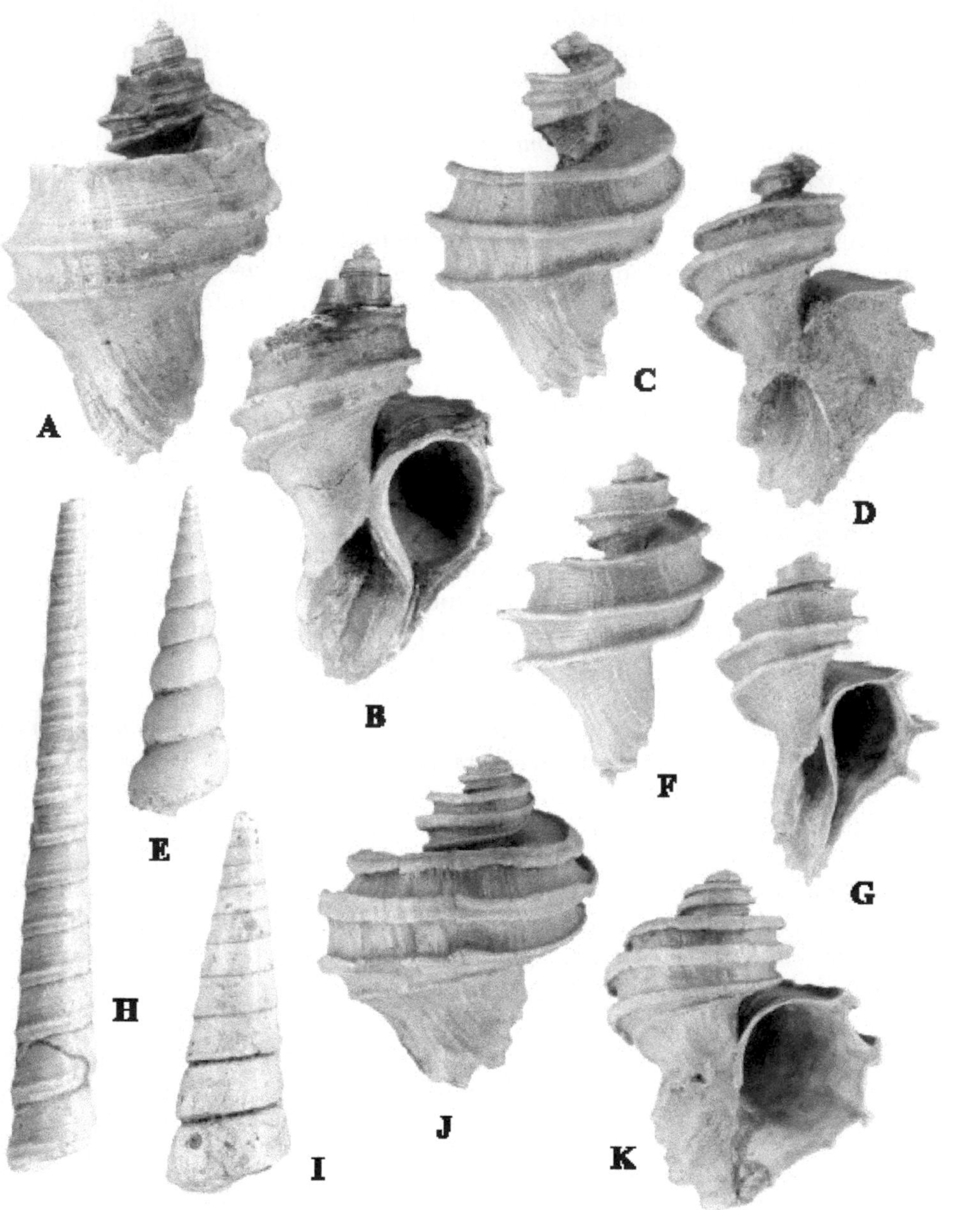

Plate 22. Index fossils and characteristic organisms of the *Torcula exaltata* Community (deep lagoon turritellid beds) of the Calvert Subsea. Included are: A,B= *Ecphorosycon pamlico* (Wilson, 1987), length 74 mm; C,D= *Trisecphora eccentrica* (Petuch, 1989), length 56 mm; E= *Mariacolpus aequistriata* (Conrad, 1863), length 31 mm; F,G= *Trisecphora prunicola* (Petuch, 1988), length 33 mm; H= *Torcula exaltata* (Conrad, 1841), length 93 mm; I= *Calvertitella indenta* (Conrad, 1841), length 41 mm; J,K= *Trisecphora tricostata* (Martin, 1904), length 58 mm. All specimens were collected in Shattuck Zone 10, Plum Point Member, Calvert Formation, at Plum Point, Calvert County, Maryland.

C. migrans (Conrad, 1843), the busyconid *Coronafulgur calvertensis* (Petuch, 1988), the ocenebrine muricid *Trisecphora schmidti* (Petuch, 1989) (Plate 23, A and B), and naticids of the genera *Euspira* and *Neverita*, and several species of large sting rays, including *Myliobatis gigas* Cope, 1867, *M. pachydon* Cope, 1867, and *M. frangens* Eastman, 1904. On sand patches between scallop beds, a rich endemic molluscan fauna was also present, including bivalves such as the venerid *Melosia staminea* (Conrad, 1839) (Plate 23, H), the glossids *Glossus markoei* Conrad, 1842) and *G. mazlea* (Glenn, 1904), the glycymerid *Glycymeris parilis* (Conrad, 1843), the crassatellid *Marvacrassatella melina* (Conrad, 1832), and the astartid *Astarte cuneiformis* (Conrad, 1840), and gastropods such as the scaphelline volutids *Scaphellopsis coronaspira* Petuch, 1988 (Plate 23, G) and *Volutifusus caricelloides* Petuch, 1988, the olivid *Oliva (Strephona) harrisi* Martin, 1904 (Plate 23, L), the small busyconids *Turrifulgur calvertensis* Petuch, 1988 and *T. prunicola* Petuch, 1993, the buccinid *Ptychosalpinx lienosa* (Conrad, 1843) (Plate 23, I), the fissurellids *Glyphis marylandicus* (Conrad, 1841) (Plate 23, M) and G. griscomi (Conrad, 1834) (both living attached to large scallops), and the vermivorous turrids *Transmariaturris calvertensis* (Martin, 1904) and *Calverturris bellacrenata* (Conrad, 1841).

The Ecphora calvertensis Community

Several major eustatic lows took place in the late Langhian (during the deposition of the Calvert Beach Member of the Calvert Formation) and these resulted in large areas of the Salisbury sea basin becoming emergent. During these low water times, severe leaching took place in the newly-deposited fossil beds and the fossil record was virtually destroyed. Only calcite-shelled organisms, such as barnacles, oysters, scallops, and ecphoras were preserved. Of these, the ecphoras are the most distinctive and abundant mollusks in Shattuck Zone 12 (Petuch, 1992). During this time, several species lived on shallow, probably intertidal, mud flats along the western half of the basin and were the major predators on small bivalves. Some of these include *Ecphorosycon kalyx* (Petuch, 1988) (Plate 24, A and B; descendant of *E. pamlico*), *Ecphora calvertensis* Petuch, 1988 (namesake of the community, Plate 24, J; descendant of *E. wardi* and the first species with four wide ribs), *Trisecphora martini* (Petuch, 1988) (Plate 24, I; descendant of the cupped-ribbed *E. schmidti*), and *Trisecphora scientistensis* (Petuch, 1992) (Plate 24, E; descendant of *T. tricostata*).

Calvert Pelagic Ecosystems

The high productivity within the Calvert upwelling systems is demonstrated by the presence of one of the richest fossil pelagic vertebrate faunas found anywhere on Earth. Here, during Calvert Formation time, over twenty-one different fish-eating toothed whales (Odontoceti) occurred (Case, 1904), some of which included the dolphins *Delphinodon mento* Cope, 1867, *D. leidyi* (Hay, 1856), *Lophocetus calvertensis* (Harlan, 1842), *Cetophis heteroclitus* Cope, 1868, *Zarhacus flagellatus* Cope, 1868, *Ixacanthus stenos* Cope, 1868, *I. atropius* Cope, 1868, and *I. coelospondylus* Cope, 1868, and the elongately rostrate *Priscodelphinus gabbi* Cope, 1868, *P. lacertosus* Cope, 1868, *P. crassangulum* Case, 1904, and *Rhabdosteus latiradix* Cope, 1867. Larger toothed whales also abounded, some of which included *Squalodon atlanticus* Leidy, 1856, *S. protervus* Cope, 1867, *Hypocetus mediatlanticus* (Cope, 1895), *Orycterocetus crocodilinus* Cope, 1867, and *O. cornutidens* (Leidy, 1856). The rich plankton resources of the Calvert Subsea also supported the first radiation of baleen whales (Mysticoceti) (Case, 1904), some of which included primitive, transitional forms such as *Ulias moratus* Cope, 1895, *Tretulias buccatus* Cope, 1895, and *Cephalotropis coronatus* Cope, 1896, and more advanced, modern-appearing giant forms such as *Cetotherium*

Plate 23. Index fossils and characteristic organisms of the *Chesapecten coccymelus* Community (deep lagoon scallop beds) of the Calvert Subsea. Included are: A,B= *Trisecphora schmidti* (Petuch, 1989), length 30 mm (first of the three-ribbed ecphoras with cupped, posterior-pointing ribs); C,D= *Ecphora wardi* Petuch, 1988, length 58 mm; E= *Chesapecten coccymelus* (Dall, 1898), length 53 mm; F= *Conradconfusus devexus* (Conrad, 1843), length 60 mm; G= *Scaphellopsis coronaspira* Petuch, 1988, length 35 mm; H= *Melosia staminea* (Conrad, 1839), length 26 mm; I= *Ptychosalpinx lienosa* (Conrad, 1843), length 50 mm; J,K= *Chesathais ecclesiastica* (Dall, 1915), length 50 mm; L= *Oliva (Strephona) harrisi* Martin, 1904, length 24 mm; M= *Glyphis marylandicus* (Conrad, 1841), length 37 mm. All specimens were collected in Shattuck Zone 10, Plum Point Member, Calvert Formation, at Plum Point, Calvert County, Maryland.

cephalum Cope, 1867, *Siphonocetus priscus* Leidy, 1851, *Balaena affinis* Owen, 1846 and *Balaenoptera sursiplana* Cope, 1895.

The plankton-feeding fish, crustacean, and mollusk fauna of the Calvert upwellings also supported a diverse elasmobranch fauna of over twenty-three species (Eastman, 1904), one of the richest known from the Neogene. Preying on the species-rich cetacean and fish faunas was a radiation of over ten species of carcharinid sharks, some of which included *Priodon egertoni* (Agassiz, 1843), *Carcharias magna* (Cope, 1867), *C. collata* Eastman, 1904, *C. laevissimus* (Cope, 1867), and *C. incidens* Eastman, 1904, and *Galeocerdo latidens* Agassiz, 1843, *G. aduncus* Agassiz, 1843, and *G. contortus* Gibbes, 1849, and *Hemipristis serra* Agassiz, 1843. The largest predator of the Calvert Subsea, the giant otodontid shark *Carcharocles megalodon* (Charlesworth, 1837), probably fed exclusively on large cetaceans. Other fish and whale-eating sharks included the hammerhead *Sphyrna prisca* Agassiz, 1843 and a large radiation of lamnid sharks, some of which included *Oxyrhina desorii* Agassiz, 1843, *O. hastalis* Agassiz, 1843, *O. sillimani* Agassiz, 1843, and *O. minuta* Agassiz, 1843 and *Odontaspis cuspidata* (Agassiz, 1843) and *O. elegans* (Agassiz, 1843). On the seafloor, the profusion of *Chesapecten* scallops and turritellid gastropods probably served as the principal prey of the giant skate *Raja dux* Cope, 1867, the previously-mentioned eagle rays of the genus *Myliobatis*, the sting rays *Trygon* sp. and *Aetobatus profundus* Cope, 1867, and the angel shark *Squatina occidentalis* Eastman 1904.

The *Ecphora chesapeakensis* Community

By latest Langhian time, large areas of the Calvert Subsea had become infilled with sediments, with the total area under marine conditions being less than half of what existed during the early Langhian. Only the intertidal mud flat environments of this dying paleosea have been preserved within Shattuck Zone 14 at the top of the Calvert Beach Member and these housed several ecphora species. Feeding on small mud flat bivalves were *Ecphora chesapeakensis* Petuch, 1992 (namesake of the community, Plate 24, C and D; descendant of *E. calvertensis* and ancestor of the Choptank Formation *E. williamsi*), *Planecphora turneri* (Petuch, 1992) (Plate 24, H; first of the thin-ribbed, four-ribbed genus *Planecphora*; see Systematic Appendix), and *Ecphora mattinglyi* Petuch, new species (see Systematic Appendix) (Plate 24, F and G; descendant of *E. wardi* and ancestor of the Choptank Formation *E. sandgatesensis* and *E. meganae*).

Communities and Environments of the Pamlico Subsea

After Aquitanian (Silverdale Subsea) time, the Albemarle Sea basin began to fill with sediments and its dimensions gradually diminished. By the Langhian, the basin was flooded as the small Pamlico Subsea and was only a fraction of its original size (Figure 17). Although within the influence of the warm Arawak Current, the Pamlico Subsea fauna was most similar to that of the cooler-water paratropical Calvert Subsea and had many taxa in common (Petuch, 1997). As indicated by the fossil beds in the Bonnerton Member of the Pungo River Formation, two marine communities existed within the deeper central areas of the Pamlico Subsea. These included the *Pecten maclellani* Community (deep lagoon scallop beds) and the *Calvertitella indenta* Community (deep lagoon turritellid beds). Like the upper beds of the Calvert formation, the Pungo River fossil beds are leached and contain only calcite shells.

The *Pecten maclellani* Community

The seafloor of the deepest areas (20-100 m) of the Pamlico Subsea was composed primarily of sand and mud and, as evidenced by phosphorite deposition, was exposed to

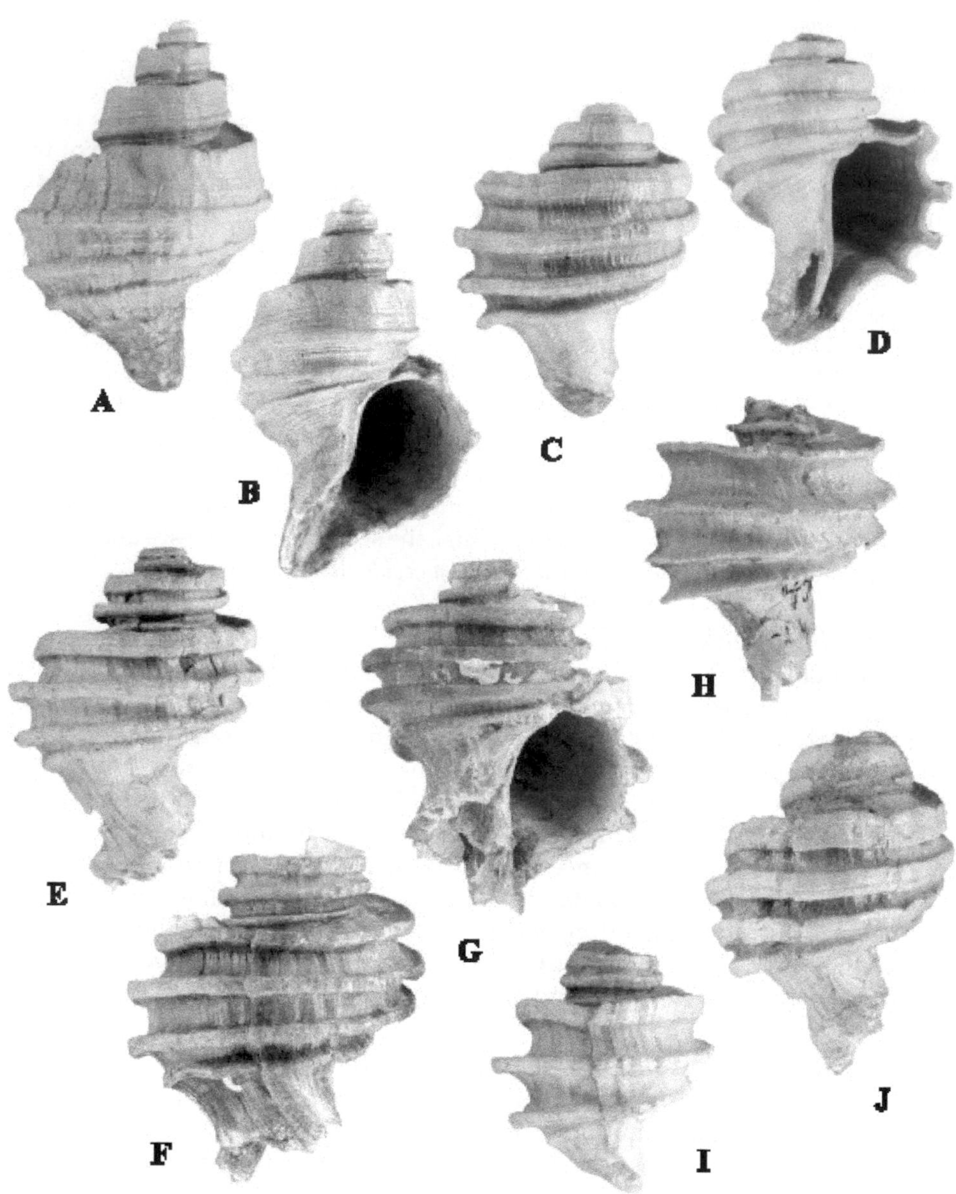

Plate 24. Index fossils and characteristic organisms of the *Ecphora calvertensis* Community (intertidal mud flats, Shattuck Zone 12) and the *Ecphora chesapeakensis* Community (intertidal mud flats, Shattuck Zone 14) of the Calvert Subsea. Included are: A,B= *Ecphorosycon kalyx* (Petuch, 1988), length 68 mm (Shattuck Zone 12); C,D= *Ecphora chesapeakensis* Petuch, 1992, length 41 mm (Shattuck Zone 14); E= *Trisecphora scientistensis* (Petuch, 1992), length 38 mm (Shattuck Zone 12); F,G= *Ecphora mattinglyi* Petuch, new species (see Systematic Appendix), holotype, length 56 mm (Shattuck Zone 14); H= *Planecphora turneri* (Petuch, 1992), length 40 mm (Shattuck Zone 14); I= *Trisecphora martini* (Petuch, 1988), length 25 mm (Shattuck Zone 12); J= *Ecphora calvertensis* Petuch, 1988, length 43 mm (Shattuck Zone 12) (first of the wide-ribbed ecphoras and ancestor of the late Miocene and Pliocene *Ecphora* species). Specimens from Shattuck Zone 12 were collected in the Calvert Beach Member (lower beds), Calvert Formation, at Governor Run, Calvert Cliffs, Calvert County, Maryland. Specimens from Shattuck Zone 14 were collected in the Calvert Beach Member (upper beds), Calvert Formation, at Scientist's Cliffs and Chesapeake Beach, Calvert County, Maryland.

almost continuous upwellings of cool, nutrient-rich water. Here, the Pamlico endemic scallop *Pecten maclellani* Gibson, 1987 (namesake of the community, Plate 25, F) and the Calvert-derived *Chesapecten coccymelus* (Dall, 1898) formed immense beds that carpeted the seafloor. These were preyed upon by a number of drilling ocenebrine muricids, some of which included *Trisecphora carolinensis* (Petuch, 1989) (Plate 25, G and H) (endemic to the Pamlico Subsea) and the Calvert-derived *Trisecphora tricostata* (Martin, 1904) (Plate 25, C and D) and *Ecphora wardi* Petuch, 1989 (Plate 25, I and J). During the earliest development of the Pamlico Subsea, in the latest Burdigalian, the ocenebrine muricids *Trisecphora chamnessi* (Petuch, 1989) (Plate 25, E) (also found in the Fairhaven Member of the Calvert Formation) and *Siphoecphora aurora* (Wilson, 1987) (Plate 25, A and B) (endemic to the Pamlico Subsea), together, occupied the niche of principal drilling predator. By the Langhian, however, these two species were replaced by *Trisecphora tricostata*, *T. carolinensis*, and *Ecphora wardi*.

The Calvertitella indenta Community

Closer to shore, in shallower water (5-20 m), the mud and sand bottom areas housed extensive beds of the Calvert-derived turritellid *Calvertitella indenta* (Conrad, 1841) (namesake of the community, Plate 26, F; preserved only as abundant molds and casts). The turritellid beds formed the basis of a molluscan ecosystem that was composed entirely of Calvert Subsea (Shattuck Zone 10, Plum Point Member) species. Some of these included the deep-burrowing hiatellid bivalve *Panopea parawhitfieldi* Gardner, 1928 (preserved only as abundant molds) and gastropods such as the epitoniids *Cirsotrema calvertensis* (Martin, 1904) (Plate 26, A and B), *Stenorhytis pachypleura* (Conrad, 1841) (Plate 26, H and I), and *Amaea prunicola* (Martin, 1904), all of which fed upon sand-dwelling zoantherians, and the drilling ocenebrine muricids *Trisecphora schmidti* (Petuch, 1989) (Plate 26, C and D), *Ecphorosycon pamlico* (Wilson, 1987) (Plate 26, E), and *Chesathais whitfieldi* Petuch, 1989 (originally described from the Kirkwood Formation of New Jersey).

Plate 25. Index fossils and characteristic organisms of the *Pecten maclellani* Community (deep lagoon scallop beds) of the Pamlico Subsea. Included are: A,B= *Siphoecphora aurora* (Wilson, 1987), juvenile specimen, length 49 mm; C,D= *Trisecphora tricostata* (Martin, 1904), length 35 mm; E= *Trisecphora chamnessi* (Petuch, 1989), length 49 mm; F= *Pecten maclellani* Gibson, 1987, length 90 mm; G,H= *Trisecphora carolinensis* (Petuch, 1989), length 26 mm; I,J= *Ecphora wardi* Petuch, 1989, length 23 mm. All specimens were collected in the Bonnerton Member, Pungo River Formation, in the Texasgulf Lee Creek Mine, Aurora, Beaufort County, North Carolina.

Plate 26. Index fossils and characteristic organisms of the *Calvertitella indenta* Community (deep lagoon turritellid beds) of the Pamlico Subsea. Included are: A,B= *Cirsotrema calvertensis* (Martin, 1904), length 28 mm; C,D= *Trisecphora schmidti* (Petuch, 1989), length 21 mm; E= *Ecphorosycon pamlico* (Wison, 1987), juvenile specimen, length 37 mm; F= *Calvertitella indenta* (Conrad, 1841), cast (steinkern), length 38 mm; G= *Chesathais whitfieldi* Petuch, 1989, length 21 mm; H,I= *Stenorhytis pachypleura* (Conrad, 1841), length 15 mm; J= *Amaea prunicola* (Martin, 1904), length 20 mm. All specimens were collected in the Bonnerton Member, Pungo River Formation, in the Texasgulf Lee Creek Mine, Aurora, Beaufort County, North Carolina.

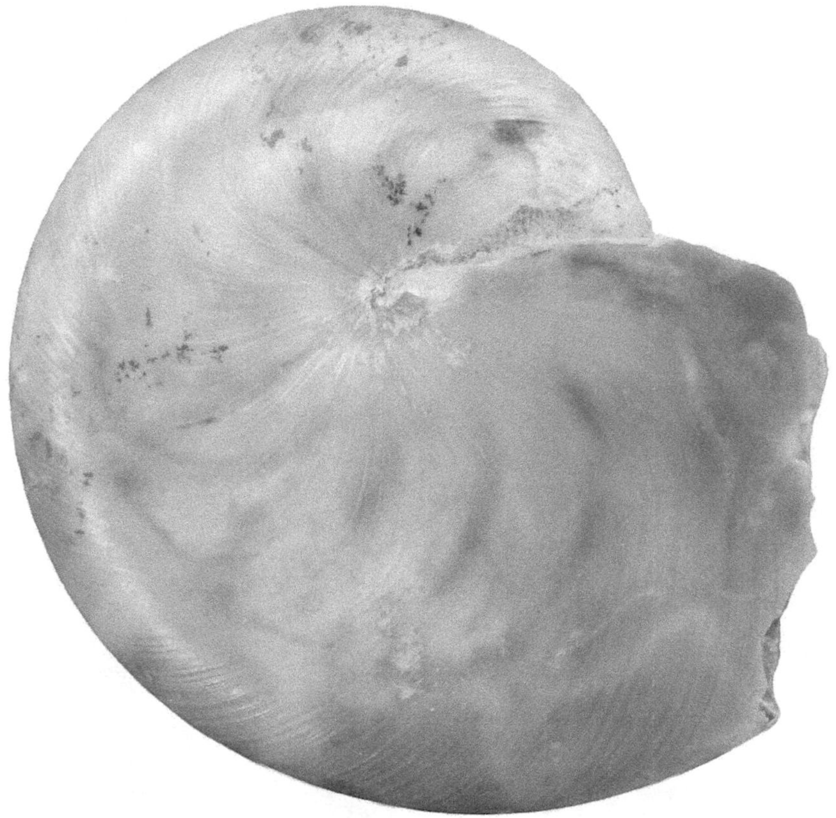

The nautiloid *Aturia* cf. *curvilineata*, Chipola Subsea, Choctaw Sea, and Burdigalian northern South America, width 76 mm; with preserved color pattern. Specimen from the Cantaure Formation of Venezuela.

Chapter 5. Middle and Late Miocene Seas

The late Middle Miocene to early late Miocene (Serravallian-early Tortonian Ages) was a time of dramatic marine climate degenerations and wildly-fluctuating sea levels (see Chapter 10). As a result of this climatic instability and the accompanying eustatic lows, many of the subsea basins were emergent for long periods of time. During the late Tortonian, a giant delta formed along the northwestern area of the Charlotte Subsea, producing a siliciclastic wedge that prograded southward to infill the entire western half of the Okeechobean Sea (Cunningham, *et. al.*, 2003). This wide platform formed the base for subsequent Plio-Pleistocene reef systems and lagoonal environments. The rate of sedimentary infilling by deltas also accelerated during the Serravallian low sea level stands, causing the Charleston Sea and Albemarle Sea basins to undergo long-term emergence (Serravallian to mid-Tortonian for the Albemarle; mid-Serravallian to early Zanclean Pliocene for the Charleston). At this time, both paleoseas existed only as remnant coastal lagoon systems. For the Middle Miocene and early Late Miocene, only the Patuxent and St. Mary's Subseas of the Salisbury Sea and the Walton Subsea of the Choctaw Sea contain fossiliferous formations that are well-preserved enough to allow for paleoecological analyses.

By early Tortonian time, the Gulf Coastal Plain Walton and Polk Subseas had ceased to exist and were replaced, in the mid-Tortonian, by the Alaqua and Charlotte Subseas. Throughout this time, the Atlantic Coastal Plain Charleston Sea basin contained only coastal lagoon systems and left little in the way of a fossil record. Of primary interest during this time was the completion of the deltaic infilling of the entire Salisbury Sea basin, with the St. Mary's Subsea representing its last subsea and "last gasp." Also of interest during this time was the reflooding and enlargement of the Albemarle Sea basin, due to the erosional removal of the Norfolk Arch and Peninsula. The subsequent Rappahannock Subsea, through "basin capture," encompassed both the southernmost remnant of the Salisbury Sea basin and an enlarged version of the Pamlico Subsea. The paleoecology and paleoceanography of the Rappahannock, Charlotte, Alaqua, Walton, Polk, Patuxent, and St. Mary's Subseas are discussed in the following sections.

Communities and Environments of the Patuxent Subsea

Although less than one-half the size of the older Calvert Subsea, the Patuxent Subsea was geomorphologically complex, having several distinctive biotopes and a wide range of bathymetries (Figure 18). Of primary importance within the Patuxent Subsea was the Millville Delta, which dominated the northern coastline and created widespread estuarine and mud flat conditions. Also of importance was the Delmarva Archipelago along the eastern side, which essentially closed off the basin and produced quiet water, lagoonal conditions across large parts of the eastern and central areas. The Patuxent Gyre, a branch of the southward-flowing, cold water Raritan Current (Petuch, 1997), entered the southeastern corner of the basin and brought open-oceanic water conditions to the deeper central and southern areas. Although not as faunistically rich as the previous Calvert Subsea, the Patuxent Subsea housed a descendant fauna that contained a large number of endemic taxa, both at the genus and species level. These are found in the three members of the Choptank Formation of the Chesapeake Group (Ward, 1992); the Drumcliff Member

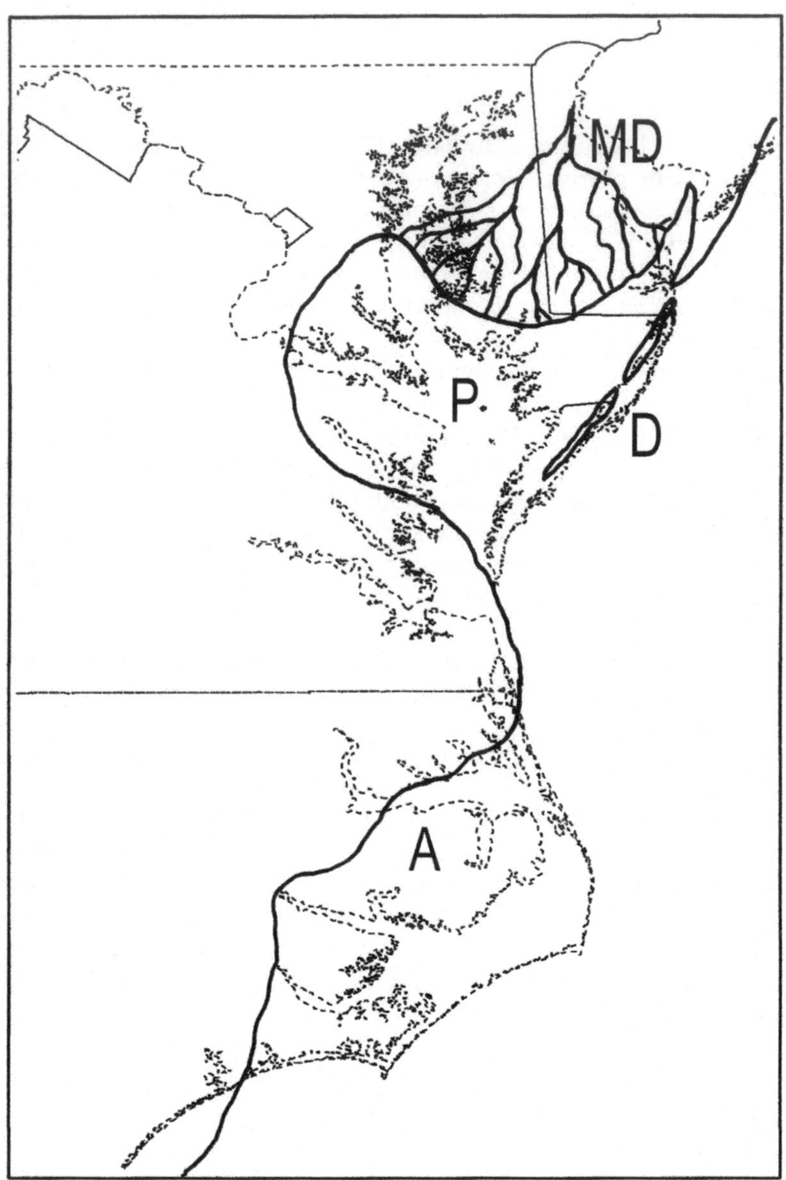

Figure 18. Approximate configuration of the coastline of the mid-Atlantic United States during early Serravallian Miocene time, superimposed upon the outlines of the Recent Delaware Bay, Chesapeake Bay, and Pamlico Sound areas. Prominent features included: P= Patuxent Subsea of the Salisbury Sea, A= Albemarle Lagoon System, MD= Millville Delta, D= Delmarva Archipelago.

(Shattuck Zone 16), the St. Leonard Member (Shattuck Zone 17), and the Boston Cliffs Member (Shattuck Zones 18 and 19) (separated by unconformities due to sea level drops). From the faunas of these three highly fossiliferous members, four distinct communities can be discerned, including the *Trisecphora smithae* Community (shallow muddy lagoon during Shattuck Zone 16 time), the *Bicorbula drumcliffensis* Community (shallow sandy lagoon during Shattuck Zone 17 time), the *Chesapecten nephrens* Community (deep lagoon scallop beds during Shattuck Zone 17 time), and the *Mariacolpus octonaria* Community (deep lagoon turritellid beds during Shattuck Zone 19 time).

The Trisecphora smithae Community

Since most of the Drumcliff Member fossil beds are heavily leached, the aragonitic component of the molluscan fauna is often removed or is poorly preserved. Only the calcitic fossils are well-preserved enough for detailed systematic and paleoecologic studies. Of these, the ecphoras (Ocenebrinae-Muricidae) are the most abundant and species-rich, and at least six different species comprising five genera occurred on the shallow muddy bottoms and mud flats of the early Patuxent Subsea. Some of these included *Ecphorosycon lindajoyceae* (Petuch, 1993) (Plate 27, A and B; last-living species of the genus *Ecphorosycon*), *Trisecphora smithae* (Petuch, 1988) (namesake of the community; Plate 27, C and D), *Ecphora sandgatesensis* Petuch, 1989 (Plate 27, F and G), *E. harasewychi* Petuch, 1989 (Plate 27, J), *Planecphora vokesi* (Petuch, 1989) (Plate 27, H and I; descendant of the Calvert *P. turneri*), and *Chesathais donaldasheri* Petuch, 1989 (Plate 27 E). These drilling muricids fed on a wide variety of shallowly-infaunal bivalves, including the arcid *Dallarca staminea* (Say, 1832), the leptonid *Aligena aequata* (Conrad, 1843), and the lucinid *Lucinoma contracta* subspecies. Competing with the ecphoras for bivalve prey, and feeding on the ecphoras themselves, was the large drilling naticid gastropod *Euspira* species (*E. heros* complex), which was often abundant on the mud flats.

The Chesapecten nephrens Community

After a short regressive interval at the end of Drumcliff time, the Patuxent Subsea again reflooded, this time to its greatest extent and development. As in the older Calvert and Pamlico Subseas, immense beds of pectinid bivalves carpeted open sandy bottoms in deeper water areas (10-100 m). In the case of the Patuxent Subsea, these beds were composed of only two large species, *Chesapecten nephrens* Ward and Blackwelder, 1975 (namesake of the community; Plate 28, E) and *Christinapecten marylandicus* (Wagner, 1839). Feeding on the scallops were several large molluscivorous gastropods, including the busyconids *Coronafulgur choptankensis* (Petuch, 1993) and *Sycopsis lindae* Petuch, 1988 (Plate 28, C), the ocenebrine muricid *Ecphora rikeri* Petuch, 1988 (Plate 29, E and F; descendant of *E. harasewychi*), and the naticid *Euspira* species (*E. heros* complex). Living on open sand patches between the scallop beds were a number of carnivorous gastropods, including the suctorial-feeding cancellariids *Trigonostoma biplicifera* (Conrad, 1841) (Plate 28, A), *Marianarona* species, *Mariasveltia patuxentia* (Martin, 1904), and *Cancellariella neritoidea* (Martin, 1904), and the vermivorous turrids *Transmariaturris choptankensis* (Martin, 1904), *Calverturris schmidti* Petuch, 1993, and *Mariaturricula rugata* (Conrad, 1862) (Plate 28, F).

The Bicorbula drumcliffensis Community

Closer to shore, on open muddy sand bottoms in shallow water (2-10 m), large colonies of infaunal bivalves formed virtual pavements. In most areas, the corbulid *Bicorbula drumcliffensis* (Oleksyshyn, 1960) (namesake of the community; Plate 28, I) formed the largest beds and occurred together with a host of larger species such as the pet-

Plate 27. Index fossils and characteristic organisms of the *Trisecphora smithae* Community (shallow muddy lagoons and mud flats during Shattuck Zone 16 time) of the Patuxent Subsea. Included are: A,B= *Ecphorosycon lindajoyceae* (Petuch, 1993), length 82 mm; C,D= *Trisecphora smithae* (Petuch, 1988), length 78 mm; E= *Chesathais donaldasheri* Petuch. 1989, length 23 mm; F,G= *Ecphora sandgatesensis* Petuch, 1989, length 40 mm; H,I= *Planecphora vokesi* (Petuch, 1989), length 62 mm; J= *Ecphora harasewychi* Petuch, 1989, length 32 mm. The specimens of *Trisecphora smithae* and *Planecphora vokesi* were collected in Shattuck Zone 16 (Drumcliff Member) of the Choptank Formation at Matoaka Cottages, St. Leonard, Calvert County, Maryland. The other species were all collected in Shattuck Zone 16 at Sandgates, St. Mary's County, Maryland, on the Patuxent River.

ricolid *Pleiorytis calvertensis* (Dall, 1900) (Plate 28, J), the dosiniinine venerid *Dosinia blackwelderi* Ward, 1992, the chionine venerid *Mercenaria cuneata* Conrad, 1869, the macomine tellinid *Florimetis biplicata* (Conrad, 1834), and the cardiid *Chesacardium blackwelderi* Ward, 1992. In cleaner sand areas within the same bathymetric range, the astartid *Astarte thisphila* Glenn, 1904 and the diplodontid *Diplodonta shilohensis* Dall, 1900 tended to dominate the infauna. Feeding on these, and many other species of small bivalves, were several drilling molluscivorous gastropods, including the muricine muricid *Panamurex (Stephanosalpinx) candelabra* (Petuch, 1988), the ocenebrine muricids *Ecphora meganae* Ward and Gilinski, 1988 (Plate 29, A and B; descendant of *E. sandgatesensis*), *Planecphora choptankensis* (Petuch, 1988) (Plate 29, G and H; descendant of *E. vokesi*), *Trisecphora patuxentia* (Petuch, 1988) (Plate 29, I and J; descendant of *T. smithae*), and *Chesathais lindae* Petuch, 1988 (Plate 29, C and D; descendant of *C. donaldasheri*), and the naticids *Euspira hemicrypta* (Gabb, 1860) (Plate 28, B) and *Euspira* species (*E. heros* complex). Also living on these open sand and mud bottoms was the predator/scavenger scaphelline volutid *Volutifusus conradiana* (Martin, 1904) (Plate 28, G and H). As is typical of all deeper water areas of the Salisbury Sea, turritellid gastropods formed scattered large beds; in this case *Torcula dianae* (Ward, 1992) and *Mariacolpus octonaria* subspecies. The ahermatypic oculinid coral *Astrohelia palmata* (Goldfuss, 1833) and the large isognomid oyster *Hippochaeta* species formed the only large-scale biohermal structures in these deeper lagoonal areas. The large fissurellid limpet *Glyphis nassula* (Conrad, 1845) (Plate 28, D) lived attached to the isognomids, as did the calyptraeid gastropod *Crucibulum multilineatum* (Conrad, 1842) and the discinid brachiopod *Discinisca lugubris* (Conrad, 1834).

The Mariacolpus octonaria Community

After another brief emergent period following Shattuck Zone 17 (St. Leonard Member) time, the Patuxent Subsea again reflooded and deposited the sediments of Shattuck Zones 18 and 19 (Boston Cliffs Member). During Zone 19 time, large sand banks accumulated in the central area of the subsea basin, presaging the subsequent complete infilling of the area in the late Serravallian. These shallow banks and clean sand-bottom lagoons supported a rich and highly endemic molluscan fauna. In deeper (10-30 m) areas, the turritellid gastropods *Mariacolpus octonaria* (Conrad, 1863) (namesake of the community; Plate 30, G) and *Torcula terebriformis* (Dall, 1892) formed immense beds and were the main food source for a wide variety of molluscivorous gastropod predators, some of which included the drilling ocenebrine muricids *Ecphora williamsi* Ward and Gilinsky, 1988 (Plate 30, A and B; first of the *E. gardnerae* complex), *E. meganae* subspecies (Plate 30, F), *Trisecphora shattucki* (Petuch, 1989) (Plate 30, E; last member of the three-ribbed genus *Trisecphora*), *Chesathais lindae drumcliffensis* Petuch, 1989 (Plate 30, H and I; last member of the genus *Chesathais*), and *Planecphora delicata* (Petuch, 1989) (Plate 30, J and K), the naticid *Euspira* species (*E. heros* complex), the busyconid *Coronafulgur choptankensis* (Petuch, 1993,) and the fasciolariid *Conradconfusus patuxentensis* (Petuch, 1993) (Plate 30, D). In shallower areas on clean sand bottoms, large beds of bivalves dominated the surface and shallow subsurface, including the small pectinid *Chesapecten monicae* Ward, 1992, the astartid *Astarte obruta* Conrad, 1834, and the arcid *Dallarca elevata* (Conrad, 1840). Larger bivalves such as the crassatellid *Marvacrassatella marylandica* (Conrad, 1832) and the cardiid *Chesacardium vostreysi* Ward, 1992 and small gastropods such as the molluscivorous trophonine muricid *Patuxentrophon patuxentensis* (Martin, 1904), the vermivorous terebrid *Laevihastula patuxentia* (Martin, 1904), the scavenger nassariid *Ilyanassa* species (*I. trivittatoides* complex), and the suctorial-feeding cancellariid *Marianarona* species (*M. alternata* complex) were also common in these shallow sandy areas.

Plate 28. Index fossils and characteristic organisms of the *Chesapecten nephrens* Community (deep lagoon pectinid beds during Shattuck Zone 17 time) and the *Bicorbula drumcliffensis* Community (shallow open muddy sand lagoons during Shattuck Zone 17 time) of the Patuxent Subsea. Included are: A= *Trigonostoma biplicifera* (Conrad, 1841), length 42 mm; B= *Euspira hemicrypta* (Gabb, 1860), length 11 mm; D= *Glyphis nassula* (Conrad, 1845), length 28 mm; E= *Chesapecten nephrens* Ward and Blackwelder, 1975, length 98 mm; F= *Mariaturricula rugata* (Conrad, 1862), length 42 mm; G,H= *Volutifusus conradiana* (Martin, 1904), length 57 mm (a species with a complex and confused nomenclatural history; also referred to as "*Scaphella virginiana* Dall, 1890", a *nomen nudum* and *dubium* and a name that should not be used, *Volutifusus choptankensis* Petuch, 1988 (a possible replacement name), and "*V. typus* Martin, 1904" and of authors); I= *Bicorbula drumcliffensis* (Oleksyshyn, 1960), length 26 mm; J= *Pleiorytis calvertensis* (Dall, 1900), length 52 mm. All specimens were collected in Shattuck Zone 17 (St. Leonard Member) of the Choptank Formation, north of Matoaka Cottages, St. Leonard, Calvert County, Maryland.

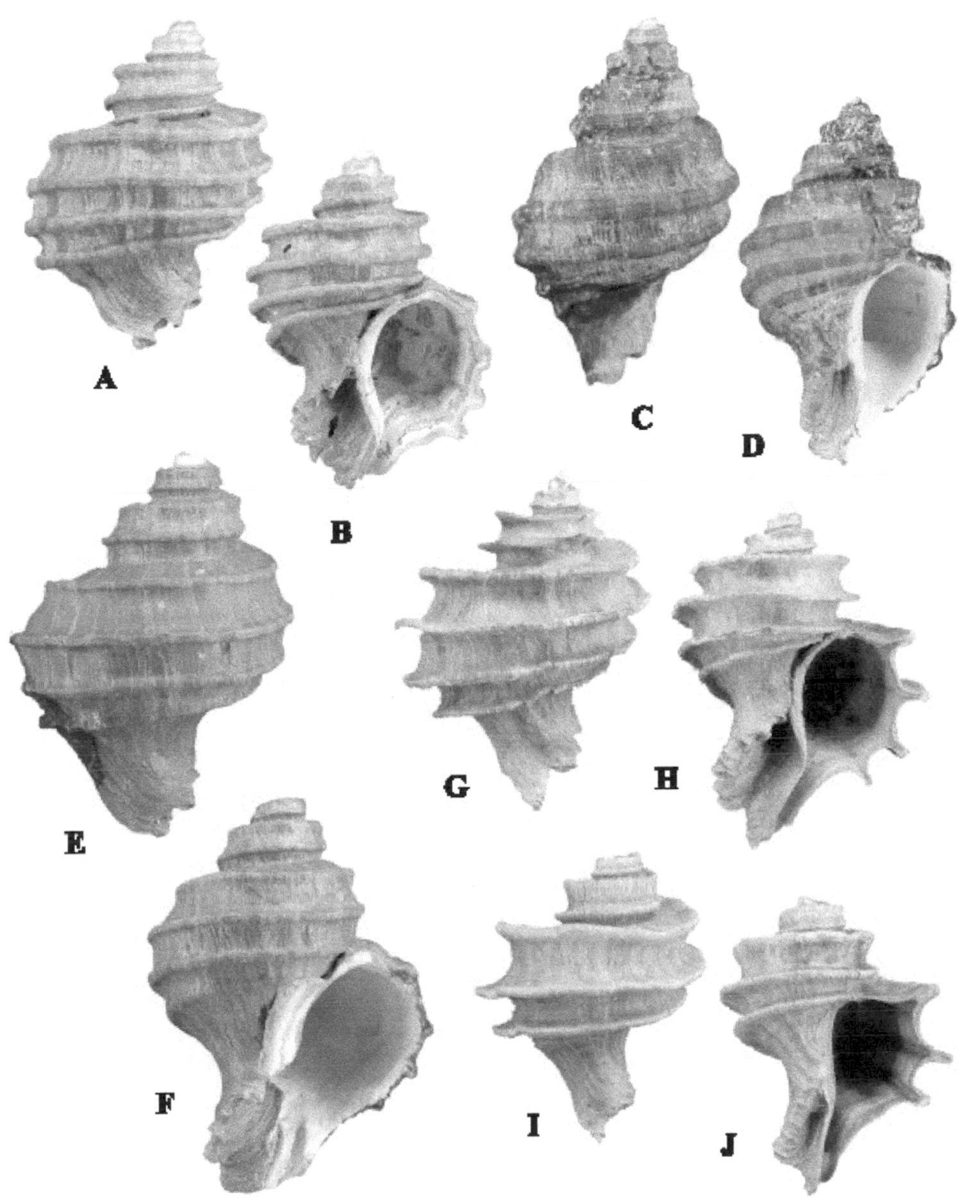

Plate 29. Ecphoras of the *Chesapecten nephrens* Community and the *Bicorbula drumcliffensis* Community, Patuxent Subsea. Includes are: A,B= *Ecphora meganae* Ward and Gilinsky, 1988, length 69 mm; C,D= *Chesathais lindae* Petuch, 1988, length 59 mm; E,F= *Ecphora rikeri* Petuch, 1988, length 80 mm; G,H= *Planecphora choptankensis* (Petuch, 1988), length 62 mm; I,J= *Trisecphora patuxentia* (Petuch, 1988), length 37 mm. The specimen of *Ecphora meganae* was collected in Shattuck Zone 17 (St. Leonard Member) of the Choptank Formation, lower bed at Drum Cliff, St. Mary's County, Maryland, on the Patuxent River. All the other specimens were collected in Shattuck Zone 17, north of Matoaka Cottages, Calvert County, Maryland.

Plate 30. Index fossils and characteristic organisms of the *Mariacolpus octonaria* Community (shallow lagoon turritellid beds during Shattuck Zone 19 time) of the Patuxent Subsea. Included are: A,B= *Ecphora williamsi* Ward and Gilinsky, 1988, length 57 mm; C= *Patuxentrophon patuxentensis* (Martin, 1904), length 18 mm; D= *Conradconfusus patuxentensis* (Petuch, 1993), length 74 mm; E= *Trisecphora shattucki* (Petuch, 1989), length 25 mm; F= *Ecphora meganae* subspecies, length 42 mm; G= *Mariacolpus octonaria* (Conrad, 1863). Length 27 mm; H,I= *Chesathais lindae drumcliffensis* Petuch, 1989, length 32 mm; J,K= *Planecphora delicata* (Petuch, 1989), length 39 mm. All specimens were collected in Shattuck Zone 19 (Boston Cliffs Member) of the Choptank Formation, upper bed at Drum Cliff, St. Mary's County, Maryland, on the Patuxent River.

Communities and Environments of the St. Mary's Subsea

By the late Serravallian, the Millville Delta and Delmarva Archipelago had filled the Salisbury Sea with sediments to such an extent that it was only one-fourth the size that it was during Calvert time (Figure 19). Also during the late Serravallian, an exceptionally severe climatic degeneration and accompanying eustatic low took place, effectively destroying the Patuxent ecosystems and leading to a regional extinction of the molluscan faunas (see Chapter 10). Particularly hard-hit were the ecphoras, where the genera *Trisecphora, Ecphorosycon,* and *Chesathais* disappeared and only one *Ecphora* lineage survived. Following this extinction event, only one species of the large four-ribbed *Ecphora* group (*E. gardnerae* complex) is found in any stratigraphic unit. All the other ecphora complexes of the Calvert and Patuxent Subseas, such as the three-ribbed (*Trisecphora*) species groups (including the cupped-ribbed *T. schmidti* complex and the thin-ribbed *T. patuxentia* complex), multiple-ribbed species (*Chesathais* and *Ecphorosycon*), and the *Ecphora rikeri* complex became extinct at this time. All subsequent late Miocene and Pliocene ecphoras (with a few five-ribbed exceptions) have four ribs and are all descended from the late Patuxent *E. williamsi* (and possibly the older *E. calvertensis*). Three and multiple-ribbed species were never seen again. The thin-ribbed *Planecphora choptankensis* complex became regionally extinct at this time, but survived farther south in Florida and the Carolinas.

The initial cold time of the late Serravallian gradually warmed to subtropical conditions during the time span of the St. Mary's Subsea. This warming pattern is dramatically illustrated in the marine paleoclimatological patterns preserved in the St. Mary's Formation. The lowest St. Mary's beds, the Conoy Member (Shattuck Zone 20), are very limited in distribution, being found only in the southernmost areas of Calvert County, Maryland (Ward, 1992). This restricted range and the impoverished nature of the molluscan fauna demonstrates that this late Serravallian transgression was small, with a relatively low water stand, had a colder marine climate, and that the basin was under marine conditions only in the south. After a brief emergent time, the St. Mary's Subsea again reflooded, this time to a greater extent and with a warmer marine climate. This second St. Mary's transgression laid down the highly fossiliferous beds of the Little Cove Point Member (Shattuck Zones 21, 22, and 23). Although having warmer marine conditions than during Conoy time, the Little Cove Point faunas were still impoverished and lacked many of the tropical faunal elements seen in the later St. Mary's beds. After a second brief emergent time in the early Tortonian, the St. Mary's Subsea reflooded for the last time, depositing the highly fossiliferous beds of the Windmill Point Member (Shattuck Zones 24 and 24A). This last transgression was the warmest, with many tropical, southern faunal elements migrating northward and appearing in the Salisbury Sea for the first time. Some of these included the gastropods *Gradiconus* (Conidae), *Mitra* s.l. (probably a new genus) (Mitridae), *Eudolium* (Tonnidae), *Cymatosyrinx* (Turridae), and *Hesperisternia* (Buccinidae). At this time, considerable erosion of the remnants of the Norfolk Arch was taking place, and the locus of deposition of the subsea was moving south into present-day Virginia (Ward, 1992).

For the entire temporal range of the St. Mary's Subsea, ten distinct marine communities can be discerned. These include the *Ecphora conoyensis* Community (shallow muddy lagoons and mud flats during Shattuck Zone 20 time), the *Ecphora asheri* Community (shallow lagoons during Shattuck Zone 21 time), the *Mariacolpus covepointensis* Community (shallow lagoon turritellid beds during Shattuck Zone 23 time), the *Chesapecten covepointensis* Community (deep lagoon pectinid beds during Shattuck Zone 23 time), the *Chesapecten santamaria* Community (deep lagoon pectinid beds during

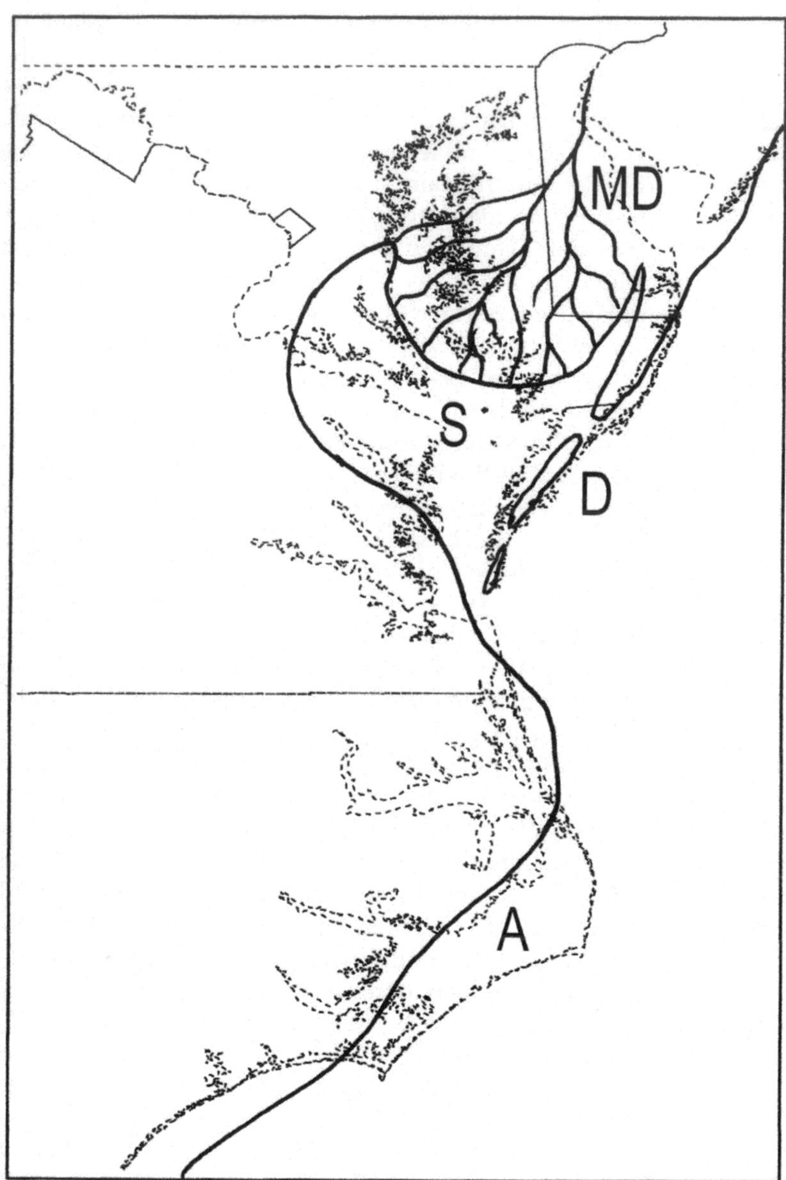

Figure 19. Approximate configuration of the coastline of the mid-Atlantic United States during late Serravallian and early Tortonian Miocene time, superimposed upon the outlines of the Recent Delaware Bay, Chesapeake Bay, and Pamlico Sound areas. Prominent features included: S= St. Mary's Subsea of the Salisbury Sea, A= Albemarle Lagoon System, MD= Millville Delta, D= Delmarva Archipelago.

Shattuck Zone 24 time), the *Lirophora alveata* Community (shallow sand bottom lagoon during Shattuck Zone 24 time), the *Mercenaria tetrica* Community (shallow sand flats during Shattuck Zone 24 time), the *Mariacolpus lindajoyceae* Community (shallow lagoon turritellid beds during Shattuck Zone 24 time), the *Bulliopsis quadrata* Community (intertidal mud flats during Shattuck Zone 24 time), and the *Gradiconus asheri* Community (shallow muddy sand lagoon during Shattuck Zone 24A time). These are synopsized in the following sections.

The Ecphora conoyensis Community

With the exception of the thick and heavily constructed *Ecphora conoyensis* Petuch, new species (see Systematic Appendix) (namesake of the community; Plate 31, D and E), few other well-preserved fossils are known from the Conoy Member. Occurring with *E. conoyensis* is a small, undescribed species of *Chesapecten*, a perfect morphological intermediate between the younger *C. covepointensis* Ward, 1992 (Shattuck Zone 23) and the older *C. monicae* Ward, 1992 (Shattuck Zone 19). The occurrence of the *Chesapecten* species indicates that this community lived in deeper parts of the lagoons behind the Delmarva Archipelago or in the open sound areas in the southern part of the basin. Badly preserved and fragmentary specimens of a *Mariacolpus* species are also present in the Conoy Member, indicating that turritellid beds occurred in the shallower areas of the earliest St. Mary's Subsea.

The Ecphora asheri Community

During Shattuck Zone 21 time (basal beds of the Little Cove Point Member), the St. Mary's Subsea still had a cool marine climate and a low sea level. Only low-diversity, shallow, muddy sand lagoonal environments have been preserved from this latest Serravallian time, with the ocenebrine muricid *Ecphora asheri* Petuch, 1988 (namesake of the community; Plate 31, A and B) being the largest and most common molluscivorous predator. This large species typically has five wide ribs and frequently is ornamented with numerous smaller, subsidiary ribs, making it the most ornately sculptured of all the Salisbury Sea ecphoras. This unusual species, the first of the large late Miocene ecphoras, was the principal predator of large sand flat-dwelling bivalves such as the cardiid *Chesacardium eshelmani* Ward, 1992 and the venerid *Mercenaria cuneata* subspecies. This ecological niche was occupied by all subsequent species of the *Ecphora gardnerae* complex. On the open sand areas between the bivalve beds were small beds of the turritellid gastropods *Torcula bohaskai* (Ward, 1992) (Plate 31, F) and *Mariacolpus covepointensis* subspecies. Living with the turritellids were several small gastropods, all survivors of the late Serravallian extinction event. Some of these included vermivores such as the turrid *Hemipleurotoma communis* (Conrad, 1830) (Plate 31, G), the suctorial-feeding cancellariid *Marianarona corbula* (Conrad, 1843) (Plate 31 C), the scavenger/predator buccinid *Celatoconus asheri* Petuch, 1988 (Plate 31, I), and the molluscivorous trophonine muricid *Scalaspira vokesae* Petuch, 1988 (Plate 31, H).

The Mariacolpus covepointensis Community

At the end of Shattuck Zone 21 time, a minor regression took place, causing much of the subsea basin to become emergent. A gradual reflooding took place during Shattuck Zone 22 time (depositing few fossils) and culminated in the high water stand and deposition of the richly fossiliferous Shattuck Zone 23 (the typical Little Cove Point beds). With deeper waters now filling the subsea, several bathymetrically-discrete communities became established. In shallow (5-10 m), sand-bottom lagoonal areas, immense beds of the

Plate 31. Index fossils and characteristic organisms of the *Ecphora conoyensis* Community (shallow muddy lagoon and mud flats during Shattuck Zone 20 time) and the *Ecphora asheri* Community (shallow sandy lagoons during Shattuck Zone 21 time) of the St. Mary's Subsea. Included are: A,B= *Ecphora asheri* Petuch, 1988, length 75 mm; C= *Marianarona corbula* (Conrad, 1843), length 13 mm; D,E= *Ecphora conoyensis* Petuch, new species (see Systematic Appendix), holotype, length 52 mm; F= *Torcula bohaskai* (Ward, 1992), length 35 mm; G= *Hemipleurotoma communis* (Conrad, 1830), length 23 mm; H= *Scalaspira vokesae* Petuch, 1988, length 18 mm; I= *Celatoconus asheri* Petuch, 1988, length 20 mm. The holotype of *Ecphora conoyensis* was collected in Shattuck Zone 20 (Conoy Member) of the St. Mary's Formation, north of Little Cove Point, Calvert County, Maryland. All other specimens were collected in Shattuck Zone 21 (lower bed of the Little Cove Point Member) of the St. Mary's Formation at Little Cove Point, Calvert County, Maryland.

turritellid gastropod *Mariacolpus covepointensis* Petuch, new species (see Systematic Appendix) (Plate 32, J) carpeted the sea floor and formed the basis of the entire community. Numerous carnivorous gastropods co-occurred with, and fed upon, the turritellids. Some of these included the busyconids *Turrifulgur covepointensis* Petuch, new species (see Systematic Appendix) (Plate 32, I) and *Sycopsis carinatum* (Conrad, 1862) (Plate 32, G), the naticid *Euspira* species (*E. heros* complex), and the fasciolariid *Conradconfusus chesapeakensis* (Petuch, 1988) (Plate 32, C and D). Living on open sand areas between the turritellid beds was a large fauna of bivalves, including the venerids *Mercenaria cuneata* subspecies, *Clementia inoceriformis* (Wagner, 1839), and *Dosinia thori* Ward, 1992, and the giant mactrid *Macrodesma subponderosa* (d'Orbigny, 1852) and many smaller species. Feeding on the larger bivalves was the drilling ocenebrine muricid *Ecphora germonae* Ward and Gilinsky, 1988 (Plate 32, E and F), the descendant of *E. asheri* and the first of the large late Miocene ecphoras with the genetically-fixed four-ribbed morphology. Also commonly occurring on these open sand patches were several other small gastropods, including the vermivorous turrids *Hemipleurotoma communis* (Conrad, 1830) and *Sediliopsis incilifera* (Conrad, 1834) and the scavenger buccinid *Bulliopsis marylandica* (Conrad, 1862) (Plate 32, H).

The Chesapecten covepointensis Community

In the deeper parts of the lagoons and sounds (10-100 m) during Shattuck Zone 23 time, the pectinid bivalve *Chesapecten covepointensis* Ward, 1992 (namesake of the community; photograph at end of chapter) formed large beds on the open sand bottoms. These were the principal food source for the large busyconid gastropod *Coronafulgur chesapeakensis* (Petuch, 1988) (Plate 32, A and B). This classic Little Cove Point index fossil occurred together with a large un-named species of the sibling genus *Busycotypus* (*B. rugosum* complex; the oldest and most primitive member). The evolutionary split between these two genera took place previously, probably during Conoy Member (Shattuck Zone 20) time (see Systematic Appendix for a discussion of the evolution of the genus *Busycotypus*). Occurring on open sand patches between the pectinid beds were numerous small gastropods, including the vermivorous terebrid *Laevihastula* cf. *simplex* (Conrad, 1830), the holothurian-feeding eulimid *Eulima migrans* Conrad, 1846, and the general carnivore scaphelline volutid *Volutifusus* species (*V. mutabilis* complex).

The Chesapecten santamaria Community

At the end of Shattuck Zone 23 time, in the early Tortonian, sea levels again dropped in a minor regression. Shortly after, sea levels rose abruptly and the St. Mary's Subsea was flooded with warm, subtropical water. By Shattuck Zone 24 (Windmill Point Member) time, the richest and most diverse molluscan fauna in the entire history of the Salisbury Sea had become established within the shrinking St. Mary's Subsea. In deeper (10-100 m) lagoonal areas, the largest Salisbury *Chesapecten* (*C. santamaria* (Tucker, 1934), namesake of the community, Plate 33, E) formed scattered beds across the open sandy bottoms. These, and numerous bivalves such as the glossid *Glossus santamaria* Ward, 1992 and the mactrid *Leptomactra delumbis* (Conrad, 1832), were the principal prey items of a number of large molluscivorous gastropods, including the busyconids *Coronafulgur coronatum* (Conrad, 1840) (Plate 33, A and B), *Busycotypus rugosum* (Conrad, 1843) (Plate 33, F), *Turrifulgur fusiforme* (Conrad, 1840) (Plate 33, C), and *Sycopsis tuberculatum* (Conrad, 1840) (Plate 33, H), and the fasciolariid *Conradconfusus parilis* (Conrad, 1832) (Plate 33, D). Living on open sand patches between the scallop beds was a large number of smaller carnivorous gastropods, including the general carnivore scaphelline volutid *Volutifusus acus* Petuch, 1988 (Plate 33, I), the scavenger/carnivore buccinids *Ptychosalpinx lindae* Petuch, 1988 (Plate 33,

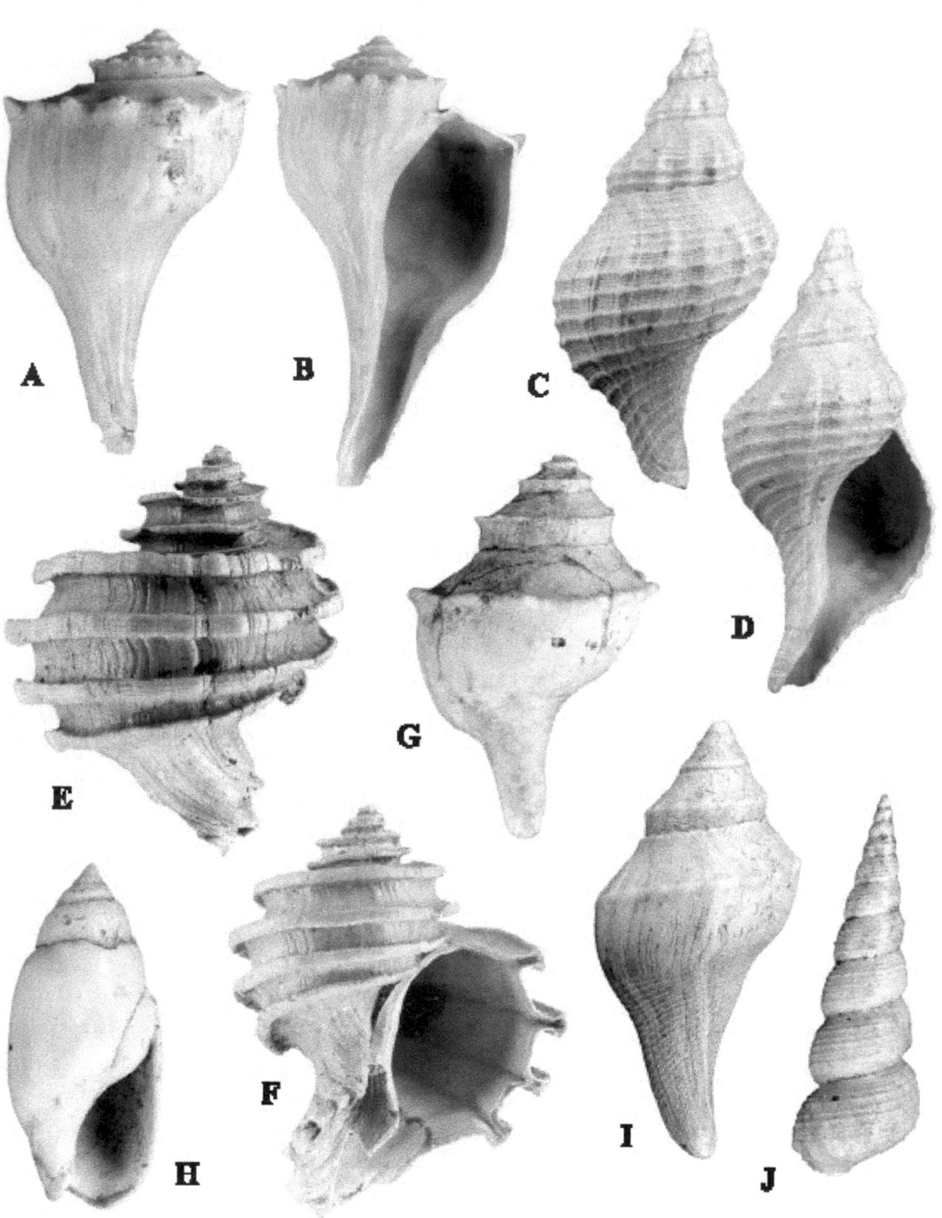

Plate 32. Index fossils and characteristic organisms of the *Mariacolpus covepointensis* Community (shallow sand lagoon turritellid beds during Shattuck Zone 23 time) and the *Chesapecten covepointensis* Community (deep lagoon pectinid beds during Shattuck Zone 23 time) of the St. Mary's Subsea. Included are: A,B= *Coronafulgur chesapeakensis* (Petuch, 1988), length 90 mm; C,D= *Conradconfusus chesapeakensis* (Petuch, 1988), length 80 mm; E,F= *Ecphora germonae* Ward and Gilinsky, 1988, length 83 mm; G= *Sycopsis carinatus* (Conrad, 1862), length 69 mm; H= *Bulliopsis marylandica* (Conrad, 1862), length 23 mm; I= *Turrifulgur covepointensis* Petuch, new species (see Systematic Appendix), holotype, length 65 mm; J= *Mariacolpus covepointensis* Petuch, new species (see Systematic Appendix), holotype, length 33 mm. All specimens were collected in Shattuck Zone 23 (upper bed of the Little Cove Point Member) of the St. Mary's Formation at Little Cove Point, Calvert County, Maryland.

G) and *P. pustulosus* Petuch, 1988, the echinoid-feeding cassid *Mariacassis caelata* (Conrad, 1830), the vermivorous turrids *Mariaturricula biscatenaria* (Conrad, 1834), *Nodisuculina engonata* (Conrad, 1862), and *Chesasyrinx rotifera* (Conrad, 1830), and the molluscivore fasciolariid *Mariafusus marylandicus* (Martin, 1904).

The Lirophora alveata Community

On open muddy sand bottom areas in shallower water (5-10 m), extensive beds of small bivalves dominated the infaunal ecosystem. Principal among these were the venerid *Lirophora alveata* (Conrad, 1830) (namesake of the community; Plate 34 L), the astartid *Astarte perplana* Conrad, 1840, and a host of small tellinids and mactrids. Feeding on the smaller bivalves was an unusually large fauna of muricids, including the trophonines *Chesatrophon laevis* (Martin, 1904) (Plate 34, D), *C. harasewychi* (Petuch, 1988), *C. chesapeakeanus* (Martin, 1904), *C. lindae* (Petuch, 1988) (Plate 34, K), *Lirosoma mariana* Petuch, 1988 (Plate 34, E), and *Scalaspira harasewychi* Petuch, 1988 (Plate 34, F), and the typhinine *Talityphis acuticosta* (Conrad, 1830). Also present were other specialized carnivorous gastropods, including the echinoid-feeding tonnid *Eudolium marylandicum* Petuch, 1988 (Plate 34, A and B), vermivorous gastropods such as the conids *Gradiconus deluvianus* (Green, 1830) (Plate 34, G and H) and *G. sanctaemariae* (Petuch, 1988) (Plate 34, I and J), suctorial-feeding cancellariids such as *Marianarona alternata* (Conrad, 1834), *M. asheri* Petuch, 1988, and *M. marylandica* Petuch, 1988, and generalized feeders such as the fasciolariid *Pseudaptyxis sanctaemariae* Petuch, 1988, the buccinid *Hesperisternia cumberlandiana* (Gabb, 1860), and the volutid *Scaphellopsis solitaria* (Conrad, 1830) (Plate 34, C).

The Mercenaria tetrica Community

Closer to shore, on shallow (0-5 m) sand banks and intertidal sand flats, extensive beds of large bivalves dominated the sea floor. Principal among these were the venerids *Mercenaria tetrica* (Conrad, 1838) (namesake of the community; Plate 35, G) and *Dosinia thori* Ward, 1992 (Plate 35, E), the mactrid *Mactrodesma subponderosa* (d'Orbigny, 1852), and the arcid *Dallarca idonea* (Conrad, 1832). Feeding on these beds of bivalves were several carnivorous gastropods, including the drilling ocenebrine muricids *Ecphora gardnerae* Wilson, 1987 (Plate 35, A and B) (specializing on large bivalves such as *Mercenaria*; see Plate 35, E and G), *Mariasalpinx emilyae* Petuch, 1988 (Plate 35, F), and *Urosalpinx subrusticus* (d'Orbigny, 1852) (Plate 35, H) (both specializing on small bivalves and oysters), and the drilling naticid *Neverita asheri* Petuch, 1988 (Plate 35, C and D). Also occurring with the drilling bivalve predators were several small vermivorous turrids, including *Oenopota parva* (Conrad, 1830), *O. marylandica* (Petuch, 1988), *Mariadrillia parvoidea* (Martin, 1904), *Chesaclava dissimilis* (Conrad, 1830), and *C. quarlesi* Petuch, 1988, the large general carnivore volutid *Volutifusus mutabilis* (Conrad, 1843) (Plate 35, I and J), and a radiation of vermivorous terebrids, including *Laevihastula simplex* (Conrad, 1830), *L. marylandica* Petuch, 1988, *L. sublirata* (Conrad, 1863), *L. chancellorensis* (Oleksyshyn, 1960), *Strioterebrum sincerum* (Dall, 1895), and *S. curviliratum* (Conrad, 1843). Large biohermal clumps of the giant barnacle *Chesaconcavus* also grew on these shallow sand banks, and these served as a food resource for the drilling ocenebrine muricids *Urosalpinx* and *Mariasalpinx*.

The Mariacolpus lindajoyceae Community

In deeper tidal channels adjacent to the sand bars, immense beds of turritellid gastropods accumulated, including *Mariacolpus lindajoyceae* Petuch, new species (see Systematic Appendix) (namesake of the community; Plate 36, H), *Torcula subvariabilis* (d'Orbigny, 1852) (Plate 36, G), and *T. chancellorensis* (Oleksyshyn, 1959) (Plate 36, F). The

Plate 33. Index fossils and characteristic organisms of the *Chesapecten santamaria* Community (deep lagoon pectinid beds during Shattuck Zone 24 time) of the St. Mary's Subsea. Included are: A,B= *Coronafulgur coronatum* (Conrad, 1840), length 98 mm; C= *Turrifulgur fusiforme* (Conrad, 1840), length 69 mm; D= *Conradconfusus parilis* (Conrad, 1832), length 106 mm; E= *Chesapecten santamaria* (Tucker, 1934), length 135 mm; F= *Busycotypus rugosum* (Conrad, 1843), length 72 mm; G= *Ptychosalpinx lindae* Petuch, 1988, length 26 mm; H= *Sycopsis tuberculatum* (Conrad, 1840), length 63 mm; I= *Volutifusus acus* Petuch, 1988, length 96 mm (compare with V. mutabilis, Plate 35 I and J; note complete lack of columellar plications and shoulder angle). All specimens were collected in Shattuck Zone 24 (Windmill Point Member) of the St. Mary's Formation at Chancellor's Point, St. Mary's County, Maryland.

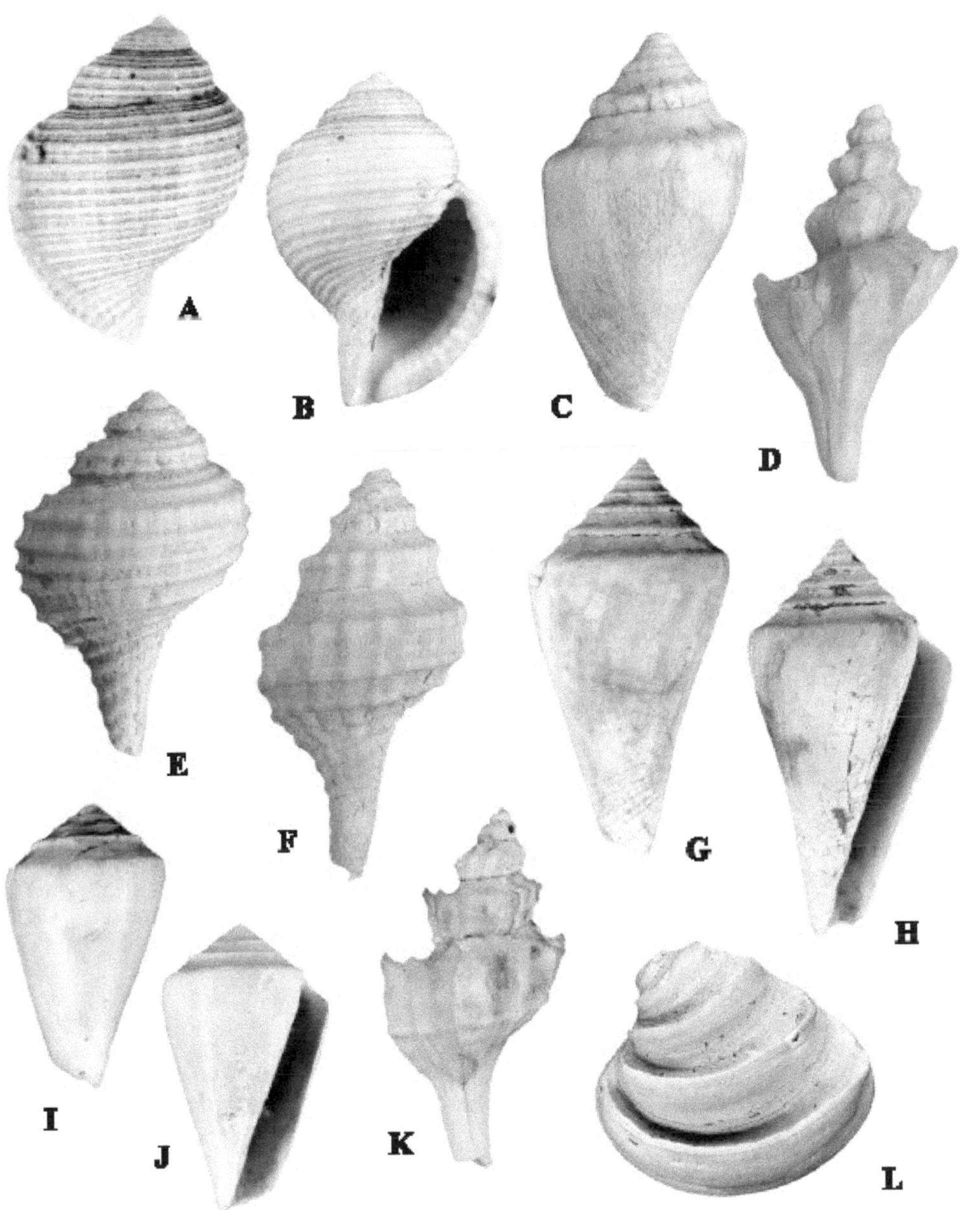

Plate 34. Index fossils and characteristic organisms of the *Lirophora alveata* Community (shallow muddy sand bottom lagoon during Shattuck Zone 24 time) of the St. Mary's Subsea. Included are: A,B= *Eudolium marylandicum* Petuch, 1988, length 44 mm; C= *Scaphellopsis solitaria* (Conrad, 1830), length 32 mm; D= *Chesatrophon laevis* (Martin, 1904), length 16 mm; E= *Lirosoma mariana* Petuch, 1988, length 25 mm; F= *Scalaspira harasewychi* Petuch, 1988, length 15 mm; G,H= *Gradiconus deluvianus* (Green, 1830), length 55 mm; I,J= *Gradiconus sanctaemariae* (Petuch, 1988), length 40 mm; K= *Chesatrophon lindae* (Petuch, 1988), length 15 mm; L= *Lirophora alveata* (Conrad, 1830), length 24 mm. All specimens were collected in Shattuck Zone 24 (Windmill Point Member) of the St. Mary's Formation at Chancellor's Point, St. Mary's County, Maryland.

Plate 35. Index fossils and characteristic organisms of the *Mercenaria tetrica* Community (shallow muddy sand lagoons during Shattuck Zone 24 time) of the St. Mary's Subsea. Included are: A,B= *Ecphora gardnerae* Wilson, 1987, length 96 mm; C,D= *Neverita asheri* Petuch, 1988, length 42 mm; E= *Dosinia thori* Ward, 1992, length 80 mm, with 4 mm diameter muricid-type hole, probably drilled by *Ecphora gardnerae*; F= *Mariasalpinx emilyae* Petuch, 1988, length 28 mm; G= *Mercenaria tetrica* (Conrad, 1838), length 95 mm, with 5 mm diameter hole drilled by *Ecphora gardnerae*; H= *Urosalpinx subrusticus* (d'Orbigny, 1852), length 25 mm; I,J= *Volutifusus mutabilis* (Conrad, 1843), length 99 mm (compare with *V. acus*; note two strong columellar plications and definite shoulder angle). All specimens were collected in Shattuck Zone 24 (Windmill Point Member), St. Mary's Formation, at Chancellor's Point, St. Mary's County, Maryland.

turritellids were the principal prey items for a large radiation of drilling naticid gastropods, some of which included *Poliniciella marylandica* Petuch, 1988 (Plate 36, A and B), *Neverita discula* Petuch, 1988 (Plate 36, D and E), *Euspira interna* (Say, 1822) (Plate 36, L), *E.* cf. *catenoides* (Conrad, 1863), *E. tuomeyi* (Whitfield, 1894) (Plate 36, J), and *Sinum fragilis* (Conrad, 1830) (molluscivorous, but may not have been a driller), and small muricids such as the trophonine *Scalaspira marylandica* Schoonover, 1941 (Plate 36, I). Living on scattered sand patches between the turritellid beds were a large number of other carnivorous gastropods, including the zoantherian-feeding epitoniids *Epitonium chancellorensis* Petuch, 1988 and *Stenorhytis expansa* (Conrad, 1842) (Plate 36, C), the generalist carnivore small volutid *Volutifusus asheri* Petuch, 1988 (Plate 36, K), suctorial-feeders such as the cancellariids *Mariasveltia lunata* (Conrad, 1830) and *Admete marylandica* (Martin, 1904) and vermivores such as the turrids *Sediliopsis distans* (Conrad, 1862), *S. gracilis* (Conrad, 1830), *Cymatosyrinx limatula* (Conrad, 1830), and *C. mariana* Petuch, 1988.

The Bulliopsis quadrata Community

The Millville Delta area, the dominant geomorphological feature of the northern part of the St. Mary's Subsea, contained several distinct environments. These ranged from pure fresh water fluvial systems to brackish water estuaries to organic-rich intertidal mud flats. This latter environment housed a rich endemic fauna of the scavenger gastropod families Buccinidae and Nassariidae, and these mud snails must have carpeted the mud flats in immense aggregations, as does the Recent Carolinian *Ilyanassa obsoleta*. Some of the buccinids included *Bulliopsis quadrata* (Conrad, 1830) (namesake of the community; Plate 37, A and B), *B. integra* (Conrad, 1842) (Plate 37 C and D), *B. subcylindrica* (Conrad, 1866) (Plate 37, E), and *B. ovata* (Conrad, 1862) (Plate 37, F and G), and some of the nassariids included *Ilyanassa marylandica* (Martin, 1904) (Plate 37, H), *I. trivittatoides* subspecies (Plate 37, I), and *I. peralta* (Conrad, 1868). The small scavenger columbellid *Zaphrona communis* (Conrad, 1862) (Plate 37, J) lived with the buccinid and nassariid colonies and competed for carrion and detritus. In the pure fresh water areas of the delta, the pleurocerid gastropod *Goniobasis marylandicus* (Martin, 1904) was abundant, particularly at the mouths of deltaic rivers. The large, razor-toothed crocodilian *Thecachampsa sicaria* Cope, 1869 was also found in these deltaic fluvial environments, and may have ventured onto the estuarine mud flats. In more brackish estuarine areas adjacent to the *Bulliopsis quadrata* Community, the oyster *Crassostrea carolinensis* subspecies formed large bars. These oyster bioherms supplied the substrate for the attached epibiont mytilid bivalves *Modiolus ducatelii* Conrad, 1840 and *Gregariella virginica* (Conrad, 1867) and a number of encrusting brackish water ectoprocts, including the chilostomes *Membranipora fistula* Ulrich and Bassler, 1904, *Cupularia denticulata* (Conrad, 1841), and *Lepralia montifera* Ulrich and Bassler, 1904 and the cyclostome *Theonoa glomerata* Ulrich and Bassler, 1904.

The Gradiconus asheri Community

At the end of Shattuck Zone 24 time, another eustatic drop took place and most of the St. Mary's Subsea was emergent. Subsequently, this emergent time was followed by another reflooding of the basin in the mid-Tortonian. The final St. Mary's depositional event was very brief, and is preserved only as a thin (approximately 0.5 m thick) shelly sandstone ledge on top of some exposures of Zone 24. This unrecognized St. Mary's unit was referred to as "Zone 24A" (Petuch, 1997) and is best developed at Chancellor's Point and along St. Inigoe's Creek, St. Mary's County, Maryland (due to its distinct indurated lithology, this uppermost St. Mary's bed is here informally referred to as the "Chancellor Point Member" of the St. Mary's Formation). The intervening emergent time between the underlying

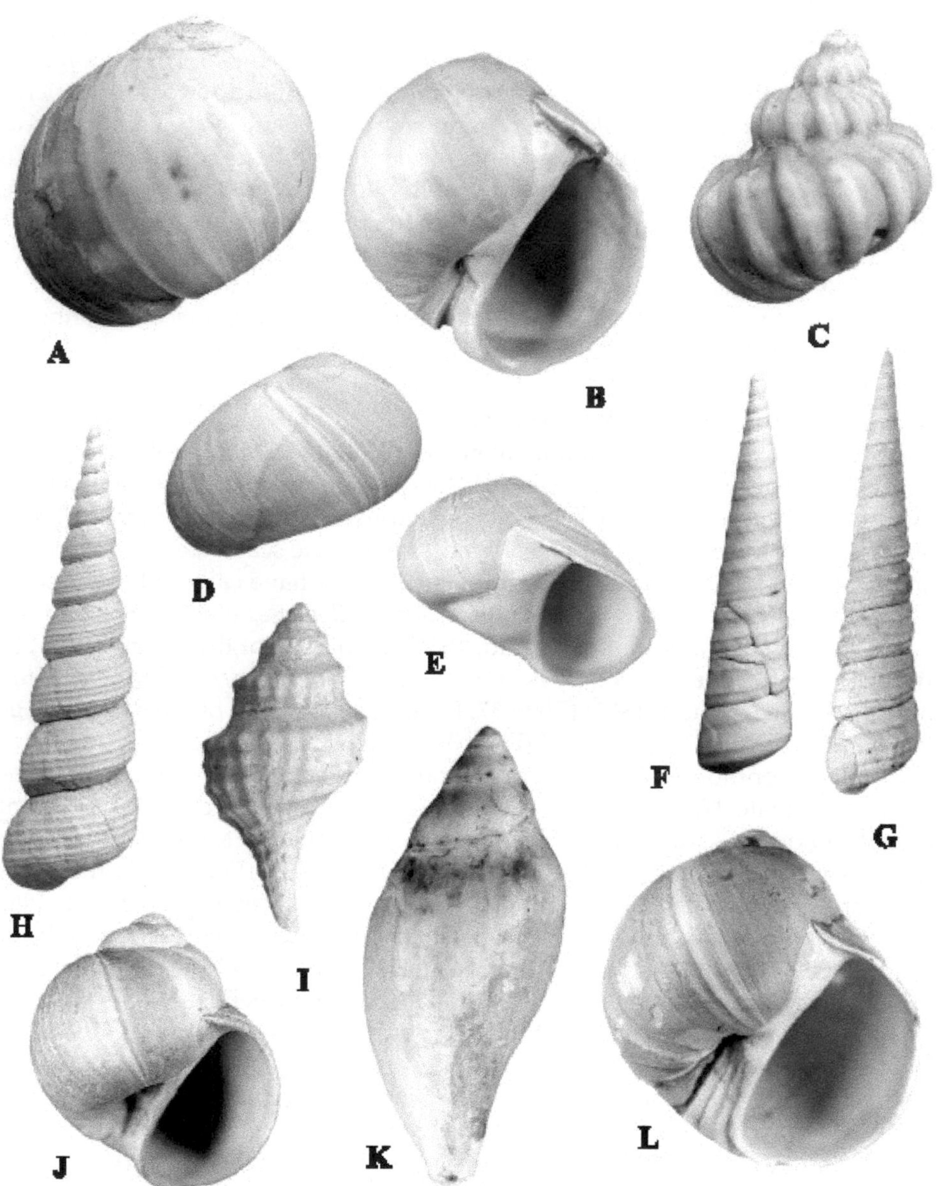

Plate 36. Index fossils and characteristic organisms of the *Mariacolpus lindajoyceae* Community (shallow lagoon turritellid beds during Shattuck Zone 24 time) of the St. Mary's Subsea. Included are: A,B= *Poliniciella marylandica* Petuch, 1988, length 20 mm; C= *Stenorhytis expansa* (Conrad, 1842), length 23 mm; D,E= *Neverita discula* Petuch, 1988, width 18 mm; F= *Torcula chancellorensis* (Oleksyshyn, 1959), length 32 mm; G= *Torcula subvariabilis* (d'Orbigny, 1852), length 48 mm; H= *Mariacolpus lindajoyceae* Petuch, new species (see Systematic Appendix), holotype, length 37 mm; I= *Scalaspira marylandica* Schoonover, 1941, length 16 mm; J= *Euspira tuomeyi* (Whitfield, 1894), length 18 mm; K= *Volutifusus asheri* Petuch, 1988, length 74 mm (compare with *V. acus* and *V. mutabilis*; a smaller, stouter, lower-spired shell than the two larger Zone 24 species); L= *Euspira interna* (Say, 1822), length 43 mm. All specimens were collected in Shattuck Zone 24 (Windmill Point Member) of the St. Mary's Formation, along Chancellor's Point, St. Mary's County. Maryland.

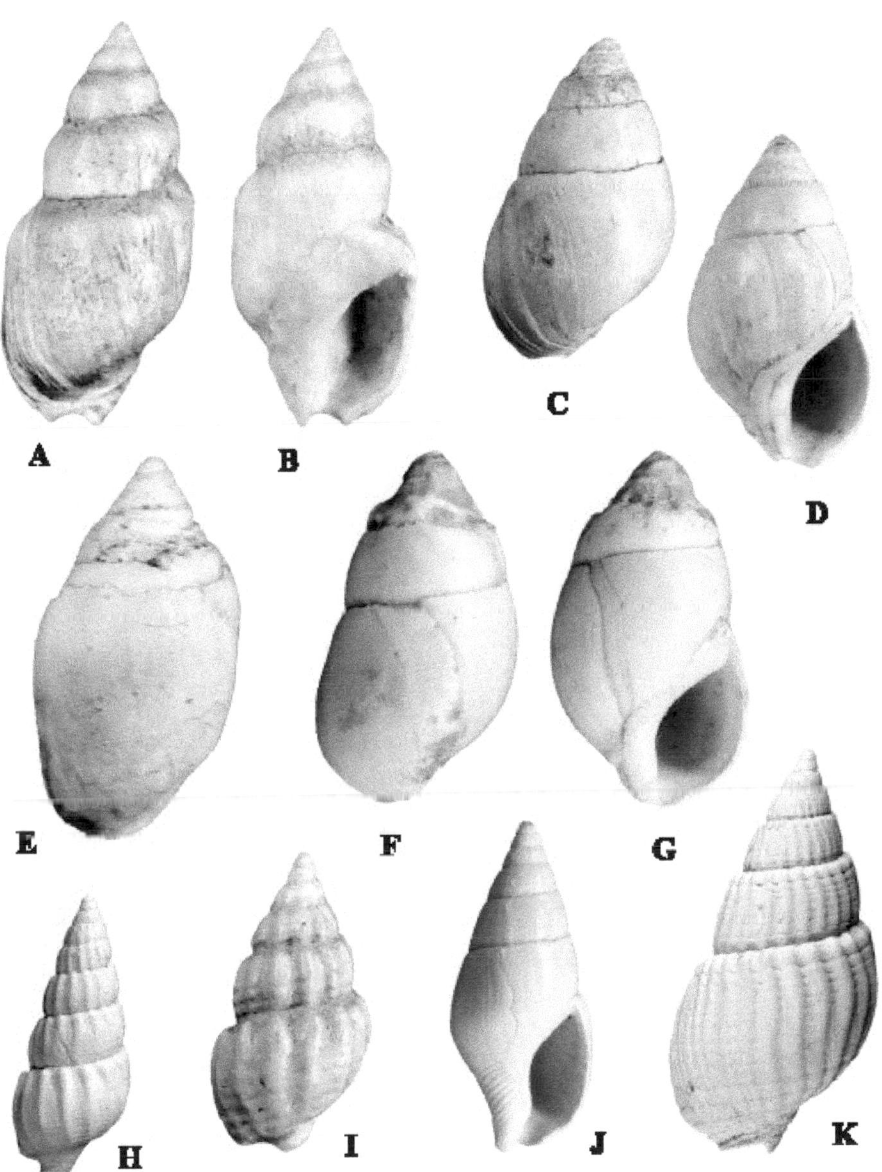

Plate 37. Index fossils and characteristic organisms of the *Bulliopsis quadrata* Community (organic-rich intertidal mud flats and estuaries during Shattuck Zone 24 time) of the St. Mary's Subsea. Included are: A,B= *Bulliopsis quadrata* (Conrad, 1830), length 23 mm; C,D= *Bulliopsis integra* (Conrad, 1842), length 18 mm; E= *Bulliopsis subcylindrica* (Conrad, 1866), length 21 mm; F,G= *Bulliopsis ovata* (Conrad, 1862), length 19 mm; H= *Ilyanassa marylandica* (Martin, 1904), length 14 mm; I= *Ilyanassa trivittatoides* subspecies, length 10 mm; J= *Zaphrona communis* (Conrad, 1862), length 9 mm; K= *Ilyanassa peralta* (Conrad, 1868), length 22 mm. All specimens were collected in a muddy-clay facies of Shattuck Zone 24 (Windmill Point Member) of the St. Mary's Formation, along the southern end of Chancellor's Point, St. Mary's County, Maryland.

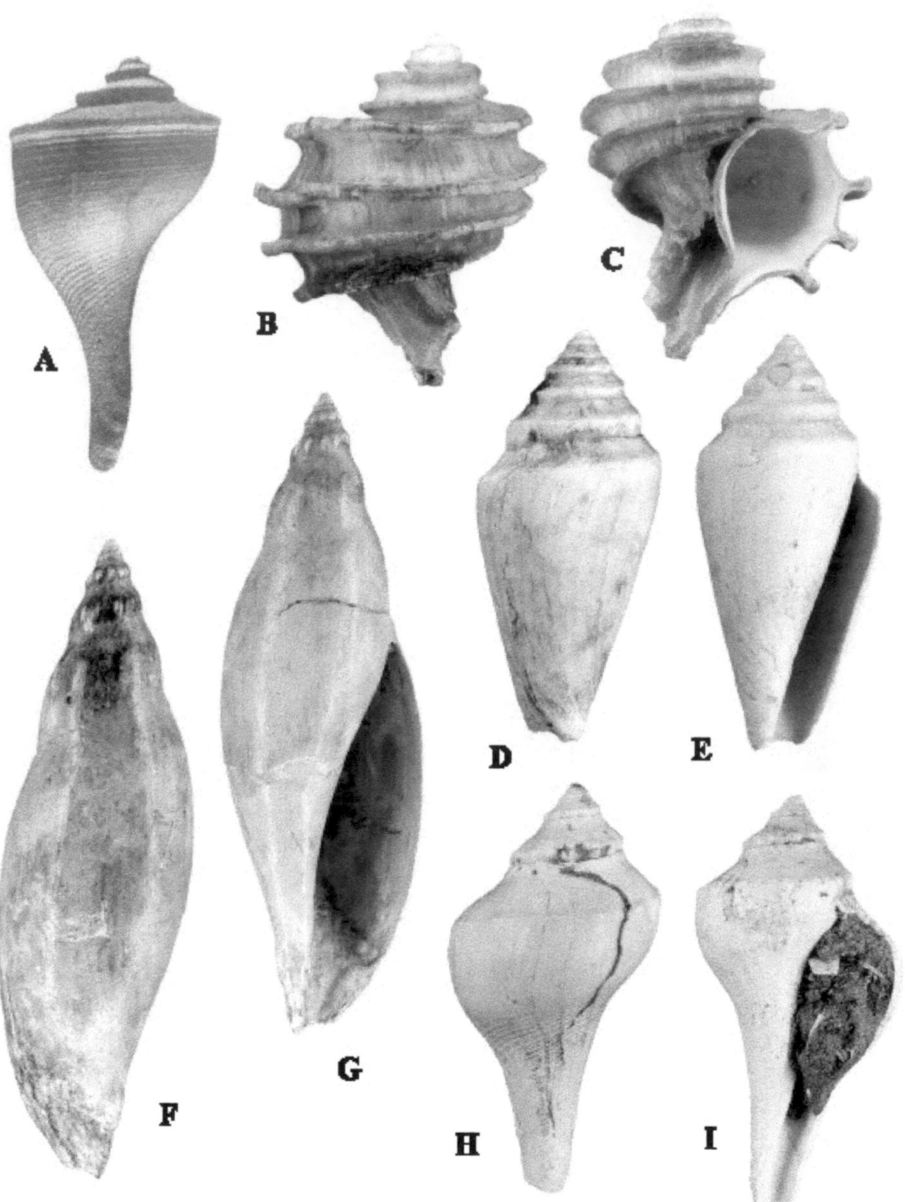

Plate 38. Index fossils and characteristic organisms of the *Gradiconus asheri* Community (shallow muddy sand lagoon during Shattuck Zone "24A" time) of the St. Mary's Subsea. Included are: A= *Busycotypus asheri* Petuch, 1988, length 54 mm; B,C= *Ecphora gardnerae angusticostata* Petuch, 1989, length 40 mm; D,E= *Gradiconus asheri* (Petuch, 1988), length 42 mm; F,G= *Volutifusus marylandicus* Petuch, 1988, length 120 mm; H,I= *Turrifulgur turriculus* Petuch, 1988, length 30 mm. All specimens were collected in Shattuck Zone "24A" ("Chancellor's Point Member") of the St. Mary's Formation, at Chancellor's Point, St. Mary's County, Maryland.

Windmill Point Member and the overlying Chancellor's Point Member must have been longer than the other St. Mary's unconformities, as most of the species had evolved into new forms. The community preserved within the Chancellor Point Member (Zone 24A) ecologically resembled the older *Lirophora alveata* Community and occurred on the same type of shallow, quiet, sand bottom areas with normal salinity. Since most specimens must be chipped or hammered from the hard sandstone, only a partial record of the ecosystem structure can be deciphered. Some of the principal organisms of this last St. Mary's lagoonal environment included a number of carnivorous gastropods, including the vermivorous conid *Gradiconus asheri* (Petuch, 1988) (namesake of the community; Plate 38, D and E), the drilling ocenebrine muricid *Ecphora gardnerae angusticostata* Petuch, 1989 (Plate 38, B and C), the bivalve-feeding busyconids *Busycotypus asheri* Petuch, 1988 (Plate 38, A) and *Turrifulgur turriculus* Petuch, 1988 (Plate 38, H and I), and the general carnivore volutids *Volutifusus marylandicus* Petuch, 1988 (Plate 38, F and G) and *Megaptygma meganucleus* (Petuch, 1988). This was the last molluscan fauna of the Salisbury Sea.

Communities and Environments of the Rappahannock Subsea

By late Tortonian time, massive erosion of the Albemarle Lagoon area and the Norfolk Arch had led to a revitalized and expanded Albemarle Sea (Figure 20), manifested as the Rappahannock Subsea. This composite paleosea, composed of the remnants of the southernmost part of the now infilled Salisbury Sea and the old Pamlico Subsea basin, had a relatively simple geomorphology. The deepest part of the Rappahannock Subsea was in the north, along the Virginia coast (Ward, 1992), and this was the center of deposition for the principal Rappahannock stratigraphic unit, the Eastover Formation. Lying behind the Delmarva Peninsula and Archipelago, this deeper water area was relatively protected from open oceanic conditions and resembled a quiet water sound. Along the western coast of the Rappahannock Subsea, a series of low barrier islands, the Bayport Archipelago (originally described as the "Bayport High" by Ward and Blackwelder, 1980), separated the eastern deep water sandy areas from the western shallow muddy estuaries. Farther south, along the North Carolina coast, there also existed extensive shallow water areas, sandy shoals (the Aurora Shoals; named for Aurora, North Carolina, site of the Lee Creek Mine), and possibly chains of barrier islands. This bathymetric and geomorphologic pattern was retained throughout the rest of the Miocene and the entire Pliocene. From the two members of the Eastover Formation (the late Tortonian Claremont Manor Member and the latest Tortonian-early Messinian Cobham Bay Member), five distinct communities can be discerned. These include the *Ecphora whiteoakensis* Community (shallow sand bottom lagoons during Claremont Manor time), the *Chesapecten ceccae* Community (deeper water sand bottom lagoons or sounds during Claremont Manor time), the *Mulinia rappahannockensis* Community (intertidal mud flats and shallow muddy lagoons during Cobham Bay time), the *Oliva idonea* Community (shallow sand bottom lagoons during Cobham Bay time), and the *Placopecten principoides* Community (deeper water sand bottom lagoons or sounds during Cobham Bay time).

The Ecphora whiteoakensis Community

The earliest Rappahannock molluscan faunas closely resembled those of the Salisbury Sea. This was particularly apparent in the shallow sandy and intertidal environments during Claremont Manor time (0-5 m). Here, the large ocenebrine muricid *Ecphora whiteoakensis* Ward and Gilinsky, 1988 (namesake of the community; Plate 39, C and D; last member of the *E. gardnerae* complex) was the principal drilling predator of a large fauna of bivalves, including the venerids *Mercenaria druidi* subspecies and *Lirophora dalli* (Olsson,

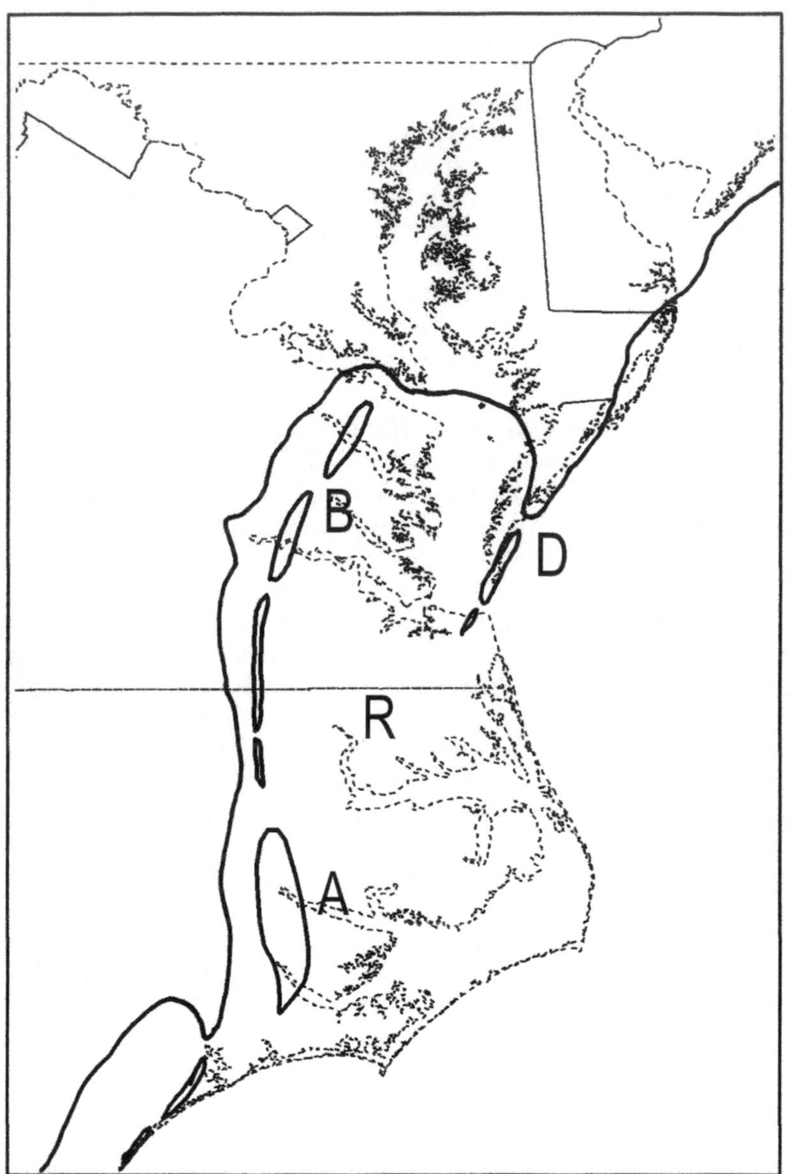

Figure 20. Approximate configuration of the coastline of the mid-Atlantic United States during late Tortonian and early Messinian Miocene time, superimposed upon the outline of the Recent Chesapeake Bay and Pamlico Sound areas. Prominent features included: R= Rappahannock Subsea of the Albemarle Sea, B= Bayport Archipelago, A= Aurora Shoals of the Rappahannock Subsea, D= Delmarva Peninsula and Archipelago.

1914), the arcids *Dallarca virginiae* (Dall, 1898) and *D. rotunda* Ward, 1992, and beds of the ostreid oyster *Ostrea brucei* Ward, 1992 and the isognomid oyster *Hippochaeta* species. Farther south, in the Aurora Shoals area, this community also included the large pycnodontine gryphaeid oyster *Gigantostrea leeana* (Wilson, 1987) (originally thought to be from the Langhian Pungo River Formation but now found to be from previously-unrecognized Eastover beds in the Lee Creek Mine, Aurora, North Carolina and from the contemporaneous Bayshore Formation in Florida). This large oyster, along with *Ostrea brucei* and *Hippochaeta* species, formed scattered biohermal structures in the intertidal areas.

The Chesapecten ceccae Community

Farther offshore, in the deeper areas of the lagoons and sounds (5-100 m), large beds of scallops dominated the sea floor. Principal among these were *Chesapecten ceccae* Ward, 1992 (namesake of the community) and *Placopecten* species (*P. principoides* complex). Occurring along with the scallops were a number of large, endemic bivalves, including the glycymerid *Costaglycymeris virginiae* (Dall, 1898), the euloxid *Euloxa latisulcata* (Conrad, 1839), and the crassatellid *Marvacrassatella surreyensis* (Mansfield, 1929). The principal gastropod predator of these shallowly infaunal bivalves was the busyconid *Coronafulgur kendrewi* subspecies (*C. coronatum* complex; see Systematic Appendix). Scattered between the pectinid beds in shallower water areas were small colonies of the Salisbury relict turritellid *Mariacolpus carinata* (Gardner, 1948).

The Mulinia rappahannockensis Community

During the latest Tortonian, the marine climate degenerated and a brief eustatic low caused much of the Rappahannock Subsea to become emergent. This was followed by a major sea level rise during the early Messinian (Cobham Bay time), when the basin was reflooded with much warmer, subtropical water; leading to the reinvasion from the south of the tropical molluscan genera *Gradiconus* (Conidae) and *Oliva* (Olividae). Along the protected shorelines, estuaries, and shallow lagoons west of the Bayport Archipelago, extensive organic-rich mud flats accumulated at this time. These intertidal mud flats and adjacent shallow muddy lagoon areas housed a rich endemic molluscan fauna that was dominated by the mactrid bivalve *Mulinia rappahannockensis* (Gardner, 1944) (Plate 39, H; namesake of the community and originally described as a *Spisula* species) (Ward, 1992). Occurring in these muddy, estuarine areas were other larger bivalves such as the venerids *Mercenaria druidi* Ward, 1992 and *Lirophora vredenburgi* Ward, 1992, the cardiid *Chesacardium blountense* (Mansfield, 1932), the arcid *Dallarca carolinensis* (Dall, 1898), the mactrid *Spisula bowlerensis* (Gardner, 1943), large bars of the ostreid oysters *Ostrea geraldjohnsoni* Ward, 1992 and *Conradostrea greeni* Ward, 1992, and small aggregations of the turritellid gastropod *Mariacolpus plebeia* (Say, 1824). The principal gastropod predators of the bivalves and turritellids were the drilling ocenebrine muricid *Urosalpinx rappahannockensis* (Gardner, 1948) (Plate 39, I), the naticid *Euspira* species (*E. tuomeyi* complex), and the busyconid *Coronafulgur kendrewi* Petuch, new species (Plate 39, A and B; see Systematic Appendix). The organic-rich open intertidal mud flats also supported large aggregations of scavenger gastropods, including the buccinid *Bulliopsis bowlerensis* Allmon, 1990 (Plate 39, G) and the nassariid *Ilyanassa* species (*I. peralta* complex).

The Oliva idonea Community

Farther offshore, on cleaner sand bottoms in quiet, shallow lagoonal areas (5-10 m), large beds of bivalves occurred. Some of these included the venerid *Dosinia* species, the astartids *Astarte cobhamensis* Ward, 1992 and *A. rappahannockensis* (Gardner, 1944), the gly-

Plate 39. Index fossils and characteristic organisms of the *Ecphora whiteoakensis* Community (shallow sand bottom lagoons during Claremont Manor Member time), the *Mulinia rappahannockensis* Community (intertidal mud flats during Cobham Bay Member time), the *Oliva idonea* Community (shallow sand bottom lagoons during Cobham Bay Member time), and the *Placopecten principoides* Community (deeper water sand bottom lagoons during Cobham Bay Member time) of the Rappahannock Subsea. Included are: A= *Coronafulgur kendrewi* Petuch, new species (see Systematic Appendix), holotype, length 146 mm; B= *Coronafulgur kendrewi* Petuch, new species, paratype, length 122 mm; C,D= *Ecphora whiteoakensis* Ward and Gilinsky, 1988, length 89 mm; E,F= *Oliva (Strephona) idonea* Conrad, 1839, length 25 mm; G= *Bulliopsis bowlerensis* Allmon, 1990, length 24 mm; H= *Mulinia rappahannockensis* (Gardner, 1944), length 24 mm; I= *Urosalpinx rappahannockensis* (Gardner, 1948), length 19 mm; J,K= *Ecphora kochi* Ward and Gilinsky, 1988, length 92 mm; L= *Placopecten principoides* (Emmons, 1858), length 135 mm. The specimens of *Coronafulgur kendrewi, Bulliopsis bowlerensis, Urosalpinx rappahannockensis*, and *Mulinia rappahannockensis* were collected in the Cobham Bay Member of the Eastover Formation, 5 km southeast of Bowlers Wharf, Essex County, Virginia, on the Rappahannock River. The specimens of *Ecphora whiteoakensis, Ecphora kochi*, and *Oliva idonea* were collected in the Eastover Formation in the Lee Creek Mine, Aurora, Beaufort County, North Carolina (previously unknown and unreported beds). The specimen of *Placopecten principoides* was collected at Murfreesboro, North Carolina.

cymerid *Costaglycymeris mixoni* Ward, 1992, the glossid *Glossus fraterna* (Say, 1824), and the arcids *Dallarca clisea* (Dall, 1898) and *Rasia arata* (Say, 1824). Feeding on the larger bivalves were several molluscivorous gastropods, including the eurytopic and eurybathic busyconid *Coronafulgur kendrewi* Petuch, the drilling ocenebrine muricids *Pterorhytis* species, *Urosalpinx* species (*Mariasalpinx* ?), and *Ecphora kochi* Ward and Gilinsky, 1988 (Plate 39, J and K), and the olivid *Oliva (Strephona) idonea* Conrad, 1839 (namesake of the community; Plate 39, E and F), which also fed on smaller, delicate bivalves and on carrion. On scattered open sand patches between the bivalve beds, small beds of the turritellid gastropod *Mariacolpus plebeia* (Say, 1824) occurred along with the vermivorous conid *Gradiconus spenceri* (Ward, 1992) and large biohermal clumps of the chamid bivalve *Pseudochama corticosa* (Conrad, 1833) and the isognomid oyster *Hippochaeta* species.

The Placopecten principoides Community

In the deep lagoon and open sound areas of the Rappahannock Subsea (10-100 m), immense beds of pectinid bivalves carpeted the sea floor. These scallop beds were made up almost entirely of three large species, including *Carolinapecten urbannaensis* (Mansfield, 1929) (first of its genus to appear in the fossil record), *Placopecten principoides* (Emmons, 1858) (namesake of the community; Plate 39, L), and *Chesapecten middlesexensis* (Mansfield, 1936). Occurring in these dense scallop beds were several other large bivalves, including the deep water arcid *Striarca centenaria* (Say, 1824), the crassatellid *Marvacrassatella urbannaensis* (Mansfield, 1929), and the hiatellid *Panopea goldfussi* Wagner, 1839. Feeding on these bivalves was the large busyconid *Busycotypus* species (*B. rugosum* complex) and the scaphelline volutid *Volutifusus* species (*V. mutabilis* complex).

Communities and Environments of the Walton Subsea

Known from only a few small exposures of the Shoal River Formation (Serravallian-mid-Tortonian) in Walton County, Florida, the Walton Subsea of the Choctaw Sea is still essentially unstudied. The little data that is available shows that the oceanographic conditions within the basin were cooler than the predecessor Chipola Subsea and the Walton fauna lacked many of the key tropical molluscan indicators, such as the Strombidae, Vasinae-Turbinellidae, Cypraeidae, Ovulidae, Modulidae, Turbinidae, Potamididae, and Lyriinae-Volutidae.

The Torcula waltonensis Community

Based upon the fossil record at Shell Bluff on the Shoal River, one distinct paratropical turritellid-based community can be discerned. This occurred in deeper lagoonal, sand bottom areas and was dominated by large beds of *Torcula waltonensis* (Gardner, 1947) (namesake of the community), *Bactrospira* species, *Eichwaldiella segmenta* (Gardner, 1947), and *Apicula jacula* (Gardner, 1947). On open sand patches between the turritellid beds, large numbers of bivalves occurred, some of which included the venerids *Lirophora funiakensis* (Gardner, 1926), *L. trimeris* (Gardner, 1926), *Hyphantosoma waltonensis* (Gardner, 1926), and *Mercenaria nannodes* (Gardner, 1926), the crassatellid *Marvacrassatella alicensis* (Mansfield, 1932), and the tellinids *Tellina piesa* (Gardner, 1926) and *Phyllodina leptalea* (Gardner, 1926). The turritellids and bivalves were the main prey items for a large drilling muricid gastropod fauna, some of which included *Microrhytis dryas* (Gardner, 1947), *Siratus nicholsi* (Gardner, 1947) (Plate 40, H), *Eupleura pterina* Gardner, 1947, *Chicoreus aldrichi* (Gardner, 1947), and *Vokesinotus tribaka* (Gardner, 1947). Also feeding on the bivalves and turritellids were several molluscivorous gastropods including the volutids *Scaphella pycnoplecta* (Gardner, 1937) and *S. florea* (Gardner, 1947), the olivid *Oliva*

(Strephona) waltoniana Gardner, 1937, and the busyconids *Busycon montforti* Aldrich, 1907, *Turrifulgur aldrichi* (Gardner, 1947), and *T. dasum* (Gardner, 1947). Echinoderm-feeders such as the personiid *Distorsio floridana* (Gardner, 1947) and the cassid *Phalium murrayi* Schmelz, 1996 also occurred on the sand patches, as did a large fauna of vermivorous conids, some of which included *Ximeniconus waltonensis* (Aldrich, 1903) (Plate 40, G), *Gradiconus drezi* (S. Hoerle, 1976) (Plate 40, J), *Lithoconus submoniliferus* (Gardner, 1937), and *Conasprella infulatus* (S. Hoerle. 1976).

Communities and Environments of the Alaqua Subsea

After a brief emergent time during the mid-Tortonian eustatic low, the Choctaw Sea basin again reflooded as the Alaqua Subsea. Due to a paucity of fossil exposures as in the preceding Walton Subsea, little is known about the oceanography and ecology of this late Tortonian-early Messinian paleosea. What little data has been assembled has come from small and ephemeral exposures of the Red Bay Formation in Walton County, Florida (Mansfield, 1935). The molluscan paleontology of these beds has shown that the Alaqua fauna was only paratropical in nature and lacked most of the key tropical index groups such as the Strombidae, Vasinae-Turbinellidae, Cypraeidae, Ovulidae, Potamididae, Turbinidae, Modulidae, and Lyriinae-Volutidae.

The *Eichwaldiella blountensis* Community

As in the Shoal River beds of the Walton Subsea, only one shallow water, sand bottom community can be discerned and, likewise, this was based on turritellid gastropods. In this case, the turritellids *Eichwaldiella blountensis* (Mansfield, 1935) (namesake of the community), *Apicula alaquaensis* (Mansfield, 1935), *A. permenteri* (Mansfield, 1935), and *Torcula vaughanensis* (Mansfield, 1935) formed immense beds that carpeted the sea floor. Feeding on the turritellids were several molluscivorous gastropods, including the muricids *Panamurex alaquaensis* (Mansfield, 1935) and *Urosalpinx vaughanensis* Mansfield, 1935, the drilling naticids *Euspira waltonensis* (Mansfield, 1935) and *E. propeinternus* (Mansfield, 1935), the busyconids *Coronafulgur propecoronatum* (Mansfield, 1935) (last-living member of the genus *Coronafulgur*) and *Pyruella blountense* (Mansfield, 1935), and the fasciolariids *Fusinus alaquaensis* Mansfield, 1935 and *Latirus* (?) *waltonensis* Mansfield, 1935. Bivalves were abundant on sand patches between the turritellid beds and some of these included the nuculanid *Yoldia waltonensis* Mansfield, 1932, the arcid *Dallarca rubisiniana* (Mansfield, 1916), the crassatellid *Marvacrassatella rubisiniana* (Mansfield, 1916), the dosiniinine venerid *Dosinia blountana* Mansfield, 1932, the cardiid *Chesacardium blountense* (Mansfield, 1932), and the large pectinids *Pecten macdonaldi* subspecies and *Lyropecten pontoni* (Mansfield, 1932). Also living on the sand patches were several carnivorous gastropods, some of which included the echinoderm-feeding cassid *Phalium waltonensis* (Mansfield, 1935) and the ranellid *Cymatium (Linatella) alaquaensis* (Mansfield, 1935) (originally described as a *Neptunea*) and a large fauna of carnivorous species, some of which included the vermivorous conids *Gradiconus vaughanensis* (Mansfield, 1935), *Ximeniconus blountensis* (Mansfield, 1935), *X. alaquaensis* (Mansfield, 1935), and *Conasprella laqua* (Mansfield, 1935), the turrids *Polystira alaquaensis* Mansfield, 1935, *Hindsiclava blountensis* (Mansfield, 1935), and *Cymatosyrinx vaughanensis* Mansfield, 1935, and the suctorial-feeding cancellariids *Cancellaria laqua* Mansfield, 1935 and *Axelella uaquala* (Mansfield, 1935).

Communities and Environments of the Polk Subsea

Due to extensive leaching by deltaic environments during emergent times in the late Miocene (see Figure 21), the paleoecological record of the older Miocene Okeechobean Sea

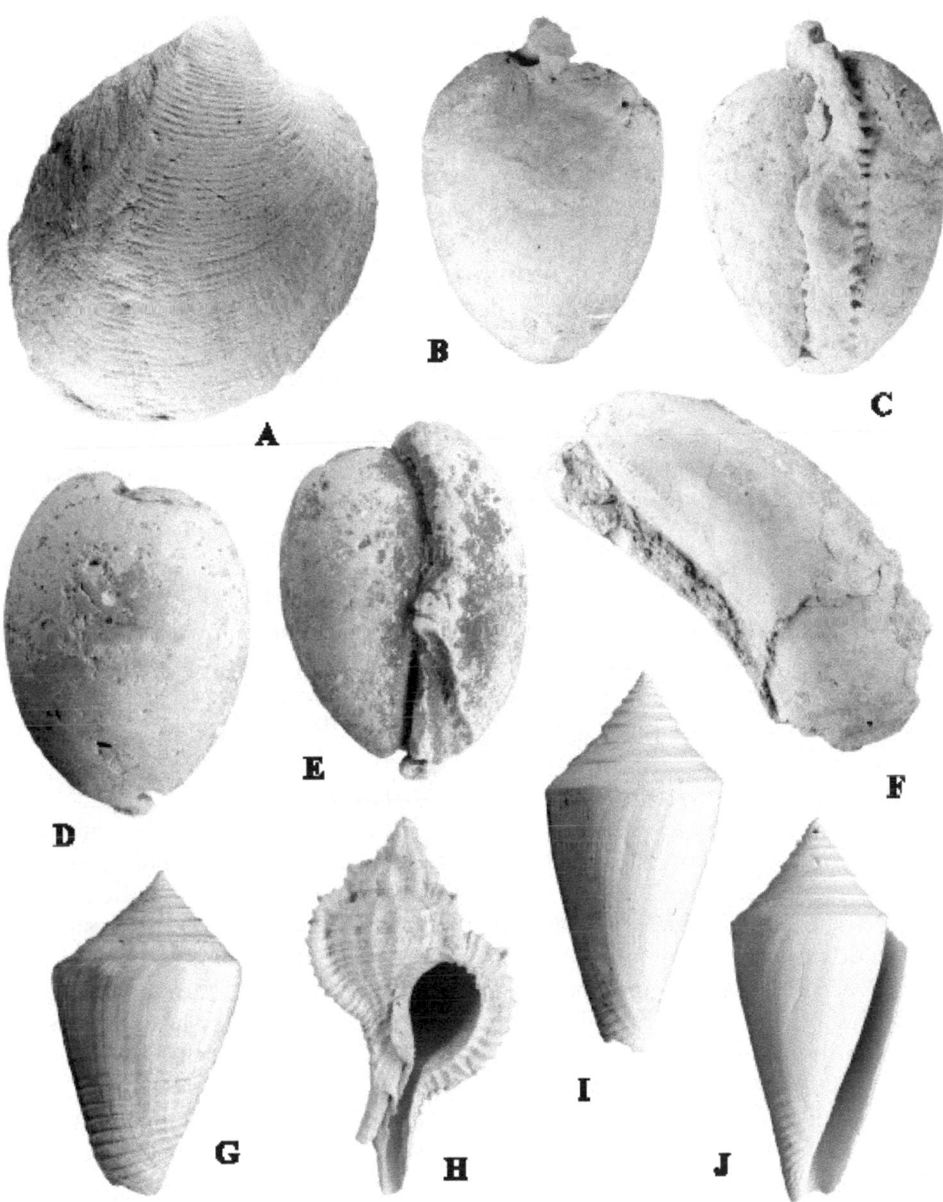

Plate 40. Index fossils and characteristic organisms of the *Clementia inoceriformis* Community (shallow muddy sand lagoon and Turtle Grass beds) of the Polk Subsea, and the *Torcula waltonensis* Community (deep lagoon turritellid beds) of the Walton Subsea. Included are: A= *Clementia inoceriformis* (Wagner, 1839), length 68 mm; B,C= *Akleistostoma* species, length 48 mm; D,E= *Calusacypraea* species, length 46 mm; F= *Perna incurvus* (Conrad, 1839), length 120 mm; G= *Ximeniconus waltonensis* (Aldrich, 1903), length 19 mm; H= *Siratus nicholsi* (Gardner, 1947), length 31 mm; I,J= *Gradiconus drezi* (S.Hoerle, 1976). The moldic fossils (A-F) were collected in the Peace River Formation in the IMC (International Minerals Corporation) quarry in extreme southwestern Polk County, Florida, along the South Prong of the Alalfia River. The specimens of *Ximeniconus waltonensis*, *Siratus nicholsi*, and *Gradiconus drezi* were collected in the Shoal River Formation at Shell Bluff, Shoal River,Walton County, Florida.

basin was all but obliterated. Only a few small stringers of moldic fossils have ever been uncovered, with most of these being located around the northern edge of the basin. One particular site, at the IMC (International Minerals Corporation) quarry in extreme southwestern Polk County, Florida, has yielded an unusually rich molluscan assemblage from the upper beds of the Peace River Formation of the Hawthorn Group (Scott, 1988). Although moldic, many of the index species were recognizable and the assemblage was found to be equivalent in age to the St. Mary's Formation of Maryland (latest Serravallian-early Tortonian).

The Clementia inoceriformis Community

The only-known Polk Subsea molluscan fauna at the IMC quarry was associated with a shallow water (0-10 m) muddy sand environment that was adjacent to Turtle Grass (*Thalassia*) beds. Bivalves were common in the assemblage and included the venerid *Clementia inoceriformis* (Wagner, 1839) (namesake of the community; Plate 40, A), the pholadid *Cyrtopleura arcuata* (Conrad, 1841), the mytilid *Perna incurvus* (Conrad, 1839) (Plate 40, F), and the mactrid *Mactra* species. Living with the bivalves on the open sand bottom were the turritellid *Torcula* species (*T. subvariabilis* complex), *Gradiconus* species (*G. deluvianus* complex), and *Oliva (Strephona)* species. Of special interest within the *Clementia inoceriformis* Community was the presence of two different genera of cowries (family Cypraeidae), including *Akleistostoma* Gardner, 1948 (Plate 40, B and C) and *Calusacypraea* Petuch, 1996 (Plate 40, D and E), both the oldest-known members of their species groups. Like their Pliocene descendants (see Chapters 6 and 7) and their living sister-group, the genus *Muracypraea* Woodring, 1959, these first *Akleistostoma* and *Calusacypraea* species lived in the Turtle Grass beds and adjacent mud flats (Petuch, 1976, 1979, 1987).

Communities and Environments of the Charlotte Subsea

During the late Tortonian and early Messinian, the Okeechobean Sea became geomorphologically more complex (Figure 21). At this time, the first coral reef systems of the Everglades Pseudoatoll (see Chapter 7) developed as two main reef tracts: the Miami-Palm Beach Reef Tracts on the eastern side and the Immokalee Reef Tract on the western side. A series of muddy coastal lagoons, the Wabasso Lagoon System (named for the Wabasso Beds of the late Peace River Formation), ran up the eastern coast of the Floridian Peninsula as far as present-day Jacksonville. The only area of the Charlotte Subsea to have well-preserved fossils was in the Peace River Lagoon System near present-day Port Charlotte, Charlotte County, Florida. Here, in the two main beds of the Bayshore Formation of the Hawthorn Group (originally described as a member of the late Pliocene Tamiami Formation; see Chapter 1), two distinct marine communities are preserved. Since the Bayshore Formation is heavily leached, only calcitic species are present. These include the *Gigantostrea leeana* Community (shallow sand bottom lagoon during early Bayshore time) and the *Chesapecten middlesexensis* Community (deep sand bottom lagoon pectinid beds during late Bayshore time). Based upon the index fossils contained in these assemblages, it can now be seen that the Bayshore Formation is chronologically equivalent to the Claremont Manor and Cobham Bay Members of the Eastover Formation. During this time, the marine climate was cool enough in southern Florida to allow Rappahannock Subsea organisms to invade the Okeechobean Sea. Interestingly enough and with few exceptions, the contemporaneous Alaqua Subsea of the Choctaw Sea had a very different molluscan fauna, lacking the characteristic Eastover-Bayshore scallops and ecphoras. This faunal disparity implies the presence of an unknown oceanographic barrier between the northwestern edge of the Okeechobean Sea and the southeastern edge of the Choctaw Sea.

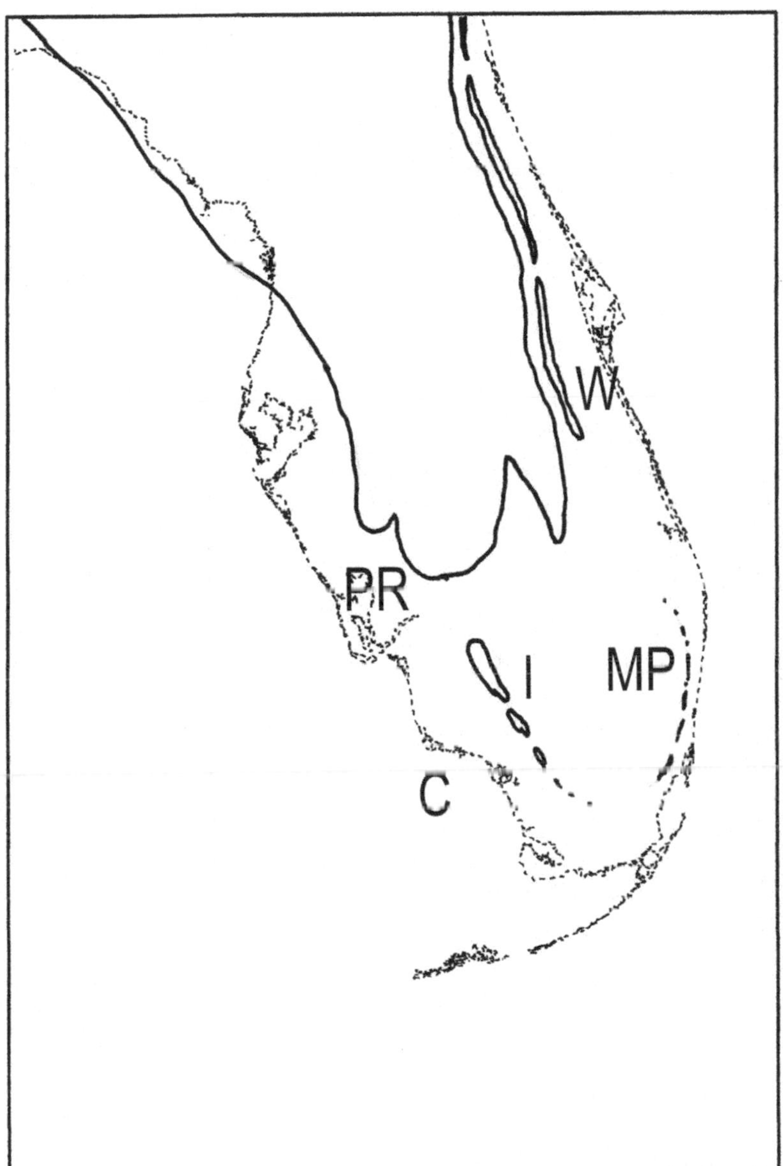

Figure 21. Approximate configuration of the coastline of southern Florida during late Tortonian and Early Messinian Miocene time, superimposed upon the outline of the Recent Florida Peninsula. Prominent features included: C= Charlotte Subsea of the Okeechobean Sea, PR= Peace River Lagoon System, I= Immokalee Reef Tract, MP= Miami-Palm Beach Reef Tract, W= Wabasso Lagoon System. At this time, a huge delta was prograding southward along the eastern side of the Immokalee Reef Tract, producing a wide, shallow platform (see Cunningham, *et. al.*, 2003).

The Gigantostrea leeana Community

During earliest Charlotte Subsea time (late Tortonian; Claremont Manor Member equivalent), barrier sand islands developed along the southwestern tip of the Floridian Peninsula. In the quiet, shallow sand bottom lagoons behind these barrier islands, immense oyster beds grew, essentially forming reeflike bioherms. At least three species produced these beds, including the pycnodontine gryphaeid *Gigantostrea leeana* (Wilson, 1987) (namesake of the community; Plate 41, E) (also found in the Eastover Formation at Aurora, North Carolina), and the ostreids *Ostrea* cf. *brucei* Ward, 1992 and *O. disparilis* subspecies. Feeding on the oysters was the large and characteristic Eastover muricid *Ecphora whiteoakensis* Ward and Gilinsky, 1988 (Plate 41, A and B; also an index fossil for the Claremont Manor Member of the Eastover Formation), which was abundant in certain localities. Large barnacles of the genus *Concavus* were also abundant on these oyster reefs and often formed large biohermal clumps.

The Chesapecten middlesexensis Community

In the latest Tortonian and early Messinian, the Charlotte Subsea was flooded with deeper and warmer water and a richer molluscan fauna became established. Because of the limited extent of the upper beds of the Bayshore Formation, only the deep lagoon pectinid bed community has been preserved. These scallop beds carpeted the Peace River Lagoon sea floor and were made up exclusively of only two species, *Carolinapecten urbannaensis* (Mansfield, 1929) (Plate 41, D) and *Chesapecten middlesexensis* (Mansfield, 1936) (Plate 41, C) (both species are index fossils for the Cobham Bay Member of the Eastover Formation). Scattered among the scallop beds were large biohermal clumps of at least two species of *Concavus* barnacles and these were the principal prey items for the Charlotte Subsea endemic ocenebine muricids *Zulloia zulloi* Petuch, 1994 (Plate 41, G) and *Z. violetae* Petuch, 1994 (Plate 41, F). Scattered among the scallop beds were small bioherms of the ostreid oysters *Ostrea geraldjohnsoni* Ward, 1992 and a small *O. disparilis* subspecies. Feeding on the scallops and oysters was the large, endemic Charlotte Subsea ocenebrine muricid *Globecphora floridana* (Petuch, 1989) (Plate 41, H, I, and J; the first and oldest member of the genus *Globecphora*).

Plate 41. Index fossils and characteristic organisms of the *Gigantostrea leeana* Community (shallow sand bottom lagoon during early Bayshore time) and Chesapecten middlesexensis Community (deep sand bottom lagoon pectinid beds during late Bayshore time) of the Charlotte Subsea. Included are: A,B= *Ecphora whiteoakensis* Ward and Gilinsky, 1988, length 101 mm; C= *Chesapecten middlesexensis* (Mansfield, 1936), width 120 mm; D= *Carolinapecten urbannaensis* (Mansfield, 1929), width 115 mm; E= *Gigantostrea leeana* (Wilson, 1987), length 138 mm; F= *Zulloia violetae* Petuch, 1994, length 34 mm; G= *Zulloia zulloi* Petuch, 1994, length 36 mm; H= *Globecphora floridana* (Petuch, 1989), highly sculptured variant, length 112 mm; I,J= *Globecphora floridana* (Petuch, 1989), typical form, length 101 mm. The specimens of *Ecphora whiteoakensis* and *Gigantostrea leeana* were collected in the lower beds of the Bayshore Formation, along the Cocoplum Canal, Port Charlotte, Charlotte County, Florida. The specimens of *Chesapecten middlesexensis*, *Carolinapecten urbannaensis*, *Zulloia zulloi*, *Zulloia violetae*, and *Globecphora floridana* were collected in the upper beds of the Bayshore Formation at a construction site in El Jobean, Port Charlotte, Charlotte County, Florida.

Chesapecten covepointensis Ward, 1992, St. Mary's Subsea, Salisbury Sea, length 102 mm. Specimen from Little Cove Point, Calvert County, Maryland.

Chapter 6. Early and Late Pliocene Seas

The late Messinian Miocene was a time of extreme climatic degeneration that resulted in repeated severe cold periods and major eustatic lows. During this catastrophic time, the eastern North American paleoceanographic basins remained emergent for over one million years, from at least 5.5 million to 4.5 million years BP. Warmer climates and sea level rises did not occur until the early Zanclean Pliocene (4.5 million years). At that time, only three paleosea basins reflooded: the Okeechobean as the Murdock Subsea, the Charleston as the Santee Subsea, and the Albemarle as the Williamsburg Subsea. These three subseas were short-lived, as sea levels dropped again during a second spate of cold climatic conditions in the late Zanclean. Based on the depression in the sea level curves shown by Krantz (1991), this short, second climatic degeneration was as severe as those of the preceding Messinian-early Zanclean times.

The beginning of the Piacenzian Pliocene saw a return to tropical and subtropical conditions for most of eastern North America. Sea levels rose to their highest since the Tortonian and reflooded four paleosea basins: the Choctaw as the Jackson Subsea, the Okeechobean as the Tamiami Subsea, the Charleston as the Duplin Subsea, and the Albemarle as the Yorktown Subsea. Because of the warmer marine climate and a wider variety of available habitats than in previous times, these Piacenzian seas, in total, contained the richest invertebrate faunas ever seen in North America. One of these, the Everglades Pseudoatoll structure of the Tamiami Subsea was so faunally rich and geomorphologically complicated that it has warranted a separate chapter in this book (Chapter 7). The paleoecology and paleoceanography of the Zanclean Okeechobean Sea (Murdock Subsea), along with the Zanclean Williamsburg and Santee Subseas, and the Piacenzian Yorktown, Duplin, and Jackson Subseas, are discussed in the following sections.

Communities and Environments of the Williamsburg Subsea

The Zanclean reflooding of the Albemarle Sea basin was not as extensive as the transgression that occurred during late Tortonian-early Messinian time, and the resultant Williamsburg Subsea (Figure 22) was only half as large as the previous Rappahannock Subsea. Of primary importance within the Williamsburg Subsea was the incipiency of the Chuckatuck Archipelago of sand barrier islands (Petuch, 1997), which sheltered the northern half of the paleosea and produced lagoonal conditions in these areas (referred to as the "Chuckatuck barrier" by Campbell, 1993). At this time, the Cape Fear Arch had developed into a large Peninsula, oceanographically separating the Charleston Sea basin (Santee Subsea) from the Albemarle Sea basin. The cold, southward-flowing Narragansett Current extended as far south as the Cape Fear Peninsula, where it veered directly eastward into the Atlantic Ocean. At the same time, the warm, northward-flowing Ciboney Current extended as far north as the Cape Fear Peninsula, where it also veered eastward in a confluence with the Narragansett Current (Petuch, 1997). Because of the Ciboney Current, the oceanographic climate of the Santee Subsea was much warmer than that of the Williamsburg Subsea. A cold water countercurrent off the Narragansett Current, the Hampton Gyre (Petuch, 1997), flooded the Williamsburg Subsea, producing oceanographic conditions similar to the Miocene Salisbury Sea.

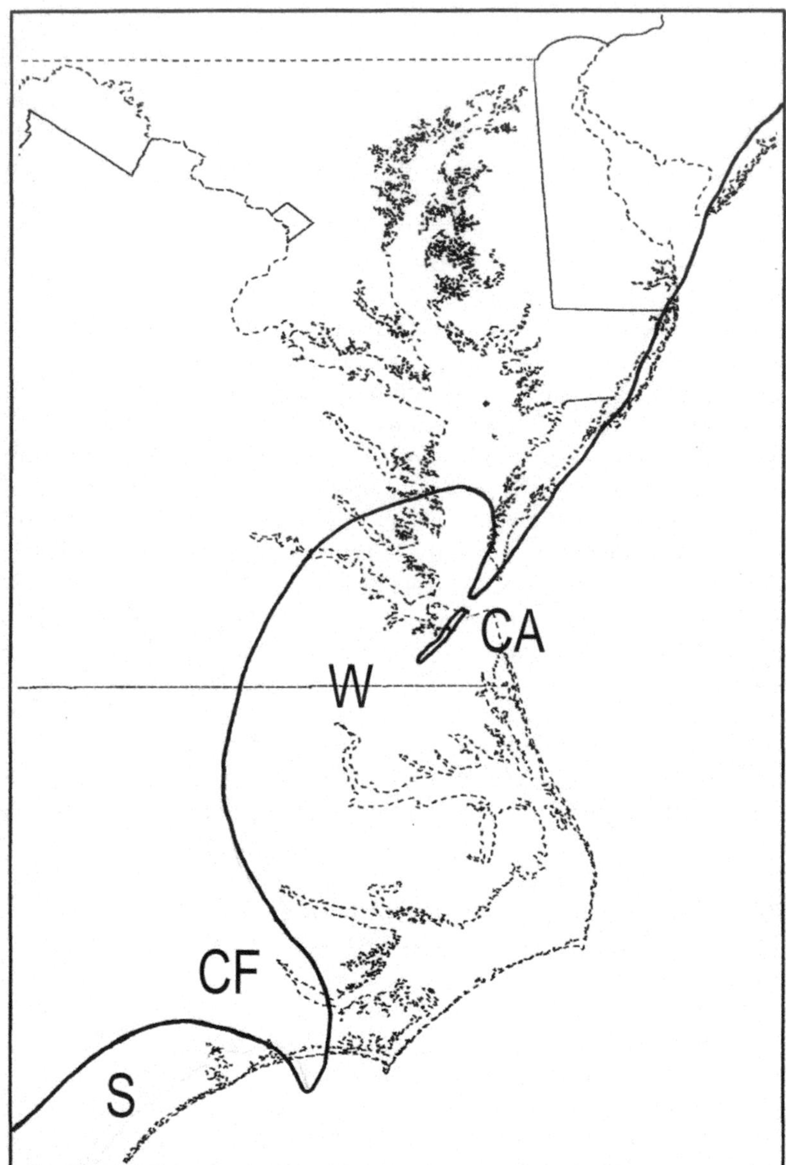

Figure 22. Approximate configuration of the coastline of the mid-Atlantic United States during Zanclean Pliocene time, superimposed upon the outlines of the Recent Cheaspeake Bay and Pamlico Sound areas. Prominent features included: W= Williamsburg Subsea of the Albemarle Sea, S= Santee Subsea of the Charleston Sea, CA= Chuckatuck Archipelago, CF= Cape Fear Arch and Peninsula.

As most of the fossiliferous beds of the Williamsburg Subsea are heavily leached, only incomplete and cursory paleoecological reconstructions can be undertaken. From specimens collected in the Sunken Meadow Member of the Yorktown Formation (Yorktown Zone 1 of Mansfield, 1928) at Williamsburg, Virginia and in the Lee Creek Mine at Aurora, North Carolina, two distinct, bathymetrically-discrete communities are known to have existed within the Williamsburg Subsea. These include the *Placopecten clintonius* Community (deep sand bottom lagoon and open sound pectinid beds) and the *Torcula bipertita* Community (shallow sand bottom lagoon turritellid beds). Having a cooler marine climate than the subseas to the south, the Williamsburg Subsea housed many Transmarian Province relictual molluscan taxa, where they survived the demise of the Salisbury Sea. Some of these included the gastropods *Mariafusus* (Fasciolariidae), *Lirosoma* and *Chesatrophon* (Trophoninae-Muricidae), and *Scaphellopsis* (Scaphellinae-Volutidae), and the bivalve *Chesacardium*.

The Placopecten clintonius Community

As in the Calvert, Patuxent, St. Mary's, and Rappahannock Subseas, the open sound and deep lagoon (20-100 m) areas of the Williamsburg Subsea supported immense beds of scallops. Several pectinid species occurred here, including *Placopecten clintonius* (Say, 1824) (namesake of the community; Plate 42, G), *Chesapecten jeffersonius* (Say, 1824) (Plate 42, C), *Carolinapecten eboreus* subspecies, and *Pecten smithi* (Olsson, 1914). At many Sunken Meadow Member exposures, such as in the Lee Creek Mine, *Placopecten clintonius* is so abundant that it makes up the majority of molluscan fossils. These dense scallop beds were the principal food source for several large carnivorous gastropods, including the busyconid *Busycotypus* species (*B. incile* complex), the ocenebrine muricid *Globecphora xenos* (Petuch, 1989) (Plate 42, D and E; a southern migrant and descendant of the Charlotte Subsea *G. floridana*), and the giant scaphelline volutid *Megaptygma* species (*M. sinuosa* complex; descendant of the St. Mary's Subsea *M. meganucleus*). The fissurellid limpet *Glyphis redimicula* (Say, 1824), and several species of calyptraeid gastropods, lived attached to the larger specimens of scallops. As in the Calvert Subsea, balaenopterid whales were also abundant in the deep sound areas (as attested by the abundance of whale bones in the Lee Creek Mine) and these were the principal prey items of the giant otodontid shark *Carcharocles megalodon* (Charlesworth, 1837) (Plate 42, F).

The Torcula bipertita Community

In shallower lagoonal areas (5-20 m) closer to shore, and behind the Chuckatuck Archipelago, extensive beds of turritellid gastropods accumulated. These were composed of just two species, the large and abundant *Torcula bipertita* (Conrad, 1844) (namesake of the community), and the smaller and less abundant *Bactrospira terstriata* (Rogers and Rogers, 1837). On open sand patches scattered among the turritellid beds, large numbers of shallowly infaunal bivalves occurred. Some of these included the crassatellid *Marvacrassatella cycloptera* (Dall, 1903), the glossid *Glossus carolina* (Dall, 1900), the margaritariid *Margaritaria abrupta* (Conrad, 1832), the cardiid *Planicardium taenopleura* (Dall, 1900), the venerids *Chione cortinaria* (Rogers and Rogers, 1837) and *Mercenaria inflata* (Dall, 1903), the cardiid *Chesacardium virginianum* (Conrad, 1839), the astartids *Astarte deltoidea* Gardner, 1944 and *A. vaginulata* Dall, 1903, and numerous small mactrids and tellinids. Feeding on both the bivalves and the turritellids were several molluscivorous gastropods, including drilling ocenebrine muricids such as the small, wide-ribbed *Ecphora pachycostata* Petuch, 1989 (Plate 42, A and B; a possible offshoot of the Rappahannock Subsea *E.*

Plate 42. Index fossils and characteristic organisms of the *Placopecten clintonius* Community (deep sand bottom lagoon and open sound pectinid beds) and the *Torcula bipertita* Community (shallow sand bottom lagoon turritellid beds) of the Williamsburg Subsea. Included are: A,B= *Ecphora pachycostata* Petuch, 1989, length 48 mm; C= *Chesapecten jeffersonius* (Say, 1824), length 130 mm; D,E= *Globecphora xenos* (Petuch, 1989), length 80 mm; F= tooth of *Carcharocles megalodon* (Charlesworth, 1837), length 95 mm; G= *Placopecten clintonius* (Say, 1824), length 122 mm; H,I= *Ecphora leecreekensis* 1989, length 42 mm (note the characteristic fine-threaded and scaly shell texture). All specimens were collected in the Sunken Meadow Member (Yorktown Zone 1) of the Yorktown Formation in the Lee Creek Texasgulf Mine, Aurora, Beaufort County, North Carolina.

whiteoakensis), the thin-ribbed, heavily-sculptured *Ecphora leecreekensis* Petuch, 1989 (Plate 42, H and I), *Pterorhytis* cf. *umbrifer* (Conrad, 1832), and *Urosalpinx* new species (elongated), the trophonine muricids *Lirosoma multicostata* Olsson, 1914 and *Chesatrophon* species (*C. lindae* complex), the drilling naticid *Euspira interna* subspecies, the fasciolariid *Mariafusus propeparilis* (Mansfield, 1929), and the scaphelline volutid *Scaphellopsis* species (*S. solitaria* complex). Also occurring in the turritellid beds were other small carnivorous gastropods, including the echinoderm-feeding *Ficus jamesi* Olsson, 1914, the sponge and hydroid-feeding trochids *Calliostoma shacklefordensis* Olsson, 1916 and *C. johnsoni* Campbell, 1993, and the vermivore conid *Ximeniconus* species (*X. marylandicus* complex). Closer to shore, in shallower water adjacent to the turritellid beds, large oyster bars developed, composed primarily of the ostreids *Ostrea compressirostra* subspecies and *Conradostrea sculpturata* subspecies. The large scallop *Chesapecten palmyrensis* (Mansfield, 1936) also occurred in shallow muddy environments, but only along the western shore of the Williamsburg Subsea.

Communities and Environments of the Murdock Subsea

Although sea levels rose during early Zanclean time, the marine climate in the Okeechobean Sea basin was still colder than during Tampa Subsea time or during the subsequent Piacenzian. Because of these cold water conditions, the Everglades Pseudoatoll coralline structure probably ceased to grow, and its eroded carbonate fines, mixed with phosphorites, filled the central lagoon of the Okeechobean Sea. The greatest geomorphological complexity of the Murdock Subsea was seen along its western coast, where a series of sand barrier islands and lagoons (the Murdock Lagoon System and Archipelago) developed (Figure 23). The cooler marine climate, and localized upwellings of cold nutrient-rich water, allowed northern Williamsburg faunal influences to invade the Murdock Subsea area. In the phosphorite-rich Murdock Station Formation (originally described as a member of the carbonate-rich Tamiami Formation; see Hunter, 1968), two distinct Williamsburg-type communities can be recognized. These included the *Concavus tamiamiensis* Community (intertidal mud flats and shallow lagoons in high-productivity waters) and the *Chesapecten palmyrensis* Community (deep lagoons in high-productivity waters). Since the Murdock Station Formation is leached, few aragonitic fossils are preserved and any paleoecological studies are skewed toward the better-preserved calcitic and apatitic fossils. The high concentration of autochthonous phosphorites in the Murdock Station Formation at Sarasota (Unit 11 in the APAC quarry; see Petuch, 1982) indicates the presence of an almost continuous upwelling system immediately offshore of the Murdock Lagoons.

The Concavus tamiamiensis Community

In the sheltered Murdock Lagoons behind the Murdock Archipelago, vast barnacle "reefs" developed in shallow intertidal areas (0-10 m). These massive cirriped bioherms were composed of at least three species, including the abundant giant *Concavus tamiamiensis* (Ross, 1965) (namesake of the community; Plate 43, A; originally incorrectly thought to have come from the younger Tamiami Formation), and the smaller *Balanus newburnensis* Weisbord, 1966 (Plate 43, C) and *Concavus glyptopoma* (Pilsbry, 1916). Scattered between the barnacle reefs were small oyster bars composed of the ostreid *Ostrea compressirostra* Say, 1824 (Plate 43, H) and the large anomiid *Placunanomia floridana* Mansfield, 1932 (Plate 43, I). These supplied the substrate for a wide variety of encrusting epibionts, including the discinid inarticulate brachiopods *Discinisca multilineata* (Conrad, 1845) (Plate 43, B) and *D. lugubris* (Conrad, 1834) (Plate 43, J) and ahermatypic corals such

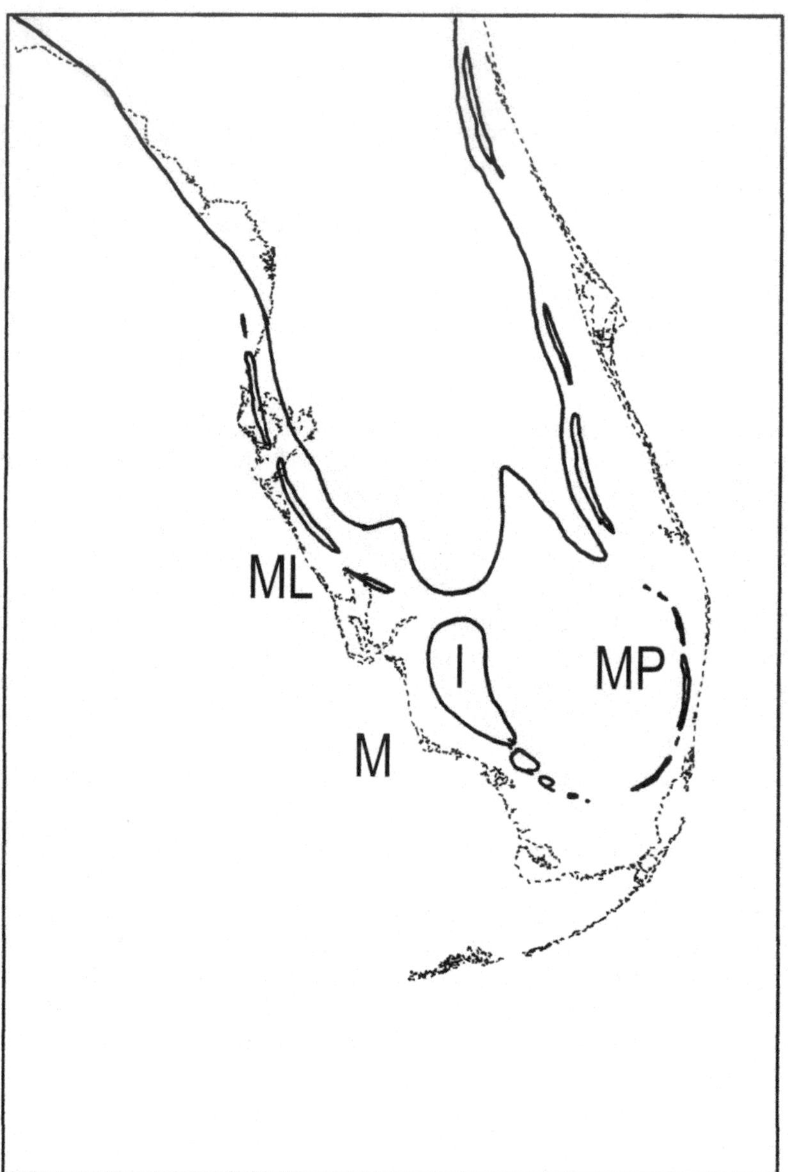

Figure 23. Approximate configuration of the coastline of the Floridian Peninsula during Zanclean Pliocene time, superimposed upon the outline of Recent Florida. Prominent features included: M= Murdock Subsea of the Okeechobean Sea, ML= Murdock Lagoon System and Archipelago, I= Immokalee Island and Reef Tract, MP= Miami-Palm Beach Reef Tract. At this time, a giant delta (originating in the Charlotte Subsea) had infilled the western half of the Okeechobean Sea and produced the large, sand-covered Immokalee Island.

as *Astrangia lineata* (Conrad, 1835) (Plate 43, K). Equipped with an extra large labial tooth for wedging open giant barnacles, the ocenebrine muricid *Pterorhytis* cf. *umbrifer* (Conrad, 1832) (Plate 43, D and E) was the primary predator of *Concavus tamiamiensis*. Feeding on the oysters, brachiopods, and interstitial bivalves were two other large common ocenebrine muricids, *Ecphora streami* (Petuch, 1994) (Plate 43, F and G) and *E. roxaneae* Petuch, 1991 (which also occurred in the Santee Subsea of the Carolinas; see Plate 49, A, B, and C).

The Chesapecten palmyrensis Community

In the deeper parts of the Murdock Lagoons, on muddy bottoms in areas influenced by upwellings, immense beds of scallops carpeted the sea floor, much like the Williamsburg Subsea. Although the pectinid beds were dominated by the large *Chesapecten jeffersonius* (Say, 1824) (Plate 44, I), several other scallop species occurred in abundance. Primary among these were the large and common *Chesapecten palmyrensis* (Mansfield, 1936) (namesake of the community; Plate 44, C; also found in the Williamsburg and Santee Subseas), *Chesapecten madisonius carolinensis* (Conrad, 1873), and *Carolinapecten eboreus* subspecies, and the smaller *Argopecten choctawhatcheensis* (Mansfield, 1932) (Plate 44, H) and *Leptopecten* cf. *leonensis* (Mansfield, 1932) (Plate 44, K). Scattered among the pectinid beds, on open muddy sand patches, were large bioherms of the ahermatypic coral *Septastrea* species (Plate 44, J), some of which grew to over 1 meter in height. In open muddy areas between the coral bioherms, several bivalves occurred, including the margaritariid *Margaritaria abrupta* (Conrad, 1832) (Plate 44, G) and the petricolid *Pleiorytis centenaria* (Conrad, 1833). Feeding on the scallops and infaunal bivalves were two species of drilling ocenebrine muricids, the thick-ribbed *Ecphora quadricostata* subspecies (Plate 44, A and B) and the inflated, thin-ribbed *Planecphora hertweckorum* (Petuch, 1988) (Plate 44, E and F; descendant of the *P. choptankensis* complex of the Salisbury Sea). The entire molluscan fauna of the *Chesapecten palmyrensis* Community was the primary food source for the Florida Walrus (*Odobenus* species; with walrus tusks being found occasionally in Unit 11 at the Sarasota quarries) and the giant sting ray *Myliobatus* (Plate 44, D).

Communities and Environments of the Santee Subsea

During Zanclean time, the Charleston Sea basin was only partially flooded and was essentially a narrow strip of coastal lagoons with some deep embayments (Figure 22). Because of the heavy leaching and bad preservation of Zanclean-aged fossil beds in southern North Carolina and northern South Carolina, little is known of the paleoecology of the Santee Subsea. The paltry information that is available, gleaned mostly from the lower beds of the Goose Creek Formation (moldic limestone) in the Martin-Marietta mine at Santee, South Carolina, shows that the Santee Subsea faunas were similar to those of the Murdock Subsea, but less species-rich and lacking many of the Murdock index species. Only one marine community is recognized at this time.

The Ecphora roxaneae Community

The molluscan fauna contained in the lowest beds of the Goose Creek Formation lived in a shallow muddy lagoon (0-10 m) behind a barrier island system. Here, large oyster bars formed, composed mostly of the ostreid *Ostrea compressirostra* Say, 1824. Scattered between the oyster bars were large clumps of the giant barnacle *Concavus proteus* (Conrad, 1834). Although forming small biohermal structures, these barnacles did not develop the extensive "reefs" that were produced by its southern Murdock congener, *C. tamiamiensis*. Feeding on the oysters and barnacles was the large ocenebrine muricid, *Ecphora roxaneae*

Plate 43. Index fossils and characteristic organisms of the *Concavus tamiamiensis* Community (intertidal mud flats and shallow lagoons) of the Murdock Subsea. Included are: A= *Concavus tamiamiensis* (Ross, 1965), clump length 152 mm; B= *Discinisca multilineata* (Conrad, 1845), length 19 mm; C= *Balanus newburnensis* Weisbord, 1966, clump length 86 mm; D,E= *Pterorhytis* cf. *umbrifer* (Conrad, 1832), length 78 mm (note the well-developed labial tooth for opening large barnacles); F,G= *Ecphora streami* (Petuch, 1994), length 102 mm; H= *Ostrea compressirostra* Say, 1824, length 99 mm; I= *Placunanomia floridana* Mansfield, 1932, length 74 mm; J= *Discinisca lugubris* (Conrad, 1834), length 21 mm; K= *Astrangia lineata* (Conrad, 1835), colony growing on back of oyster, length 110 mm. All specimens were collected in the Murdock Station Formation of the Hawthorn Group, in Quality Aggregates pit 6, Sarasota, Sarasota County, Florida.

Plate 44. Index fossils and characteristic organisms of the *Chesapecten palmyrensis* Community of the Murdock Subsea. Included are: A,B= *Ecphora quadricostata* subspecies, length 60 mm; C= *Chesapecten palmyrensis* (Mansfield, 1936), length 83 mm; D= tail spine of *Myliobatus* sting ray, length 62 mm; E,F= *Planecphora hertweckorum* (Petuch, 1988), length (broken) 81 mm; G= *Margaritaria abrupta* (Conrad, 1832), length 64 mm (note the healed shell damage caused by attempted sting ray predation); H= *Argopecten choctawhatcheensis* (Mansfield, 1932), length 36 mm; I= *Chesapecten jeffersonius* (Say, 1824), length 142 mm; J= *Septastrea* species, length 162 mm; K= *Leptopecten* cf. *leonensis* (Mansfield, 1932), length 21 mm. All specimens were collected in the Murdock Station Formation of the Hawthorn Group, in Quality Aggregates pit 6, Sarasota, Sarasota County, Florida.

Petuch, 1991 (namesake of the community; Plate 49, A, B, and C; also found in the Murdock Subsea). This is the single most abundant fossil in many beds of the lower Goose Creek Formation, and may have been the ancestor of the "T"-shaped ribbed genus *Latecphora* of the Duplin and Tamiami Subseas. In these same lowermost beds, fragments of the large scallops *Chesapecten jeffersonius* (Say, 1824), *C. madisonius carolinensis* (Conrad, 1973), and *C. palmyrensis* (Mansfield, 1936) are also present, indicating that a deeper water lagoonal environment existed in the near vicinity at that time.

Communities and Environments of the Yorktown Subsea

In late Zanclean time, another period of severe climatic degenerations and sea level drops took place and the paleosea basins of eastern North America were emergent for over 0.5 million years (Ward and Huddlestun, 1988). At the beginning of Piacenzian time, the climate became much warmer and the resultant sea level rise flooded the paleosea basins to their maximum extent. In the north, the Albemarle Sea basin (as the Yorktown Subsea) flooded all the way to present-day Petersburg, Virginia (Figure 24). At this time, the Chuckatuck Archipelago reached its maximum development, essentially closing off the northern half of the Yorktown Subsea and producing extensive shoals and lagoons along its western side (the "Chuckatuck barrier" of Campbell, 1993). Due to the sheltering effect of the Chuckatuck Archipelago and the stronger influence of the warm water Ciboney Current, the cool water intrabasinal Hampton Gyre had weakened or dissipated by Piacenzian time. This allowed many warm water organisms to invade the Pliocene Albemarle Sea for the first time. Some of these southern migrants included the gastropods *Akleistostoma* (Cypraeidae), *Contraconus* (Conidae), *Sconsia* (Cassidae), *Fasciolaria (Cinctura), Terebraspira,* and *Triplofusus* (all Fasciolariidae) and *Cariboliva* (Olividae). Although the Yorktown Zone 1 cold water scallop genus *Placopecten* (*P. clintonius*) disappeared from the Albemarle Sea basin at this time, other warm-tolerant northern elements became established. Some of these included the gastropods *Bulbus* (Naticidae), *Buccinum* (Buccinidae) (see Campbell, 1993), *Trophonopsis* (Muricidae), *Solariella* (Trochidae), and the bivalves *Mytilus* (*M. edulis* complex, Mytilidae) and *Hiatella* (Hiatellidae).

From the highly fossiliferous beds of the Rushmere, Mogarts Beach, and Moore House Members of the Yorktown Formation (collectively Yorktown Zone 2 of Mansfield, 1928; of Virginia and northern North Carolina), four distinct marine communities can be discerned. These included the *Chesapecten septenarius* Community (deep lagoon pectinid beds), the *Torcula alticostata* Community (shallow lagoon turritellid beds), and the *Marvacrassatella virginica* Community (shallow muddy sand lagoons), and the *Mytilus edulis alaeformis* Community (estuarine mud flats). These communities, and their accompanying biotopes and lithofacies, interfingered along the western edge of the Chuckatuck Archipelago. Although containing many tropically-derived elements, these communities were still only paratropical in nature, lacking key tropical indicators such as the Strombidae, Melongenidae, Potamididae, Modulidae, and Turbinellidae (and Vasinae).

The Chesapecten septenarius Community

In the sheltered deeper lagoon and sound areas (10-100 m) behind the Chuckatuck Archipelago during Rushmere Member time, pectinid bivalves formed extensive beds on the open sand bottoms. These were composed almost exclusively of four large species, including *Chesapecten septenarius* (Say, 1824) (namesake of the community; Plate 45, E), *C. madisonius* (Say, 1824) (Plate 45, F), *Christinapecten decemnarius* (Conrad, 1834) and variety *virginianus* (Conrad, 1840), and *Carolinapecten eboreus* (Conrad, 1833) (Plate 45, A). Other, less abundant, pectinids also occurred in these beds, including the bay scallop *Argopecten*

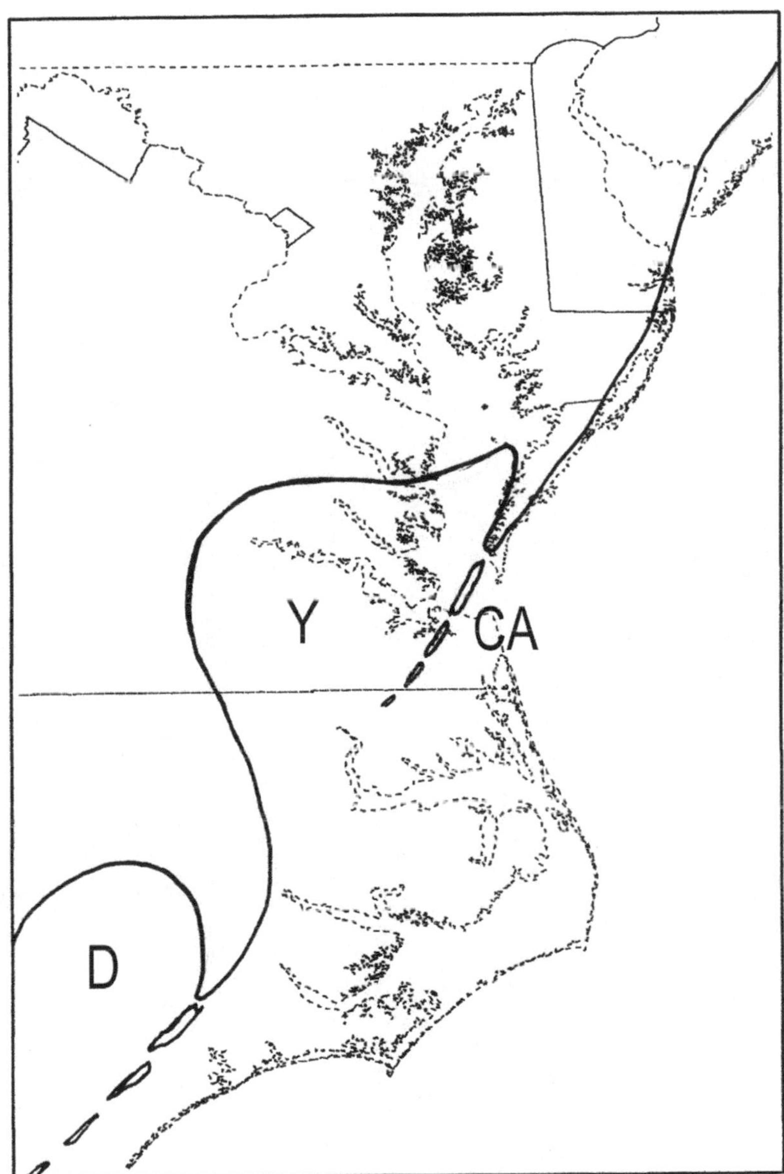

Figure 24. Approximate configuration of the coastline of the mid-Atlantic United States during Piacenzian Pliocen time, superimposed upon the outline of the Recent Chesapeake Bay and Pamlico Sound areas. Prominent features included: Y= Yorktown Subsea of the Albemarle Sea, D=Duplin Subsea of the Charleston Sea, CA= Chuckatuck Archipelago.

Plate 45. Index fossils and characteristic organisms of the *Chesapecten septenarius* Community (deep lagoon pectinid beds) of the Yorktown Subsea. Included are: A= *Carolinapecten eboreus* (Conrad, 1833), length 112 mm; B,C= *Ecphora quadricostata* (Say, 1824), length 46 mm; D= *Busycotypus incile* (Conrad, 1833), length 168 mm; E= *Chesapecten septenarius* (Say, 1824), length 82 mm; F= *Chesapecten madisonius* (Say, 1824), length 129 mm; G= *Glyphis oblonga* (H. Lea, 1843), length 31 mm; H= *Volutifusus obtusus* (Emmons, 1858), length 99 mm; I= *Volutifusus spengleri* Petuch, 1991, length 160 mm; J= *Busycon maximus* (Conrad, 1839), length 149 mm; K= *Megaptygma sinuosa* (Gabb, 1862), juvenile specimen, length 93 mm. All specimens were collected in the Rushmere Member (lower Zone 2) of the Yorktown Formation, in the Lone Star Cement Company pit, Chuckatuck, Virginia.

comparilis (Tuomey and Holmes, 1855), the extremely flattened glass scallop *Amusium mortoni* (Ravenel, 1844), and *Stralopecten rogersii* (Conrad, 1834). Smaller bivalves lived on sand patches between the scallop beds and some of these included the periplomatid *Cyathodonta dalli* (Mansfield, 1929), the thraciids *Thracia transversa* H. Lea, 1843 and *T. maddelysensis* Gardner, 1944, the venerid *Macrocallista reposta* (Conrad, 1834), and several species of small mactrids and tellinids. Feeding on these smaller bivalves and on the larger scallops were several large molluscivorous gastropods, including the busyconids *Busycotypus incile* (Conrad, 1833) (Plate 45, D), *Busycon maximus* (Conrad, 1839), and *Sinistrofulgur contrarium* (Conrad, 1840), the buccinid *Atractodon curvirostra* (Conrad, 1864), the drilling ocenebrine muricid *Ecphora quadricostata* (Say. 1824) (Rushmere Member only; replaced by stratigraphic subspecies in the Mogarts Beach and Moore House Members), and the scaphelline volutids *Volutifusus obtusus* (Emmons, 1858) (Plate 45, H), *V. spengleri* Petuch, 1991 (Plate 45, I), *V. emmonsi* Petuch, 1994, and *Megaptygma sinuosa* (Gabb, 1862) (Plate 45, K). Other carnivorous gastropods also occurred within the scallop beds, including the small echinoderm-feeding cassid *Sconsia hodgii* (Conrad, 1841), the ficid *Ficus holmesi* (Conrad, 1867), and the large vermivorous conid *Contraconus adversarius* (Conrad, 1840). The large fissurellid limpet, *Glyphis oblonga* (H. Lea, 1843) (Plate 45, A), lived attached to the larger scallops.

An ecological structure very similar to the *Chesapecten septenarius* Community is seen in the scallop beds off Recent North Carolina. Here, large busyconids (*Busycotypus canaliculatum*, *Sinistrofulgur laeostomum*, and *Busycon carica eliceans*) and scaphelline volutids (*Scaphella junonia* and *Volutifusus georgianus*) prey upon the deep water bay scallop *Argopecten gibbus carolinensis* and the venerid *Macrocallista maculata*. Also occurring in these beds, as in the Yorktown community, is a small echinoderm-feeding cassid (*Phalium granulatum*), a ficid (*Ficus communis*), a large vermivorous conid (*Gradiconus delessertii*), and a large drilling muricid (*Hexaplex fulvescens*).

The *Torcula alticostata* Community

Closer to shore, in quiet, shallow water lagoonal areas (5-10 m) during Rushmere Member time, immense beds of turritellid gastropods carpeted the sandy sea floor. These were composed of at least three species, some of which included *Torcula alticostata* (Conrad, 1834) (namesake of the community; Plate 46, K; extremely abundant in Zone 2), *Torculoidella fluxionalis* (Rogers and Rogers, 1837), and *Archimediella virginica* (Campbell, 1993). Living on open sand patches between the turritellid beds were numerous small bivalves, some of which included the venerids *Chione cribraria* (Conrad, 1843) and *Lirophora latilirata* subspecies, the glossid *Glossus fraternus* (Say, 1824), the macomid *Macoma cookei* Gardner, 1944, the astartids *Astarte concentrica* Conrad, 1834, *A. roanokensis* Gardner, 1944, and *A. undulata* Say, 1824, and the nuculanid *Yoldia laevis* (Say, 1824). Feeding on the turritellids and small bivalves were numerous molluscivorous gastropods, some of which included the trophonine muricids *Trophonopsis petiti* Campbell, 1993, *Lirosoma sulcosa* (Conrad, 1830) (Plate 46, A), *Chesatrophon tetricus* (Conrad, 1832) (Plate 46, D), and *Scalaspira strumosa* (Conrad, 1830) (Plate 46, H) (the last three being Salisbury Transmarian relicts), the eurytopic and eurybathic muricid *Ecphora quadricostata* (Say, 1824) (Plate 48, G and H), a radiation of the buccinid genus *Ptychosalpinx*, including *P. fossulata* (Conrad, 1843) (Plate 46, B), *P. altile* (Conrad, 1832) (Plate 46, C), *P. multirugata* (Conrad, 1841) (Plate 46, G), and *P. laqueata* (Conrad, 1832) (Plate 46, J), and the fasciolariids *Fusinus burnsi* (Dall, 1890) (Plate 46, F) and *F. exilis* (Conrad, 1832). On the scattered open sand patches, a large fauna of other carnivorous gastropods occurred, some of which included the vermivorous terebrids *Laevihastula hamptonensis* (Campbell, 1993) (Plate 46,

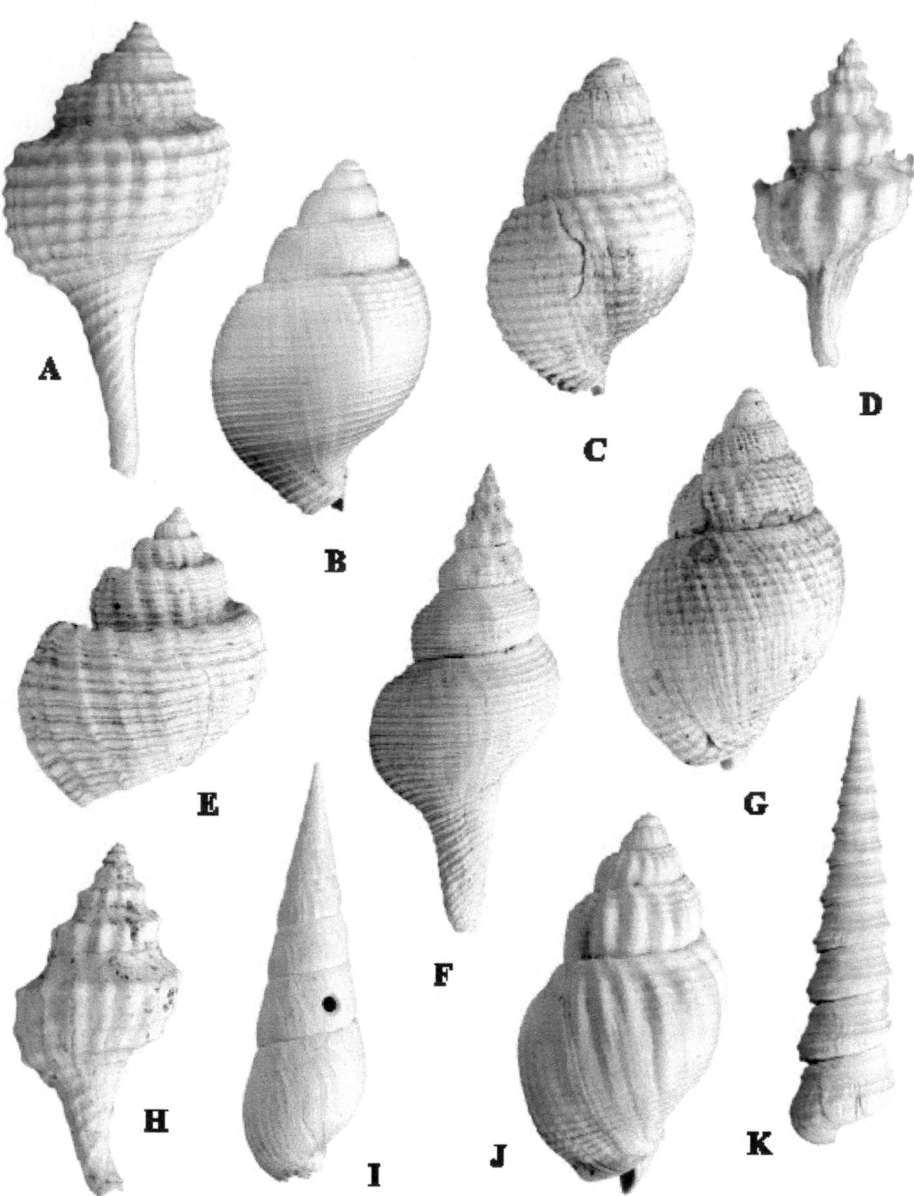

Plate 46. Index fossils and characteristic organisms of the *Torcula alticostata* Community (shallow lagoon turritellid beds) of the Yorktown Subsea. Included are: A= *Lirosoma sulcosa* (Conrad, 1830), length 34 mm; B= *Ptychosalpinx fossulata* (Conrad, 1843), length 36 mm; C= *Ptychosalpinx altile* (Conrad, 1832), length 34 mm; D= *Chesatrophon tetricus* (Conrad, 1832), length 28 mm; E= *Ventrilia perspectiva* (Conrad, 1834), length 23 mm; F= *Fusinus burnsi* (Dall, 1890), length 87 mm; G= *Ptychosalpinx multirugata* (Conrad, 1841), length 37 mm; H= *Scalaspira strumosa* (Conrad, 1830), length 26 mm; I= *Laevihastula hamptonensis* (Campbell, 1993), length 43 mm; J= *Ptychosalpinx laqueata* (Conrad, 1832), length 39 mm; K= *Torcula alticostata* (Conrad, 1834), length 62 mm. All specimens were collected in the Rushmere Member (lower Zone 2) of the Yorktown Formation, in the Lone Star Cement Company pit, Chuckatuck, Virginia.

I), *Strioterebrum. indenta* (Conrad, 1836), *S. emmonsii* (Dall, 1896), and *S. grayi* (Gardner, 1948), and the suctorial-feeding cancellariids *Ventrilia perspectiva* (Conrad, 1834) (Plate 46, E), *V. smithfieldensis* (Oleksyshyn, 1960), and *Axelella williamsi* (Petit, 1976), and the eurytopic vermivorous conid *Ximeniconus marylandicus* (Green, 1830) (also found in the *Marvacrassatella virginica* Community).

The Marvacrassatella virginica Community

In shallow intertidal areas (0-5 m) behind the Chuckatuck Archipelago and along the sheltered northwestern coast during Rushmere Member time, large expanses of organic-rich muddy sand bottoms supported an amazingly large and rich bivalve fauna. Some of these, which must have literally paved the sea floor, included larger species such as the crassatellid *Marvacrassatella virginica* (Gmelin, 1791) (namesake of the community; Plate 47, J), the venerids *Mercenaria tridacnoides* (Lamarck, 1818), *M. rileyi* (Conrad, 1838), and *M. nucea* Dall, 1903, the cardiids *Chesacardium acutilaqueatum* (Conrad, 1839) (Plate 47, A) and *Planicardium virginianum* (Conrad, 1839), the hiatellid *Panopea reflexa* (Say, 1824), the arcids *Sectiarca lienosa* (Say, 1831), *Dallarca callicestosa* (Dall, 1898), *Granoarca propatula* (Conrad, 1844), and *Cunearca staminea* (Say, 1831), and the margaritariid *Margaritaria abrupta* (Conrad, 1832), and smaller species such as the glycymerid *Costaglycymeris subovata* (Say, 1824) (Plate 47, C), the carditid *Cyclocardia granulata* (Say, 1824), and numerous small mactrids and tellinids. Feeding on this incredibly rich bivalve fauna was an equally diverse molluscivorous gastropod fauna, some of which included the ocenebrine muricids *Pterorhytis umbrifer* (Conrad, 1832) (Plate 47, B), *Urosalpinx cannabinis* (Conrad, 1844) (Plate 47, G), *U. phrikna* (Gardner and Aldrich, 1919), *Vokesinotus lamellosus* (Emmons, 1858), *Trossulasalpinx trossulus* (Conrad, 1832), *Ecphora quadricostata* (Say, 1824) (Plate 48, G and H), the thin-ribbed *Planecphora mansfieldi* (Petuch, 1989) (Plate 48, C and D; descendant of the Murdock Subsea *P. hertweckorum*), and the inflated *Globecphora parvicostata* (Pilsbry, 1911) (Plate 48, A and B; descendant of the Williamsburg Subsea *G. xenos*), the fasciolariids *Fasciolaria (Cinctura) rhomboidea* Rogers, 1839 and *Terebraspira sparrowi* (Emmons, 1858), and the naticids *Euspira perspectiva* (Rogers and Rogers, 1837) (Plate 47, E), *Euspira sayana* (Campbell, 1993) (Plate 47 F), and *Bulbus chesapeakensis* Campbell, 1993. Also living on the open sand bottom were numerous smaller carnivorous gastropods, including the generalist carnivore olivid *Cariboliva canaliculata* (H. Lea, 1843) (Plate 47, K), the generalist carnivore marginellids *Bullata antiqua* (Redfield, 1852) and *Microspira oviformis* (Conrad, 1868), the vermivore conid *Ximeniconus marylandicus* (Green, 1830) (Plate 47, D), and the sponge and hydroid-feeding trochids *Calliostoma mitchelli* (Conrad, 1834) (Plate 47, I) and *Solariella ricei* Campbell, 1993 (Plate 47, H). In small, scattered patches of sea grasses (possibly *Zostera*) growing on the mud flats and bivalve beds, the only Yorktown cowrie, *Akleistostoma carolinensis* (Conrad, 1841), occurred. This small cypraeid probably had the same ecological preferences as the closely related and morphologically similar *Muracypraea mus* (Linnaeus, 1758), which lives in muddy intertidal sea grass beds in the Recent Gulf of Venezuela and feeds on both the grass blades and the epifauna (Petuch, 1976; 1988) (see Chapter 10).

It is interesting to note that most of the taxa discussed in the above-mentioned Yorktown communities existed, with little morphological change, throughout the time span of the Rushmere, Mogarts Beach, and Moore House Members of the Yorktown Formation. Some exceptions do occur, however, and these include rapidly evolving groups (such as genera in the bivalve families Astartidae and Pectinidae and the gastropod subfamily Ocenebrinae) that generate new species in each consecutive member. These rapidly-evolving species are ideal biostratigraphic index fossils and can be essen-

Plate 47. Index fossils and characteristic organisms of the *Marvacrassatella virginica* Community (shallow muddy sand lagoons) of the Yorktown Subsea. Included are: A= *Chesacardium acutilaqueatum* (Conrad, 1839), length 108 mm; B= *Pterorhytis umbrifer* (Conrad, 1832), length 37 mm; C= *Costaglycymeris subovata* (Say, 1824), length 35 mm; D= *Ximeniconus marylandicus* (Green, 1830), length 28 mm; E= *Euspira perspectiva* (Rogers and Rogers, 1837), length 31 mm; F= *Euspira sayana* (Campbell, 1993), length 46 mm; G= *Urosalpinx cannabinis* (Conrad, 1844), length 35 mm; H= *Solariella ricei* Campbell, 1993, length 9 mm; I= *Calliostoma mitchelli* (Conrad, 1834), length 24 mm; J= *Marvacrassatella virginica* (Gmelin, 1791), length 72 mm; K= *Cariboliva canaliculata* (H. Lea, 1843), length 42 mm. All specimens were collected in the Rushmere Member (lower Zone 2) of the Yorktown Formation, in the Lone Star Cement Company pit, Chuckatuck, Virginia.

Plate 48. Ecphoras of the Yorktown Formation (Zone 2; Rushmere, Mogarts Beach, and Moore House Members). A,B= *Globecphora parvicostata* (Pilsbry, 1911), length 70 mm, Rushmere Member, Lee Creek Mine, Aurora, North Carolina; C,D= *Planecphora mansfieldi* (Petuch, 1989), length 66 mm, Rushmere Member, Lee Creek Mine, Aurora, North Carolina; E,F= *Ecphora quadricostata umbilicata* (Wagner, 1839), length 68 mm, Moore House Member, Rice's Pit, Hampton, Virginia; G,H= *Ecphora quadricostata* (Say, 1824), length 41 mm, Rushmere Member, Lee Creek Mine, Aurora, North Carolina; I,J= *Ecphora quadricostata rachelae* Petuch, 1989, length 42 mm, Mogarts Beach Member, upper Yorktown beds in the Lee Creek Mine, Aurora, North Carolina.

tial for fine-tuned regional correlations. Of particular importance are the large and common Yorktown ecphoras, which undergo distinct shell morphology changes in every chronostratigraphic increment. In the Rushmere Member, three species occur together in most assemblages: *Ecphora quadricostata* (Plate 48, G and H), *Globecphora parvicostata* (Plate 48, A and B), and *Planecphora mansfieldi* (Plate 48, C and D). All three also occur together in the contemporaneous Duplin Subsea of the Charleston Sea and the Tamiami Subsea of the Okeechobean Sea. By Mogarts Beach Member time, the *Globecphora* and *Planecphora* species were extinct, and only one ecphora had survived: the high-spired Mogarts Beach indicator *Ecphora quadricostata rachelae* Petuch, 1989 (Plate 48, I and J). This subspecies is also found, though rarely, in the Pinecrest Member (Unit 7 at Sarasota) of the Tamiami Formation. Likewise, in the Moore House Member (basal beds), only one species existed, this being the high-spired and widely flaring *Ecphora quadricostata umbilicata* (Wagner, 1839) (Plate 48, E and F). This last-living ecphora occurred only in the Yorktown Subsea.

The Mytilus edulis alaeformis Community

Along the western and southern shores of the Yorktown Subsea during Rushmere Member time, the *Marvacrassatella virginica* Community also extended into estuarine areas at the mouths of rivers. Here, it was replaced by large beds of the mytilid mussels *Mytilus edulis alaeformis* Sowerby, 1821 (namesake of the community) and *Modiolus gigas* (Dall, 1898) interfingered with the mud flat bivalve beds. Farther upstream, in brackish water, the estuarine mactrid bivalves *Rangia clathrodon* (Conrad, 1833) and *Mulinia congesta* (Conrad, 1833) formed virtual monoculture beds. On these intertidal estuarine mud flats, large aggregations of nassariid gastropods scavenged carrion and detritus, much like the *Ilyanassa* colonies in the Recent Long Island Sound area. Some of these included *Ilyanassa granifera* (Conrad, 1868), *I. sexdentata* (Conrad, 1844), *Paranassa porcina* (Say, 1824), *Scalanassa harpuloidea* (Conrad, 1844), *S. scalaspira* (Conrad, 1868), and *Uzita quadrulatus* (H. Lea, 1843). Competing with the nassariids for carrion were several species of small columbellids, including *Zafrona communis* subspecies, *Astyris edenensis* (Richards and Harbison, 1947), and several species of the Yorktown endemic genus *Juliamitrella, including J. chesapeakensis* Campbell, 1993, *J. gardnerae* (Ward and Blackwelder, 1987), and *J. smithfieldensis* (Mansfield, 1929). In some areas of the estuaries and river mouths, large oyster bars formed massive biohermal structures. These were composed entirely of the ostreids *Ostrea compressirostra* Say, 1824 and *Conradostrea sculpturata* (Conrad, 1840).

Communities and Environments of the Duplin Subsea

During Piacenzian time, the Charleston Sea basin flooded to its maximum extent and was geomorphologically more complex than during previous transgressions (Figure 25). At this time, the Cape Fear Arch had eroded and regressed inland, separating the Charleston Sea basin into two large embayments: the northern Magnolia Embayment (named for the fossil beds at Natural Well, Magnolia, Duplin County, North Carolina) and the southern Raysor Embayment (named for the Raysor Formation of South Carolina). The Magnolia Embayment was partially enclosed by a chain of sand barrier islands, the Onslow Archipelago (named for Onslow Bay, North Carolina), which produced lagoonal and sheltered open sound conditions in the northern and central parts of the Duplin Subsea. The Raysor Embayment contained large areas of shallow muddy sand shoals and sand flats and these intertidal environments graded into the tropical estuaries of the Satilla Lagoon System to the south. The northward-flowing warm water Ciboney Current extended into the Duplin Subsea basin, producing warm-temperate and subtropical climatic conditions (Petuch, 1997). Because of the warmer marine climate, the Duplin Subsea shared

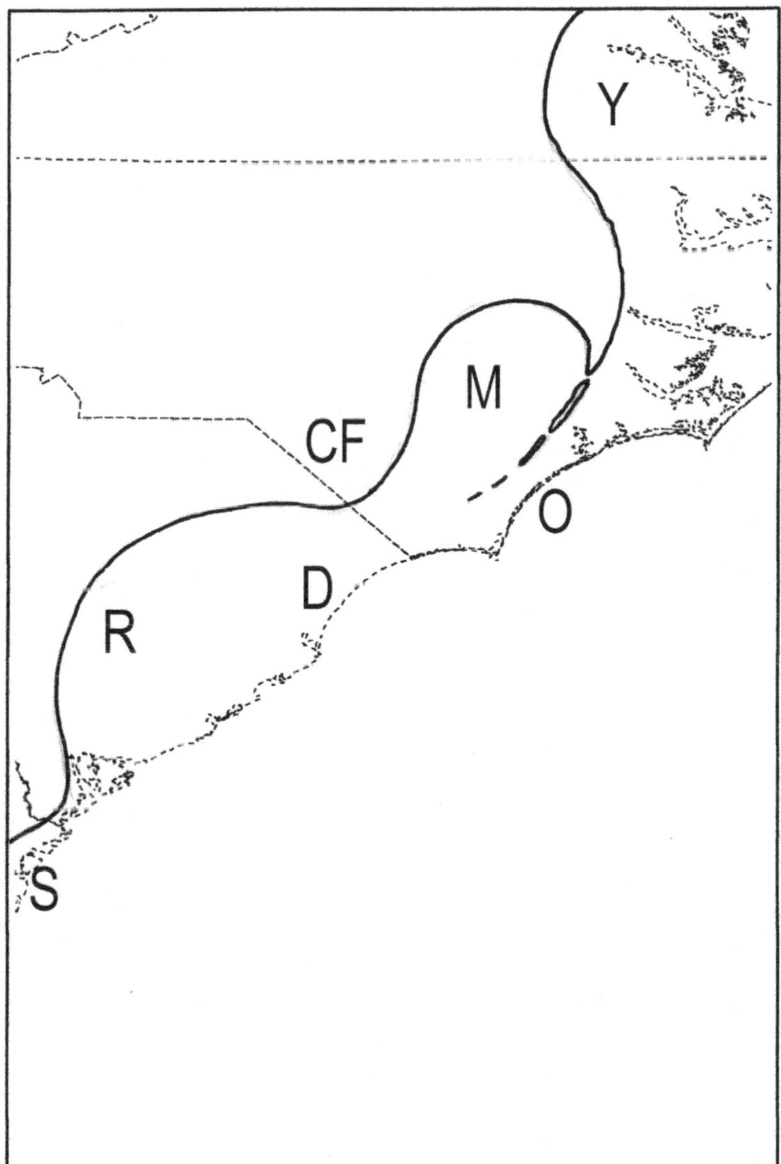

Figure 25. Approximate configuration of the coastline of the Carolinas during Piacenzian Piocene time, superimposed upon the outline of the coasts of Recent North and South Carolina. Prominent features included: D= Duplin Subsea of the Charleston Sea, Y= Yorktown Subsea of the Albemarle Sea, M= Magnolia Embayment, R= Raysor Embayment, S= Satilla Lagoon System, O= Onslow Archipelago, CF= Cape Fear Arch.

many faunal elements with the southern Floridian Tamiami Subsea and its molluscan fauna was essentially transitional between those of the Albemarle and Okeechobean Seas. Even though under the influence of the Ciboney Current, the tropical indicator families Strombidae, Turbinellidae (and Vasinae), Melongenidae, and Modulidae were absent from the Duplin Subsea, indicating possible colder winter temperatures. Many of the Duplin Subsea taxa are illustrated in Chapter 7 (Tamiami Subsea faunas).

The fauna of the Duplin Subsea is known primarily from small exposures of the type Duplin Formation in Duplin County, North Carolina, the Duplin Formation in Sumter County (Muldrow), South Carolina, the Raysor and Goose Creek (upper beds) Formations of southern South Carolina, and the Satilla River beds of Georgia. For the most part, the Raysor and Goose Creek Formations are leached and their fossils are poorly preserved, making paleoecological reconstructions difficult. From all these units, four main communities can be discerned: the *Carolinapecten walkerensis* Community (deep lagoon pectinid beds), the *Torcula etiwanensis* Community (shallow lagoon turritellid beds), the *Costaglycymeris tuomeyi* Community (shallow sand lagoon and sand flats), and the *Rangia solida* Community (subtropical estuaries). As in other paleoseas, the lithofacies associated with the turritellid beds and the bivalve-rich sand flat environments interfinger, both laterally and bathymetrically.

The Carolinapecten walkerensis Community

During the early Piacenzian (equivalent to the Rushmere Member of the Yorktown Formation and the Buckingham Member of the Tamiami Formation of Florida), the sand bottom deep lagoon areas (20-100 m) behind the Onslow Archipelago and along the Raysor Embayment housed immense scallop beds. These were composed of the large pectinids *Carolinapecten walkerensis* (Tucker, 1934) (namesake of the community) and *Chesapecten septenarius* (Say, 1824), smaller species such as *Argopecten comparilis* (Tuomey and Holmes, 1855) and *Christinapecten decemnarius* (Conrad, 1834), the endemic glass scallop *Amusium holmesi* (Dall, 1898), and the lions paw scallop *Nodipecten peedeensis* (Tuomey and Holmes, 1856) (also found in the Buckingham Member of the Tamiami Formation). Living among the scallops and on scattered open sand patches were numerous smaller bivalves, including the venerids *Macrocallista reposta* (Conrad, 1834), *Lirophora athleta* (Conrad, 1863), and *Transenella carolinensis* Dall, 1903, the astartid *Astarte conradi* Gardner, 1943, and the glossid *Glossus carolina* (Dall, 1900). Feeding on the scallops and smaller bivalves were several large gastropods, including the busyconids *Busycon filosum* Conrad, 1862, *B. tritonis* (Conrad, 1867), *Busycotypus conradi* (Tuomey and Holmes, 1856), and *Sinistrofulgur contrarium* subspecies, the fasciolariid *Triplofusus duplinensis* (B. Smith, 1940), the eurytopic and eurybathic ocenebrine muricid *Ecphora quadricostata* (Say, 1824), the pleioptygmatid *Pleioptygma carolinensis* (Conrad, 1841), and the scaphelline volutid *Scaphella trenholmi* (Tuomey and Holmes, 1856). Several carnivorous gastropods were also present on the scallop beds, and some of these included echinoderm-feeders such as the cassid *Sconsia hodgii* (Conrad, 1841) and the ficid *Ficus holmesi* Conrad, 1867 and the sponge and hydroid-feeding trochids *Calliostoma cyclum* Dall, 1892 and *C. hertfordense* Gardner, 1948. An extremely rich vermivorous gastropod fauna also occurred on the open sand patches, attesting to the large resident biomass of polychaete annelid worms. Some of these included the turrids *Mangelia emissaria* Olsson, 1916, *Cymatosyrinx tiara* (Gardner, 1948), *C. ziczac* (Gardner, 1948), and *C. lunata* (H. Lea, 1843), the terebrids *Strioterebrum grayi* (Gardner, 1948), *S. carolinensis* (Conrad, 1841, *S. neglecta* (Emmons, 1858), and *Myurellina unilineata* (Conrad, 1841), and the conid *Contraconus adversarius* (Conrad, 1840). A large fauna of suctorial-feeding cancellariids also occurred on the open sand bottoms, some of which included *Massyla rapella* subspecies, *M. venusta* (Tuomey and Holmes, 1856), *Cancellaria rotunda* Dall, 1892, *C. depressa* Tuomey and Holmes, 1857, and *C. plagiostoma* Conrad, 1834.

The Torcula etiwanensis Community

In shallower sand bottom lagoonal areas (5-20 m) during the early Piacenzian (Rushmere and Buckingham time), extensive turritellid beds accumulated behind the Onslow Archipelago and along sheltered western coastal areas. These were made up primarily of the large *Torcula etiwanensis* (Tuomey and Holmes, 1856) (namesake of the community) and *Eichwaldiella holmesi* (Dall, 1892), and the smaller species *Torculoidella fluxionalis duplinensis* (Gardner and Aldrich, 1919). Feeding on the turritellids were several drilling naticid gastropods, including *Polinices carolinensis* (Conrad, 1841), *Euspira propeinternus* (Mansfield, 1935), *E. perspectiva* subspecies, and *Neverita emmonsi* (Conrad, 1856), and the ocenebrine muricids, *Ecphora quadricostata* (Say, 1824) (Plate 49, G and H) and the "T"-shaped ribbed *Latecphora violetae* (Petuch, 1988) (Plate 49, I). This last species, which also occurs in the Buckingham Member of the Tamiami Formation of Florida, appears to have been a specialized feeder on turritellid gastropods, preying on both free-living genera and on the sessile gregarious genus *Vermicularia*. The wide, "T"-shaped ribs of the genus *Latecphora* were an amazing biomechanical adaptation against peeling crab predation (Vermeij, 1978), where the shell thickness was essentially doubled (as in an "I" beam) without doubling the shell weight (Petuch, 1989).

The Costaglycymeris tuomeyi Community

In shallow water muddy sand lagoons and intertidal sand flats (0-5 m) during the early Piacenzian (Rushmere and Buckingham time), dense beds of shallowly infaunal bivalves dominated the sea floor. Some of the component species included the venerids *Mercenaria rileyi* (Conrad, 1838), *M. tridacnoides* (Lamarck, 1818), *Costacallista emmonsi* (Gardner, 1943), and *Chione cribraria* (Conrad, 1843), the endemic glycymerid *Glycymeris tuomeyi* Dall, 1898 (namesake of the community), the cardiid *Planicardium virginianum* (Conrad, 1839), and the mactrid *Hemimactra alta* (Gardner, 1943). Primary gastropod predators on this rich bivalve fauna included the busyconid *Fulguropsis carolinensis* (Tuomey and Holmes, 1856), the fasciolariids *Heilprinia carolinensis* (Dall, 1892), *Fasciolaria (Cinctura) rhomboidea* Rogers, 1839, *Terebraspira sparrowi* (Emmons, 1858), *T. nodulosa* (Emmons, 1858), and *T. acuta* (Emmons, 1858), the naticid *Naticarius* cf. *plicatella* (Conrad, 1863), and the large muricids *Phyllonotus globosus* (Emmons, 1858), *Globecphora parvicostata* (Pilsbry, 1911) (Plate 49, D, E, and F), *Planecphora mansfieldi* (Petuch, 1989), and the eurytopic and eurybathic *Ecphora quadricostata* (Say, 1824) (Plate 49, G and H). Also occurring on the intertidal sand flats were several general carnivore/scavenger gastropods, some of which included the olivids *Oliva (Strephona) carolinensis* (Conrad, 1863) and *O. (Strephona) robesonensis* Gardner, 1948, the marginellids *Microspira oliviformis* (Tuomey and Holmes, 1856) and *Bullata taylori* (Olsson, 1916), and the large nassariid *Globinassa gastrophila* (Olsson, 1916). Small scattered patches of sea grass (possibly *Thalassia*) also grew near the intertidal sand flats and these supported two species of weed-bed cowries, the small *Akleistostoma pilsbryi* (Ingram, 1947) and the larger, widespread *A. carolinensis* (Conrad, 1841). Also scattered between the small sea grass patches and bivalve beds were biohermal clumps of the oysters *Ostrea compressirostra* Say, 1824 and *Conradostrea sculpturata* (Conrad, 1840). Living on these oysters was the large, flattened fissurellid limpet, *Diodora carolinensis* (Conrad, 1875).

Plate 49. Ecphoras of the *Ecphora roxaneae* Community (shallow muddy lagoon and mud flats) of the Santee Subsea and the *Torcula etiwanensis* ommunity (shallow lagoon turritellid beds) and *Costaglycymeris tuomeyi* Community (shallow sand areas and intertidal sand flats) of the Duplin Subsea. Included are: A,B= *Ecphora roxaneae* Petuch, 1991, length 110 mm; C= *Ecphora roxaneae* Petuch, 1991, length 67 mm; D,E= *Globecphora parvicostata* (Pilsbry, 1911), length 52 mm; F= *Globecphora parvicostata* (Pilsbry, 101), length 57 mm; G,H= *Ecphora quadricostata* (Say, 1824), length 45 mm; I= *Latecphora violetae* (Petuch, 1988), length 84 mm. The specimens of *E. roxaneae* are from the lower beds of the Goose Creek Formation and the specimens of *G. parvicostata*, *E. quadricostata*, and *L. violetae* are from the upper beds of the Goose Creek Formation; all from the Martin-Marietta Mine, Santee, South Carolina.

The Rangia solida Community

During the late Piacenzian, contemporaneous with the Moore House Member of the Yorktown Formation and the Fruitville Member of the Tamiami Formation, the southern part of the Raysor Embayment and the Satilla Lagoon System had developed into a continuous extensive subtropical estuary. This brackish water area housed a rich molluscan fauna that had many species in common with the even richer Myakka Lagoon System of the Tamiami Subsea (see Chapter 7) and the Pliocene estuaries of the Louisiana and Texas coasts (Dall, 1913). Here, at the mouths and deltas of several major paleorivers, extensive organic-rich mud flats developed. Farther inland, these intertidal flats graded into dense mangrove forests (probably of the Black Mangrove *Avicennia*) and oligohaline water conditions. Literally paving the mud flats in immense monoculture beds were the mactrid bivalves *Rangia solida* Dall, 1913 (namesake of the community) and *Mulinia congesta magnoliana* (Dall, 1898). Feeding on algal and bacterial films on the surface of these mactrid beds were large aggregations of gastropods, including the potamidids *Pachycheilus cancelloides* (Aldrich, 1911) and *P. satillensis* (Aldrich, 1911) and a radiation of rissoideans, including *Paludestrina satillensis* (Aldrich, 1911), *P. georgiensis* (Aldrich, 1911), *P. plana* (Aldrich, 1911), and *P. expansilabris* (Aldrich, 1911). The algivorous neritid gastropod *Neritina sparsilineata* (Dall, 1913) was also abundant on the mactrid beds and in the mangrove forests. Several small scavenger and carrion-feeding gastropods were also present on the mud flats, including the nassariid *Uzita smithiana* (Olsson, 1916) and the columbellid *Zafrona matsoni* (Mansfield, 1924). Farther upstream in the paleorivers, the planorbid fresh water gastropod *Planorbis antiquitus* Aldich, 1911 was abundant on emergent vegetation, and its dead shells commonly washed onto the estuarine mud flats.

Communities and Environments of the Jackson Subsea

After a long emergent period during the late Messinian Miocene and the entire Zanclean Pliocene, the Choctaw Sea basin finally reflooded at the beginning of the Piacenzian as the Jackson Subsea. Although much smaller in area than previous Choctaw subseas, this early Pliocene paleosea was geomorphologically complex (Figure 26). The principal structural feature within the Jackson Subsea was the Alum Bluff Archipelago (named for Alum Bluff, Liberty County, Florida, on the Apalachicola River), a chain of low muddy-sand islands that formed on top of the topographic highs produced by the older Chipola Subsea coral reefs and carbonate sand banks. These islands, and the surrounding shallow banks and lagoons were the center of deposition for the highly-fossiliferous Jackson Bluff Formation of the Alum Bluff Group. By the late Piacenzian, the islands of the Alum Bluff Archipelago contained large evaporate hypersaline lakes, and these were the sources of the gypsum and "aluminous clay" beds (evaporites) seen in the upper part of the Jackson Bluff Formation at Alum Bluff (Mansfield, 1932). Subsequent heavy erosion by the proto-Apalachicola, proto-Choctawhatchee, and proto-Ochlockonee Rivers during the Pleistocene has removed most of the Jackson Subsea Islands, and only part of one large island remains until today. This last-remaining paleoisland can be seen, in bisected form, as the high cliffs of Alum Bluff. The Apalachicola River is presently rapidly eroding this last, small insular remnant and, in the near geological future, there will be no record of these large topographic highs having ever existed.

In the early Piacenzian, contemporaneous with the Rushmere Member of the Yorktown Formation and the Buckingham member of the Tamiami Formation, the Jackson Subsea was under the influence of the Apalachee Gyre (Petuch, 1997), a cool-water gyre off the Gulf Loop Current, and its marine climate was at best warm-temperate.

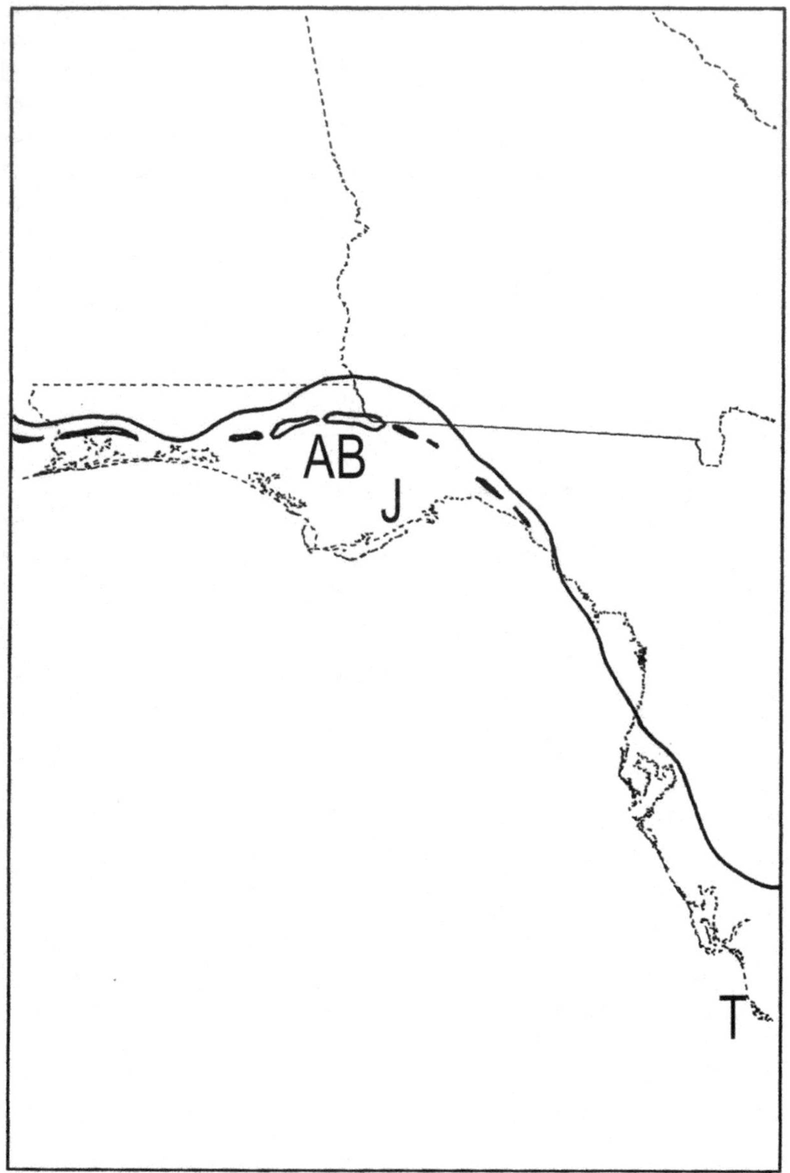

Figure 26. Approximate configuration of the coastline of the northeastern Gulf of Mexico during the Piacenzian Pliocene, superimposed upon the outline of northwestern Florida. Prominent features included: J= Jackson Subsea of the Choctaw Sea, AB= Alum Bluff Archipelago, T= Tamiami Subsea of the Okeechobean Sea.

In these cooler conditions, the lower beds of the Jackson Bluff Formation were deposited in the sheltered lagoons near and behind the island system (the "*Ecphora* Zone" of Mansfield, 1930; 1932) and the component fauna can be seen to be only paratropical in nature. The flow of the Apalachee Gyre across the narrow Jackson Subsea shelf also produced upwelling systems of cool, nutrient-rich water and this added to the high productivity of the coastal waters. At this time, the Jackson Subsea fauna was only paratropical in nature and lacked examples of the tropical index families Strombidae, Cypraeidae, Modulidae, Turbinellidae, and Turbinidae. From the highly-fossiliferous lower beds of the Jackson Bluff Formation ("*Ecphora* Zone"), two distinct marine communities can be discerned. These include the *Torcula alumensis* Community (shallow lagoonal turritellid beds) and the *Dallarca alumensis* Community (muddy-sand banks and intertidal sand flats). Later in the Piacenzian, contemporaneous with the Mogarts Beach and Moore House Members of the Yorktown Formation and the Pinecrest and Fruitville Members of the Tamiami Formation, the Jackson Subsea area began to fill with deltaic sediments. This extensive shoaling caused the formation of immense sand flats that filled the entire Choctaw Sea basin. Due to the reduced water depth, the Apalachee Gyre ceased to circulate within the basin and this resulted in a warmer marine climate during the deposition of the upper beds of the Jackson Bluff Formation (the "*Cancellaria* Zone" of Mansfield, 1930; 1932). The molluscan fauna of this last remnant of the Jackson Subsea was eutropical in nature and contained species of *Strombus* (Strombidae), *Turbo* (Turbinidae), *Modulus* (Modulidae), *Hystrivasum* (Vasinae-Turbinellidae), and *Akleistostoma* Cypraeidae). This late Jackson fauna essentially resembled an impoverished version of the Pinecrest fauna of the Tamiami Subsea and contained no endemic taxa. By the latest Piacenzian, the Jackson Subsea was completely infilled and the Choctaw Sea ceased to exist.

The Torcula alumensis Community

In deeper areas around and behind the Alum Bluff Archipelago (5-20 m) and on adjacent shallow banks, immense beds of suspension-feeding turritellid gastropods accumulated. Here, the plankton-rich waters of the Apalachee Gyre and upwelling systems supported the richest turritellid fauna ever found in a single North American paleosea. The plankton-fed turritellid biomass of these beds was impressive, attested to by the co-existence of several very large species. Some of these included *Torcula alumensis* (Mansfield, 1930) (namesake of the community; Plate 50, G), *T. gardnerae* (Mansfield, 1930), *T. etiwanensis* (Tuomey and Holmes, 1856), *Apicula cookei* (Mansfield, 1930), and *A. clarksvillensis* (Mansfield, 1930). Smaller turritellids also occurred in abundance in these beds, and some of these included *Torculoidella fluxionalis duplinensis* (Gardner and Aldrich, 1919) and *T. jacksonensis* (Mansfield, 1930). On open sand patches between the turritellids, large numbers of small bivalves occurred, some of which included the astartid *Astarte floridana* Dall, 1903, the macomids *Macoma gardnerae* (Mansfield, 1932) and *M. alumensis* Dall, 1900, the nuculanids *Nuculana trochila* (Dall, 1898) and *Yoldia tarpaeia* (Dall, 1898), and the pandorid *Pandora crassidens* (Conrad, 1838). Feeding on the turritellids and interstitial bivalves was a large number of molluscivorous gastropods, including the large naticid *Euspira perspectiva* (Rogers, 1837), the eurytopic and eurybathic ocenebrine muricids *Ecphora quadricostata striatula* (Petuch, 1986) (Plate 50, D and E) and *Planecphora mansfieldi* (Petuch, 1989) (Plate 50, A and B) (both the namesake of the "*Ecphora* Zone"), the busyconids *Busycoarctum tudiculatum* (Dall, 1890) (Plate 50, C), *Busycotypus libertiense* (Mansfield, 1930) (Plate 50, F), *Busycon alumense* Mansfield, 1930 (Plate 50, H), *Brachysycon propeincile* (Mansfield, 1930) (Plate 50, I), and *Sinistrofulgur hollisteri* Petuch, 1994, and the scaphelline volutids *Volutifusus emmonsi* Petuch, 1994 (Plate 50, K) and *Scaphella mansfieldi* Petuch, 1994. Small bioherms of the ahermatypic branching coral *Oculina* species also occurred on

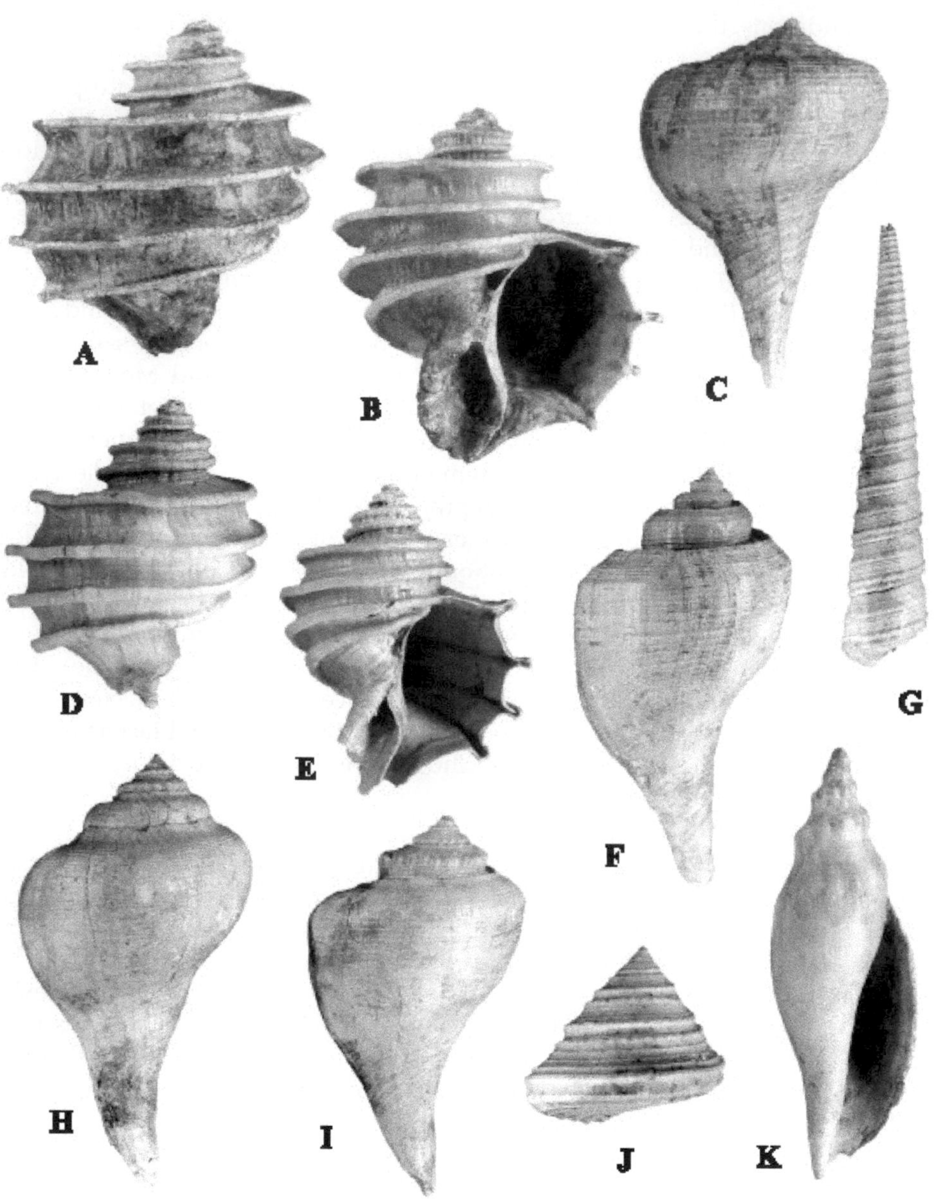

Plate 50. Index fossils and characteristic organisms of the *Torcula alumensis* Community (shallow lagoon turritellid beds) of the Jackson Subsea. Included were: A,B= *Planecphora mansfieldi* (Petuch, 1989), length 47 mm; C= *Busycoarctum tudiculatum* (Dall, 1890), length 46 mm; D,E= *Ecphora quadricostata striatula* Petuch, 1986, length 49 mm; F= *Busycotypus libertiense* (Mansfield, 1930), length 128 mm; G= *Torcula alumensis* (Mansfield, 1930), length 141 mm; H= *Busycon alumense* Mansfield, 1930, length 74 mm; I= *Brachysycon propeincile* (Mansfield, 1930), length 66 mm; J= *Calliostoma aluminium* Dall, 1892, length 11 mm; K= *Volutifusus emmonsi* Petuch, 1994, length 92 mm. All specimens were collected in the lower bed of the Jackson Bluff Formation ("*Ecphora* Zone") at Alum Bluff, Liberty County, Florida, Apalachicola River.

Plate 51. Index fossils and characteristic organisms of the *Dallarca alumensis* Community (shallow muddy sand areas and intertidal sand flats) of the Jackson Subsea. Included were: A= *Mercenaria rileyi* (Conrad, 1838), length 103 mm (note large hole drilled by *Ecphora*); B= *Heilprinia dalli* (Mansfield, 1930), length 125 mm; C= *Sinum chesapeakensis* Campbell, 1993, length 33 mm; D= *Rasia campsa* ((Dall, 1898), length 35 mm; E= *Dallarca alumensis* (Mansfield, 1032), length 58 mm; F= *Eupleura miocenica* Dall, 1890, length 20 mm; G= *Ptychosalpinx duerri* Petuch, 1994, length 39 mm; H= *Calotrophon libertiense* (Mansfield, 1930), 23 mm; I= *Lirophora xesta* (Dall, 1903), length 28 mm; J= *Contraconus lindajoyceae* Petuch, 1991, length 82 mm; K= *Heilprinia gunteri* (Mansfield, 1930), length 83 mm. All specimens were collected in the lower bed of the Jackson Bluff Formation ("*Ecphora* Zone") at Alum Bluff, Liberty County, Florida, Apalachicola River.

sand patches between the turritellid beds and these provided the habitat for the large fissurellid limpet *Glyphis alumensis* (Mansfield, 1930) and the sponge and hydroid-feeding endemic trochid *Calliostoma aluminium* Dall, 1892.

The Dallarca alumensis Community

The organic-rich intertidal sand flats and shallow (0-5 m) muddy sand areas adjacent to the Alum Bluff Archipelago both supported a rich fauna of bivalves, some of which formed extensive beds. Primary among these were the arcids, including the endemic *Dallarca alumensis* (Mansfield, 1932) (namesake of the community; Plate 51, E), the endemic *Rasia campsa* (Dall, 1898) (Plate 51, D), and *Cunearca scalaris* (Conrad, 1843), the arcoid noetiid *Noetia incile* (Say, 1824), and venerids, including *Mercenaria rileyi* (Conrad, 1838) (Plate 51, A), *Lirophora ulocyma* (Dall, 1900), *L. athleta* (Conrad, 1863), the endemic *L. xesta* (Dall, 1903) (Plate 51, I), *Chione erosa* Dall, 1903 and *C. cortinaria* (Rogers, 1837), and the endemic *Dosinia obliqua* Dall, 1903. Other common bivalves included the endemic semelid *Semele alumensis* Dall, 1900, the crassatellid *Marvacrassatella meridionalis* (Dall, 1900), and the cardiids *Planicardium virginianum* (Conrad, 1839) and *Protocardia jacksonense* Mansfield, 1932. Unlike the other Miocene and Pliocene North American paleoseas, scallops occurred only sparsely and were uncommon in the Jackson Subsea during "*Ecphora*" Zone time. The most often-encountered pectinid species were *Pecten ochlockoneensis* Mansfield, 1932, the bay scallop *Argopecten jacksonensis* (Mansfield, 1932) and the small *Leptopecten leonensis* (Mansfield, 1932). Feeding on the bivalve beds was a large fauna of molluscivorous gastropods, including the naticids *Euspira propeinternus* (Mansfield, 1935) and *N.* cf. *waltonensis* (Mansfield, 1935), and *Sinum chesapeakensis* Campbell, 1993 (Plate 51, C), the muricids *Eupleura miocenica* Dall, 1890 (Plate 51, F), *Pterorhytis marshalli* (Mansfield, 1930), *Calotrophon libertiense* (Mansfield, 1930) (Plate 51, H), *Phyllonotus leonensis* (E. Vokes, 1967), and the ubiquitous eurytopic *Ecphora quadricostata striatula* Petuch, 1986 and *Planecphora mansfieldi* (Petuch, 1989), the small busyconid *Busycotypus aepynotum* (Dall, 1890), the buccinid *Ptychosalpinx duerri* Petuch, 1994 (Plate 51, G), and the fasciolariids *Heilprinia dalli* (Mansfield, 1930) (Plate 51, B) and *H. gunteri* (Mansfield, 1930) (Plate 51, K). Also occurring in the shallow muddy sand lagoonal areas were numerous specialized carnivorous gastropods, including vermivores such as the conids *Contraconus lindajoyceae* Petuch, 1991 (Plate 51, K) and *C. adversarius* (Conrad, 1840), the suctorial-feeding cancellariid *Cancellaria depressa* subspecies, and general carnivore/scavengers such as the olivids *Oliva (Strephona) duerri* Petuch, 1994 and *O. (Strephona) alumensis* Petuch, 1994, and the buccinid *Hesperisternia filicata* subspecies.

Chapter 7. The Everglades Pseudoatoll

Probably the single most spectacular marine geomorphological feature in Cenozoic eastern North America was the Everglades Pseudoatoll of the Okeechobean Sea (Tamiami Subsea) (named for the Everglades basin of southern Florida; Figure 27). Having begun its growth in the early Oligocene (Chapter 3 of this book; Petuch, 1987), this immense coral reef structure reached its maximum development in the late Piacenzian and formed the template for the island tracts of the Pleistocene Okeechobean Sea and, ultimately, the modern Everglades (Petuch, 1986, 1992, 1994, 1997). Being essentially a "U"-shaped system of reefs, the Everglades Pseudoatoll surrounded the central area of the Okeechobean Sea in much the same way as the reefs tracts of the South Pacific Kwajalein, Eniwetock, and Jaluit Atolls surround their central lagoons (Figure 28). Unlike these true atolls, which formed on sinking volcanic islands, the Okeechobean reef system is classified as a pseudoatoll, in that it formed around a late Eocene-aged paleodepression on a tectonically stable carbonate platform (Petuch, 1987; 1997). This paleodepression, which may have formed as a remnant collapse feature around a late Eocene submarine asteroid impact site, is located under the southern part of the present-day Everglades National Park (see Petuch, 1987 for some of the anomalous geological features associated with the depression). At its greatest development, the "U"-shaped pseudoatoll extended over 330 linear kilometers, making it longer than the Recent Great Barrier Reef of Belize. Today, the buried remnants of this gigantic reef complex are the cores of the higher-elevation areas that surround the low, flat Everglades. Along the southeastern coast of Florida, this rim-like feature forms the base upon which all the coastal cities, from Palm Beach to Miami, are built.

The main pseudoatoll coral reef structures and their associated faunas are preserved in the upper part of the Tamiami Formation, in both the reefal limestone of the Golden Gate Member and in the lagoonal deposits of the Fruitville Member, and in the older Pinecrest Member (including an undescribed reefal facies) (see Chapter 1). Data gleaned from these units has shown that the Everglades Pseudoatoll was structurally complex and was asymmetrical in shape (Figure 28). The western side was better developed, with a wide, zonated reef system (the Immokalee Reef Tract), a large chain of carbonate sediment islands (the Immokalee Archipelago), and a wide, shallow carbonate-siliciclastic platform behind the reef tract (the Hendry Platform). Along the eastern side, the geomorphology was simpler, with a long, narrow, zonated reef system (the Miami-Palm Beach Reef Tract) and a narrow back-reef area. A wide, deep lagoon (the Loxahatchee Trough; over 100 m in depth) had developed between the eastern edge of the Hendry Platform and the Miami-Palm Beach Reef Tract, resembling a miniature version of the Tongue of the Ocean embayment of the Recent Great Bahama Bank. The northern part of the Tamiami Subsea intrareefal lagoon tapered into an elongated, shallow embayment (the Kissimmee Embayment) that was lined with mud flats and mangrove jungles. These mangrove environments extended along the southwestern Floridian coast, northwest of the pseudoatoll, and were incorporated into a large and complex tropical estuary and brackish water lagoon system (the Myakka Lagoon System). Altogether, these areas, and their associated environments, contained the richest coral and molluscan faunas found anywhere in the Pliocene Americas and were comparable in species diversity to the Recent Southwest Pacific.

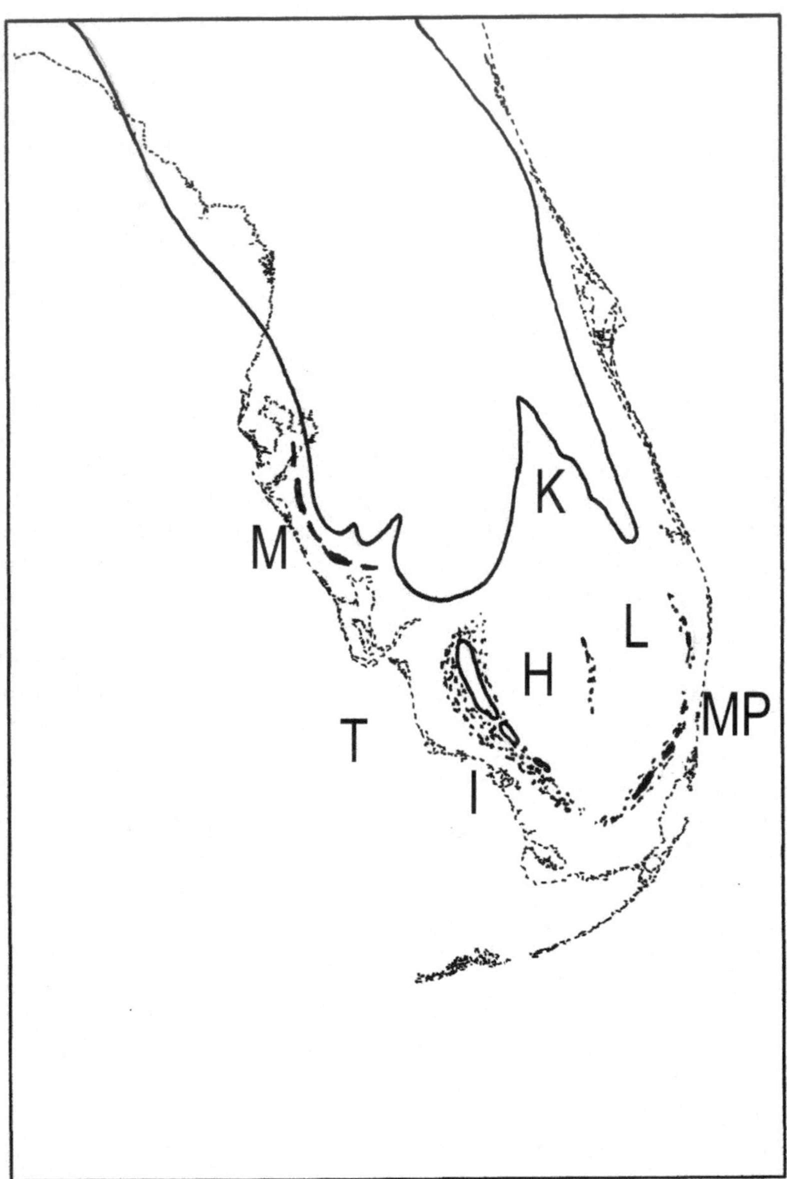

Figure 27. Approximate configuration of the coastline of Florida during Piacenzian Pliocene time, superimposed upon the outline of the Recent Floridian Peninsula. Prominent features included: T= Tamiami Subsea, M= Myakka Lagoon System, K= Kissimmee Embayment, I= Immokalee Island and Reef Tract, H= Hendry Platform, L= Loxahatchee Trough, MP= Miami and Palm Beach Reef Tracts. The Immokalee, Miami, and Palm Beach Reef Tracts, together, make up the Everglades Pseudoatoll. The Hendry Platform formed on top of the siliciclastic wedge of the older Charlotte-Murdock Subseas delta.

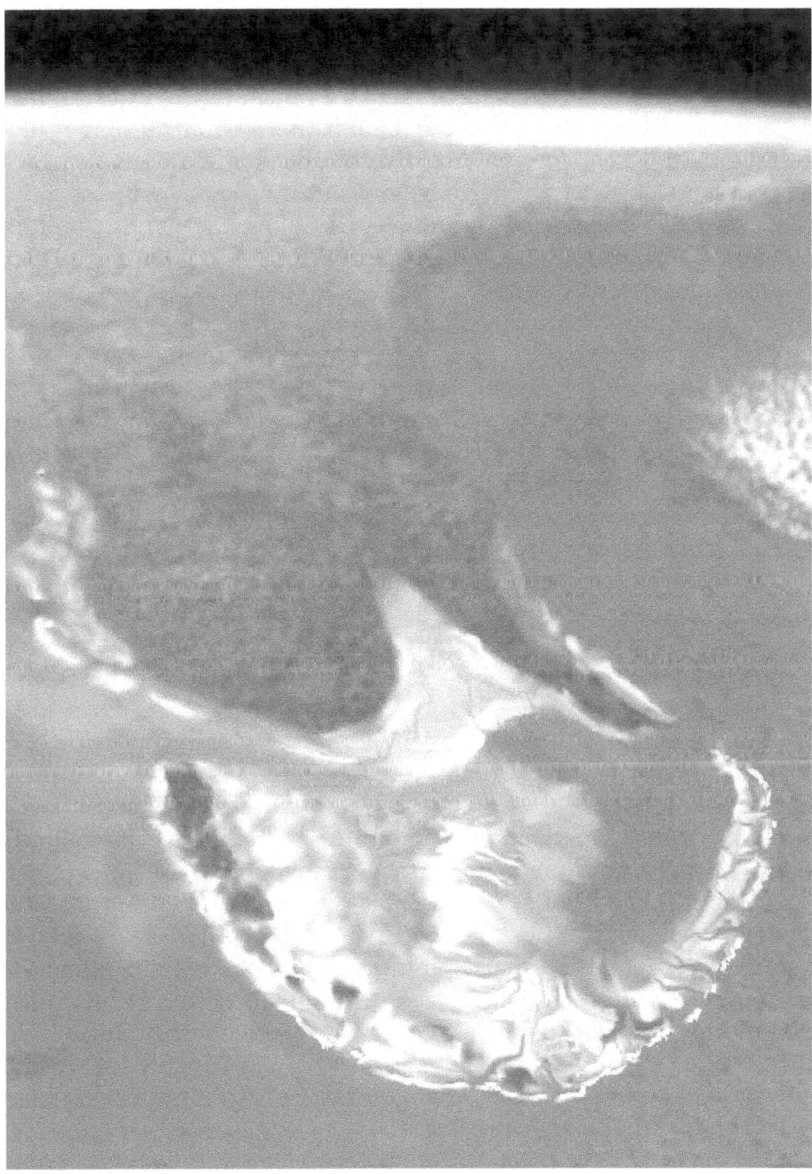

Figure 28. Simulated satellite image of the Floridian Peninsula during the Piacenzian Pliocene, showing the possible appearance of the Tamiami Subsea (looking north). The geomorphological features are the same as those shown on Figure 27. Especially prominent are the coral reefs of the "U"-shaped Everglades Pseudoatoll, the wide carbonate banks of the Hendry Platform and Kissimmee Embayment, and the deep lagoon of the Loxahatchee Trough. The GIS imagery was done by Charles Roberts, Ph.D., Department of Geography and Geology, Florida Atlantic University, using paleoenvironmental and paleobathymetric data supplied by the author, enhancing a NASA space shuttle photograph of Florida.

Chronologically, the development of the Everglades Pseudoatoll during the Pliocene breaks into three separate time frames and these correlate directly with the three sets of members of the Tamiami Formation. For time-equivalencies, the numbered and dated Sarasota Pliocene units (Units 10-2) associated with these members (Petuch, 1982, 1994; Chapter 1 of this book) will be used throughout this chapter as chronostratigraphic standards. The first and oldest of these developmental events occurred during the deposition of the early Piacenzian Buckingham Member (Unit 10 at Sarasota). From this time, four separate marine communities can be recognized within the Tamiami Subsea. These included the *Mercenaria tridacnoides* Community (shallow muddy-sand areas and intertidal sand flats), the *Apicula buckinghamensis* Community (shallow lagoon turritellid beds), the *Nodipecten peedeensis* Community (deep lagoonal pectinid beds), and the *Phyllangia blakei* Community (lagoonal coral bioherms). The second developmental event took place during the deposition of the mid-Piacenzian Pinecrest and Ochopee Members (Units 9-5 at Sarasota, and equivalents). From this time, four main communities can be discerned, and these included the *Strombus floridanus* Community (shallow muddy sand areas and intertidal mud flats), the *Apicula gladeensis* Community (shallow lagoon turritellid beds), the *Nodipecten floridensis* Community (deep lagoon pectinid beds), and the *Hyotissa meridionalis* Community (shallow lagoon oyster beds).

The third and most complex developmental stage of the Everglades Pseudoatoll system took place during the deposition of the late Piacenzian Golden Gate and Fruitville Members (Units 4-2 at Sarasota, and equivalents). At this time, five separate communities occurred on the coral reef tracts of the pseudoatoll, and these correlated with wave action-related coral zonational patterns. On a transect running from the open oceanic seaward edge to the sheltered inner Tamiami basin, the main communities included the *Pocillopora crassoramosa* Community (high energy reef crest), the *Dichocoenia tuberosa* Community (lower energy front reef platform), the *Isophyllia desotoensis* Community (lower energy back reef platform), the *Stylophora affinis* Community (sheltered back reef), and the *Antillia bilobata* Community (sheltered reef lagoon). During Unit 4 time, the Myakka Lagoon estuary was at its maximum development and two brackish water ecosystems occurred, including the *Pyrazisinus scalinus* Community (mangrove forests and mud flats) and the *Mulinia sapotilla* Community (shallow muddy lagoon bivalve beds). Later, during Units 3-2 time, large shoals developed within the Tamiami basin and the Myakka Lagoon System and these areas supported several unusual communities, two of which included the *Perna conradiana* Community (shallow lagoon mussel beds during Unit 3 time) and the *Chama emmonsii* Community (shallow lagoon chamid bivalve bioherms during Unit 2 time). During Units 4-2 time, the Kissimmee Embayment contained its own set of endemic communities and these included the *Pyrazisinus kissimmeensis* Community (mangrove forests and mud flats during Unit 3 time), the *Siphocypraea kissimmeensis* Community (sand flats and *Thalassia* beds during Unit 4 time), and the *Siphocypraea penningtonorum* Community (sand flats and *Thalassia* beds during Unit 2 time). In total, twenty main communities are recognized as having occurred within the Pliocene Everglades Pseudoatoll and these are described in the following sections.

The Early Piacenzian Communities of the Everglades Pseudoatoll

In the earliest Piacenzian, during Buckingham time (Unit 10 at Sarasota, and equivalent to the Rushmere Member of the Yorktown Formation), the Tamiami Subsea was under the influence of colder water conditions similar to those in the Duplin and Jackson Subseas. Because of the cooler marine climate and lower sea level stands, the older reef tracts of Zanclean origin had ceased in their development and were emergent as low carbonate islands. Most of the carbonate fines from these eroding, exposed reefs filled the western side of the Tamiami basin and laid the foundation for the subsequent Hendry Platform and carbonate banks of the Ochopee Member. No new major reef growth took place at this time. Although sharing many molluscan faunal elements with the northern Jackson and Duplin Subseas, the Tamiami marine communities during Buckingham time still contained a high degree of endemism and were the richest known in eastern North America. Interestingly, no *Pyrazisinus*-based tropical mud flat or mangrove environments are preserved from Buckingham time. These communities may have been spatially restricted due to the colder water conditions and survived in localized refugia within the Myakka Lagoon System or the Kissimmee Embayment.

The *Mercenaria tridacnoides* Community

Within the Tamiami Subsea area, the shallow intrabasinal sand banks (0-5 m) and the coastal intertidal muddy-sand bars housed a rich bivalve fauna, comprising an immense biomass. Here, the large venerids *Mercenaria tridacnoides* (Lamarck, 1818) (namesake of the community; Plate 52, C), *M. ochlockoneensis* (Mansfield, 1932) (Plate 52, B), and *M. rileyi* (Conrad, 1838) formed dense pavements and dominated the entire sea floor. Smaller venerids, such as *Lirophora ulocyma* (Dall, 1895) (Plate 52, E), *L. leonensis* (Mansfield, 1932), and *Chione erosa* Dall, 1903 and the crassatellid *Marvacrassatella meridionalis* Dall, 1900) also formed dense beds that intercollated with the larger species. These shallowly-infaunal bivalves were the principal prey items of numerous large molluscivorous gastropods, some of which included the largest busyconid fauna ever found in a single community, with *Busycon filosum* (Conrad, 1862) (Plate 52, D) and *B. pachyus* Petuch, 1994, *Busycoarctum tudiculatum* (Dall, 1890) and *B.* species (Plate 52, F), *Sinistrofulgur hollisteri* Petuch, 1994 (Plate 52, A), *S. grabaui* Petuch, 1994, *Busycotypus libertiense* (Mansfield, 1930), *B. mansfieldi* Petuch, 1994, and *Brachysycon propeincile* (Mansfield, 1930), the melongenids *Tropochasca petiti* Olsson, 1967 (Plate 52, I) and *Echinofulgur cannoni* Petuch, 1994, and the large muricid *Phyllonotus leonensis* (E. Vokes, 1967). Scattered between the bivalve beds were patches of sea grass (probably *Thalassia*) and these supported a small fauna of weed-associated species, including the cowries *Akleistostoma carolinensis* (Conrad, 1841) (Plate 52, G and H) and *A. pilsbryi* (Ingram, 1947) and the modulid *Modulus woodringi* Mansfield, 1930. On open muddy bottoms within the Myakka Lagoon System, this same assemblage also contained a representative of the endemic estuarine-restricted cowrie genus *Calusacypraea* (*C. duerri* (Petuch, 1996); Plate 52, J and K).

The *Apicula buckinghamensis* Community

In deeper areas within the lagoons (5-10 m) during Buckingham time, extensive beds of turritellid gastropods accumulated on the open sea floor. These were composed primarily of *Apicula buckinghamensis* (Mansfield, 1939) (namesake of the community; Plate 53, L), *A. cookei* (Mansfield, 1930), *Eichwaldiella holmesi* (Dall, 1892) (Plate 53, G), and *Torcula* cf. *gardnerae* (Mansfield, 1930). On scattered open sand patches between the turritellid beds, small beds of bivalves also occurred and these included the astarid *Astarte floridana* Dall, 1903, the macomid *Macoma alumensis* Dall, 1900, the chamid *Arcinella* species, the carditid *Carditamera vaughani* (Dall, 1903), and the arcid *Granoarca propatula* (Conrad, 1843). The turritellids and bivalves were the primary prey items of several molluscivorous gastropods, some of which included the busyconids *Fulguropsis carolinensis* (Tuomey and Holmes, 1856) (Plate 53, E), *Pyruella harasewychi* Petuch, 1982 (Plate 53, D), and *P. waltfrancei* Petuch, 1994, the fasciolariids *Heilprinia dalli* (Mansfield, 1930), *H. gunteri* (Mansfield, 1930), *Fusinus dianeae* Petuch, 1994 (Plate 53, H), *Terebraspira* species (*T. spar-*

Plate 52. Index fossils and characteristic organisms of the *Mercenaria tridacnoides* Community (shallow muddy sand areas and intertidal mud flats during Buckingham (Unit 10) time) of the Tamiami Subsea. Included are: A= *Sinistrofulgur hollisteri* Petuch, 1994, length 187 mm; B= *Mercenaria ochlockoneensis* (Mansfield, 1932), length 86 mm; C= *Mercenaria tridacnoides* (Lamarck, 1818), length 122 mm; D= *Busycon filosum* (Conrad, 1862), length 133 mm; E= *Lirophora ulocyma* (Dall, 1895), length 46 mm; F= *Busycoarctum* species, length 190 mm; G,H= *Akleistostoma carolinensis* (Conrad, 1841), length 60 mm; I= *Tropochasca petiti* Olsson, 1967, length 85 mm; J,K= *Calusacypraea duerri* (Petuch, 1996), length 67 mm. All specimens were collected in the Buckingham Member (Unit 10) of the Tamiami Formation in Quality Aggregates Phase 6 pit, Sarasota, Sarasota County, Florida.

rowi complex), and *Fasciolaria (Cinctura)* species (*F. rhomboidea* complex), and the specialized turritellid-feeding ocenebrine muricid *Latecphora violetae* (Petuch, 1988) (Plate 53, A and B). The open sand patches also contained several other carnivorous gastropods, including vermivores such as the conids *Contraconus lindajoyceae* Petuch, 1991 (Plate 53, H), *C. adversarius* (Conrad, 1840), and *Seminoleconus violetae* (Petuch, 1988) (Plate 53, F), suctorial-feeders such as the cancellariids *Cancellaria leonensis* Mansfield, 1930 and *Ventrilia alumensis* (Mansfield, 1930), and general carnivores/scavengers such as the olivids *Oliva (Strephona) keatoni* Petuch, 1994 and *O. (Strephona) duerri* Petuch, 1994, and the marginellids *Bullata taylori* (Olsson, 1916) and *B. antiqua* (Redfield, 1852). The deeper water cowrie, *Akleistostoma crocodila* (Petuch, 1994) (Plate 5, J and K; Chapter 2) also occurred on the open sand patches, possibly feeding on algae or sponges.

The Nodipecten peedeensis Community

In the deepest parts of the lagoons (10-100 m) during Buckingham time, several species of pectinid bivalves formed immense beds that covered the sea floor. Several species made up these dense beds, with the most prominent being *Carolinapecten walkerensis* (Tucker, 1934) (*C. buckinghamensis* (Mansfield, 1939) is a synonym) (Plate 53, J), *Nodipecten peedeensis* (Tuomey and Holmes, 1856) (namesake of the community; Plate 53, C), *Argopecten jacksonensis* (Mansfield, 1932), and *Pecten ochlockoneensis* Mansfield, 1932. Feeding on the scallops were several molluscivorous gastropods, some of which included the deeper water busyconid *Lindafulgur lindajoyceae* (Petuch, 1991), and the scaphelline volutids *Scaphella martinshugari* Petuch, 1994, *S. mansfieldi* Petuch, 1994, and *Volutifusus emmonsi* Petuch, 1994. On open sand patches between the scallop beds, other specialized carnivorous gastropods occurred, including echinoderm-feeders such as the cassid *Sconsia metae* Petuch, 1994 and the ficid *Ficus holmesi* Conrad, 1867, vermivores such as the conid *Jaspidiconus harveyensis* (Mansfield, 1930), the terebrid *Myurellina unilineata* (Conrad, 1841), and the turbinellid *Hystrivasum barkleyae* Petuch, 1994 (Plate 53, K), and general carnivore/scavengers such as the buccinids *Ptychosalpinx perprotractus* Petuch, 1994, and *Calophos* cf. *wilsoni* Allmon, 1990.

The Phyllangia blakei Community

Buckingham time, the deeper, cleaner water areas of the Tamiami lagoons (10-20 m) also housed a rich fauna of ahermatypic corals. These cooler water scleractinians formed large biohermal structures that often reached 1 meter in height and covered large areas of the sea floor. Several genera of astrangioid corals grew together to make up these bioherms, with the principal species being *Phyllangia blakei* Wells, 1947 (namesake of the community; Plate 54, F), *Astrangia leonensis* Weisbord, 1971 (Plate 54, E), *A. talquinensis* Weisbord, 1971, *Septastrea* species (*S. crassa* complex), *Paracyathus vaughani* Gane, 1895 (Plate 54, G), and *Oculina* species (*O. sarasotana* complex). Small heads of the hermatypic coral *Montastrea* species (*M. costata* complex) also grew in conjunction with the astrangioids. Several bivalves nestled within these bioherms, some of which included the encrusting plicatulid *Plicatula rudis* H. Lea, 1845, the chamid *Chama congregata* Conrad, 1833, and the spondylid *Spondylus bostrychites* Guppy, 1867, and the coral-nesting limid *Lima* cf. *carolinensis* Dall, 1898. The principal predators of these interstitial and encrusting bivalves were the ocenebrine muricids *Pterorhytis marshalli* (Mansfield, 1930) (Plate 54, H and I), *Planecphora mansfieldi* (Petuch, 1989) (Plate 54, A and B), and *Ecphora quadricostata striatula* Petuch, 1986 (Plate 54, C and D). Both ecphoras, as eurytopic and eurybathic species, also occurred on the adjacent turritellid and scallop beds, but appeared to prefer the coral bioherm environments, as they were often encrusted with *Astrangia* coralla.

Plate 53. Index fossils and characteristic organisms of the *Apicula buckinghamensis* Community (shallow lagoon turritellid beds during Buckingham (Unit 10) time) and the *Nodipecten peedeensis* Community (deep lagoon pectinid beds during Buckingham (Unit 10) time) of the Tamiami Subsea. Included are: A,B= *Latecphora violetae* (Petuch, 1988), length 98 mm; C= *Nodipecten peedeensis* (Tuomey and Holmes, 1856), length 118 mm; D= *Pyruella harasewychi* Petuch, length 73 mm; E= *Fulguropsis carolinensis* (Tuomey and Holmes, 1856), length 121 mm; F= *Seminoleconus violetae* (Petuch, 1991), length 53 mm; G= *Eichwaldiella holmesi* (Dall, 1900), length 109 mm; H= *Fusinus dianeae* Petuch, 1994, length 108 mm; I= *Lindafulgur lindajoyceae* (Petuch, 1991), length 131 mm; J= *Carolinapecten walkerensis* (Tucker, 1934), length 96 mm; K= *Hystrivasum barkleyae* Petuch, 1994, length 88 mm; L= *Apicula buckinghamensis* (Mansfield, 1939), length 65 mm. All specimens were collected in the Buckingham Member (Unit 10) of the Tamiami Formation in Quality Aggregates Phase 6 pit, Sarasota, Sarasota County, Florida.

Feeding on the corals themselves was the coralliophilid gastropod *Babelomurex leonensis* (Mansfield, 1937). The algae growing on dead coral colonies supported the only cerithiid gastropod in the community, *Cerithioclava cannoni* Petuch, 1994. Also associated with the coral bioherms was the largest of the Buckingham cowries, *Akleistostoma erici* (Petuch, 1999) (Plate 54, J and K). This large cypraeid probably fed on algae, sponges, and hydroids, much like the omnivorous Recent cowrie *Macrocypraea cervus*, which occurs on coral bioherms off North Carolina (Petuch, 1972).

The Middle Piacenzian Communities of the Everglades Pseudoatoll

The mid-Piacenzian, during Pinecrest Member time (equivalent to the Mogarts Beach Member of the Yorktown Formation), saw a return to a warmer marine climate (Campbell, 1993) and the beginning of the maximum development of the pseudoatoll. At this time, the Hendry Platform had also developed to its fullest extent, covering the entire western half of the intra-atoll lagoon. This shallow carbonate area, which resembled the Recent Bahama Banks, housed several unique communities, each with a highly endemic molluscan fauna.

The *Strombus floridanus* Community

On the shallow banks (0-5 m) of the Hendry Platform and Myakka Lagoon System, huge aggregations of the strombid gastropod *Strombus floridanus* Mansfield, 1930 (namesake of the community; Plate 55, F) literally covered the open sand areas, feeding on surficial algal films. This was the first time since the early Chipola Subsea that strombids had dominated any marine ecosystem in eastern North America. Beds of this small algivorous species also occurred in the latest development of the Jackson Subsea (in the upper beds of the Jackson Bluff Formation), where it was so abundant that an entire chronostratigraphic zone was named for it (the "*Strombus* Zone;" Mansfield, 1930). Also occurring in abundance with the strombids were several large venerid bivalves, including *Mercenaria carolinensis* (Conrad, 1873) (Plate 55, C), *Chione procancellata* Mansfield, 1932 (Plate 55, K), and *Macrocallista reposta* (Conrad, 1834). Preying upon the strombids and bivalves were several large molluscivorous gastropods, some of which included the melongenid *Melongena taurus* Petuch, 1994 (Plate 55, A), the fasciolariids *Fasciolaria (Cinctura) rhomboidea* Rogers, 1839 and *Triplofusus duplinensis* (B. Smith, 1940) (Plate 55, B), the busyconids *Busycoarctum tropicalis* (Petuch, 1994) and *Sinistrofulgur contrarium* (Conrad, 1840), and the ocenebrine muricid *Ecphora quadricostata rachelae* Petuch, 1989 (Plate 55, I and J). In areas adjacent to river mouths, with muddier substrates and lowered salinities, the estuarine cowries *Calusacypraea (Myakkacypraea) briani* (Petuch, 1996) (Plate 55, D and E) and *C. globulina* Petuch, new species (Plate 5, F and G, Chapter 2; see Systematic Appendix) occurred and formed large aggregations. Small beds of Turtle Grass (*Thalassia*) were scattered between the open sand areas and these supported large beds of the weed-loving cowries *Akleistostoma floridana* (Mansfield, 1931) (Plate 55, G and H) and *Siphocypraea trippeana* Parodiz, 1988 (Plate 56, E and F). The vermivorous turbinellid *Turbinella streami* Petuch, 1991 and the large strombid *Macrostrombus hertweckorum* (Petuch, 1991) also occurred on, or adjacent to, the Turtle Grass patches.

The *Apicula gladeensis* Community

On deeper parts of the Hendry Platform carbonate sand banks and in the more nutrient-rich areas of the Myakka Lagoon System (5-10 m), immense beds of turritellid gastropods dominated the sea floor. These were made up of several species, with the most abundant being *Apicula gladeensis* (Mansfield, 1931) (namesake of the community; Plate

Plate 54. Index fossils and characteristic organisms of the *Phyllangia blakei* Community (lagoonal coral bioherms during Buckingham (Unit 10) time) of the Tamiami Subsea. Included are: A,B= *Planecphora mansfieldi* (Petuch, 1989), length 85 mm; C,D= *Ecphora quadricostata striatula* Petuch, 1986, length 57 mm; E= *Astrangia leonensis* Weisbord, 1971, length 33 mm; F= *Phyllangia blakei* Wells, 1947, length 54 mm; G= *Paracyathus vaughani* Gane, 1895, length 38 mm; H,I= *Pterorhytis marshalli* (Mansfield, 1930), length 42 mm; J,K= *Akleistostoma erici* (Petuch, 1999), length 74 mm. All specimens were collected in the Buckingham Member (Unit 10) of the Tamiami Formation in Quality Aggregates Phase 6 pit, Sarasota, Sarasota County, Florida.

Plate 55. Index fossils and characteristic organisms of the *Strombus floridanus* Community (shallow muddy sand areas and intertidal mud flats during Pinecrest (Unit 7) time) of the Tamiami Subsea. Included are: A= *Melongena taurus* Petuch, 1994, length 210 mm; B= *Triplofusus duplinensis* (B. Smith, 1940), length 356 mm; C= *Mercenaria carolinensis* (Conrad, 1873), length 131 mm; D,E= *Calusacypraea briani* (Petuch, 1996), length 74 mm; F= *Strombus floridanus* Mansfield, 1930, length 66 mm; G,H= *Akleistostoma floridana* (Mansfield, 1931), length 90 mm; I,J= *Ecphora quadricostata rachelae* Petuch, 1989, length 63 mm; K= *Chione procancellata* Mansfield, 1932, length 54 mm. All specimens were collected in the Pinecrest Member (Unit 7) of the Tamiami Formation in Quality Aggregates Phase 6 pit, Sarasota, Sarasota County, Florida.

56, G), *A. seminole* Petuch, 1994, *Eichwaldiella pontoni* (Mansfield, 1931) (Plate 56, J), *E. mansfieldi* (Olsson, 1967), and *E. magnasulcus* Petuch, 1991 (Plate 56, H). The density of the individual turritellids, and the large size of the individuals themselves (particularly *Eichwaldiella pontoni*; see Petuch, 1986), prevented sympatric bivalves from becoming established within the colonies. This same pattern of competitive exclusion is seen in the monoculture turritellid beds of the Recent Gulf of Venezuela (Petuch, 1976). Feeding on the turritellids were several molluscivorous gastropods, including the wide-ribbed ecphora *Latecphora bradleyae* (Petuch, 1988) (Plate 56, K and L; a specialized turritellid feeder), the naticids *Dallitesta coensis* (Dall, 1903) and *Neverita emmonsi* (Conrad, 1858), the melongenid *Echinofulgur dalli* Petuch, 1994, the eurytopic muricid *Phyllonotus globosus* (Emmons, 1858) (Plate 57, J; also on oyster and *Vermicularia* beds), the fasciolariids *Terebraspira sparrowi* (Emmons, 1858) and *T. nodulosa* (Emmons, 1858), the busyconids *Fulguropsis excavatum* (Conrad, 1840), *Busycotypus bicoronatum* (Tripp, 1988), and *Pyruella rugosicostata* Petuch, 1982, and the buccinids *Cymatophos lindae* Petuch, 1991 and *Calophos wilsoni* Allmon, 1990.

The *Nodipecten floridensis* Community

In the deepest areas of the Everglades Pseudoatoll lagoon and Myakka Lagoon System (10-100 m), huge swarms of pectinid bivalves carpeted the sea floor. These immense, extensive beds were composed of several large species, with the most commonly encountered being *Carolinapecten eboreus* (Conrad, 1833), *Nodipecten floridensis* (Tucker and Wilson, 1932) (namesake of the community; Plate 56, B), and *Pecten hemicyclus* (Ravenel, 1834). Living on sand patches scattered between the scallop beds were numerous shallowly infaunal bivalves, some of which included the arcids *Larkinia sellardsi* (Mansfield, 1932) and *Caloosarca notoflorida* (H. Vokes, 1969), the cardiid *Venericardia olga* (Mansfield, 1939), the semelid *Semele harveyensis* Mansfield, 1932, and the tellinid *Cymatoica marcottae* Olsson and Petit, 1964. Feeding on the scallops and infaunal bivalves were several molluscivorous gastropods, some of which included the scaphelline volutids *Volutifusus spengleri* Petuch, 1991, *V. obtusus* (Emmons, 1858), and *Scaphella* species (*S. mansfieldi* complex), the giant busyconid *Busycon titan* Petuch, 1994, and the fasciolariids *Heilprinia carolinensis* (Dall, 1892), *H. miamiensis* Petuch, 1994, and *H. diegelae* Petuch, 1994. Occurring with these bivalve predators was a large fauna of vermivores, attesting to the abundance of polychaete worms in this deeper water ecosystem. Some of these included the conids *Contraconus adversarius* (Conrad, 1840) (Plate 56, A), *Spuriconus cherokus* (Olsson and Petit, 1964) (Plate 56, C), *Jaspidiconus marymansfieldae* (Petuch, 1994), and *Lithoconus druidi* (Olsson, 1967) (Plate 56, D), and the turrids *Cymatosyrinx aclinica* Tucker and Wilson, 1933 and *Hindsiclava antealesidota* (Mansfield, 1930). A large fauna of suctorial-feeding cancellariids was also present, and included *Cancellaria rotunda* Dall, 1892, *C. miamiensis* Petuch, 1994, *Trigonostoma druidi* Olsson and Petit, 1964, *Ventrilia carolinensis* (Emmons, 1858), *V. senarium* (Petit and Hoerle, 1976), *Extractrix hoerlei* (Olsson, 1967), and *Massyla propevenusta* (Mansfield, 1930). Other carnivorous gastropods included echinoderm feeders such as the cassids *Cassis floridensis* Tucker and Wilson, 1932 and *Sconsia hodgii* (Conrad, 1841), the ranellid *Cymatium (Linatella) floridanum* (Mansfield, 1930) (also possibly molluscivorous), and the ficid *Ficus jacksonensis* Olsson and Harbison, 1953, and general carnivores/scavengers such as the buccinids *Ptychosalpinx multirugatus* (Conrad, 1841) and *Hesperisternia filicata* (1843), the olivid *Oliva (Strephona) carolinensis* (Conrad, 1863), and the marginellid *Bullata popenoei* (Mansfield, 1930). The deeper water cypraeid *Akleistostoma mansfieldi* (Petuch, 1999) also lived in this ecosystem and was probably a major predator on sponges.

On the clean carbonate banks at the extreme southwestern edge of the pseudoatoll lagoon, a similar community occurred in deeper areas (10-20 m). Preserved in the Ochopee

Plate 56. Index fossils and characteristic organisms of the *Strombus floridanus* Community (shallow muddy sand areas and intertidal mud flats during Pinecrest (Unit 7) time), the *Apicula gladeensis* Community (shallow lagoon turritellid beds during Pinecrest (Unit 7) time), and the *Nodipecten floridensis* Community (deep lagoonal pectinid beds during Pinecrest (Unit 7) time) of the Tamiami Subsea. Included are: A= *Contraconus adversarius* (Conrad, 1840), length 98 mm; B= *Nodipecten floridensis* (Tucker and Wilson, 1932), length 92 mm; C= *Spuriconus cherokus* (Olsson and Petit, 1964), length 73 mm; D= *Lithoconus druidi* (Olsson, 1967), length 147 mm; E,F= *Siphocypraea trippeana* Parodiz, 1988, length 57 mm; G= *Apicula gladeensis* (Mansfield, 1931), length 103 mm; H= *Eichwaldiella magnasulcus* Petuch, 1991, length 113 mm; I= *Cymatophos lindae* Petuch, 1991, length 69 mm; J= *Eichwaldiella pontoni* (Mansfield, 1931), length 165 mm; K,L= *Latecphora bradleyae* (Petuch, 1988), length 68 mm. All specimens were collected in the Pinecrest Member (Unit 7) of the Tamiami Formation in Quality Aggragates Phase 6 pit, Sarasota, Sarasota County, Florida.

Member of the Tamiami Formation, this ecosystem was essentially a biofacies of the *Nodipecten floridensis* Community and contained many of the same species. Here, a larger pectinid fauna occurred, including the endemics *Nodipecten collierensis* (Mansfield, 1931), *Argopecten evergladesensis* (Mansfield, 1931), *A. tamiamiensis* (Mansfield, 1931), and *Carolinapecten gladeensis* (Mansfield, 1931), which lived together with *Nodipecten floridensis* and *Pecten hemicyclus*. The eccentric echinoid *Encope tamiamiensis* Mansfield, 1931 was also abundant on these clean carbonate sea floors and formed dense beds on open areas between the scallop beds. Unfortunately, the Ochopee Member is highly leached, with only calcitic fossils and molds being preserved, so the true nature of this ecosystem may never be known.

The Hyotissa meridionalis Community

In quiet, sheltered shallow areas (0-2 m) within the pseudoatoll and Myakka Lagoon System, the gryphaeid oyster *Hyotissa meridionalis* (Heilprin, 1886) (namesake of the community; Plate 57, K) formed extensive monoculture bioherms. These were so all-pervasive in some areas that entire stratigraphic units of the Pinecrest Member are made up completely of single-species oyster bars (Units 2, 6, and 9 at Sarasota; Petuch, 1982). In previous works on the systematics of the Tamiami mollusks (ie. Olsson and Petit, 1964; Petuch, 1982, 1994), this large, ornate pycnodontine gryphaeid was referred to the taxon *Hyotissa* (or *Pycnodonta*) *haitensis* (Sowerby, 1850). That species, however, is an older, Miocene form that is ancestral to the Pliocene *H. meridionalis* and its ecophenotypic varieties *tamiamiensis* and *monroensis* (Mansfield, 1932). The biohermal structures formed by *H. meridionalis* supplied the biotope for a large and diverse molluscivorous gastropod fauna, some of which included the drilling muricids *Pterorhytis fluviana* (Dall, 1903) (Plate 57, A and B), *Eupleura metae* Petuch, 1994, *Trossulasalpinx maryae* Petuch, 1994, *Vokesinotus lamellosus* (Emmons, 1858), *Chicoreus floridanus* E. Vokes, 1965 (Plate 57, E), *C. xestos* E. Vokes, 1974 (Plate 57, F), *C. stephensae* Petuch, 1994 (Plate 57, H), *Phyllonotus globosus* (Emmons, 1858) (Plate 57, J), *Acantholabia sarasotaensis* Petuch, 1991, and the giant *Hexaplex hertweckorum* (Petuch, 1988) (Plate 57, G; the largest-known American fossil muricid), the thaidid *Thais (Stramonita) caloosana* (Tucker and Wilson, 1933), and the echinofulgurine melongenid *Tropochasca lindae* Petuch, 1991 (Plate 57, D). The turbinellid *Hystrivasum locklini* (Olsson and Harbison, 1953) (Plate 57, C) also occurred in abundance on these oyster beds, feeding on interstitial and encrusting polychaete worms. In slightly deeper water adjacent to the oyster bars (2-5 m), large bioherms of the sessile gregarious turritellid (worm shell) *Vermicularia recta* Olsson and Harbison, 1953 (Plate 57, I) grew in abundance. The large turritellid-feeding ecphora *Latecphora bradleyae* (Petuch, 1988) also lived on these *Vermicularia* bioherms and was the main predator (note: several specimens of *L. bradleyae* were found, in place and in life-position, on the top of the worm shell layers at Sarasota). Like *Hyotissa meridionalis*, the worm shells also make up entire stratigraphic units within the Pinecrest Member (Units 5 and 8 at Sarasota; Petuch, 1982).

The Late Piacenzian Communities of the Everglades Pseudoatoll

During the late Piacenzian (Fruitville Member time; equivalent to the Moore House Member of the Yorktown Formation) three areas of the Tamiami Subsea reached their greatest development, produced their greatest number of environments, and contained their most species-rich invertebrate faunas (Petuch, 1992). These included the coral reef tracts of the pseudoatoll edge (found in the Golden Gate Member), the tropical estuary of the Myakka Lagoon System (found in the Fruitville Member), and the shallow banks and estuaries of the Kissimmee Embayment (found in the Kissimmee facies of the Fruitville

Plate 57. Index fossils and characteristic organisms of the *Hyotissa meridionalis* Community (shallow lagoon oyster beds during Pinecrest (Units 9-5) time) of the Tamiami Subsea. Included are: A,B= *Pterorhytis fluviana* (Dall, 1903), length 53 mm; C= *Hystrivasum locklini* (Olsson and Harbison, 1953), length 96 mm; D= *Tropochasca lindae* Petuch, 1991, length 107 mm; E= *Chicoreus floridanus* E. Vokes, 1965; F= *Chicoreus xestos* E. Vokes, 1974, length 38 mm; G= *Hexaplex hertweckorum* (Petuch, 1988), length 80 mm; H= *Chicoreus stephensae* Petuch, 1994, length 42 mm; I= *Vermicularia recta* Olsson and Harbison, 1953, clump length 226 mm; J= *Phyllonotus globosus* (Emmons, 1858), length 79 mm; K= *Hyotissa meridionalis* (Heilprin, 1886), length 116 mm. All specimens were collected in the Pinecrest Member (Units 9-5) of the Tamiami Formation in Quality Aggregates Phase 6 pit, Sarasota, Sarasota County, Florida, with the exception of *Pterorhytis fluviana*, which came from the Pinecrest Member exposures at the Bird Road dig, Miami, Dade County, Florida (Petuch, 1986).

Member). From these three areas, alone, twelve distinct communities can be recognized. As pointed out in Chapter 2, most of the coral fauna of the pseudoatoll reef tracts also occurs in the Gatunian Province to the south, demonstrating the provinciatonal nature of the Tamiami Subsea. On the other hand, most of the mollusks from the pseudoatoll are unique to southern Florida and are classic examples of provinciatonal endemics. With a combination of both Gatunian and endemic Caloosahatchian species, the pseudoatoll reef tracts contained the richest scleractinian coral fauna ever found in the western Atlantic, with over seventy species known from the Golden Gate Member alone.

The Pocillopora crassoramosa Community

On the outer, seaward edge of the pseudoatoll reef tracts (the reef crest), heavy surf pounded incessantly and produced a high energy surge environment. These high energy conditions were best developed along the Miami-Palm Beach Reef Tract (Petuch, 1986), where large trans-Atlantic swell rolled in directly from West Africa, unimpeded by the still-undeveloped and essentially nonexistent deep Bahama Banks. In these high energy reef crest environments, only a few hardy scleractinians, hydrocorals, and coralline algae could exist and thrive. These included massive, low, rounded scleractinian corals such as the seriatoporids *Pocillopora crassoramosa* Duncan, 1863 (namesake of the community; Plate 58, A) and *P. barracoaensis* Vaughan, 1919, the poritid *Goniopora jacobiana* Vaughan, 1919 (Plate 58, K), and the astrocoeniid *Astrocoenia meinzeri* Vaughan, 1919 (Plate 58, B), and the ramose hydrocoral *Millepora alcicornis* Linnaeus, 1758 (variety *ramosa*) (Plate 58, D and G). The coral-dwelling barnacle *Ceratoconcha prefloridana* (Brooks and Ross, 1960) (Plate 58, G) was abundant within the large hydrocoral masses and often formed large aggregations. Clinging to exposed limestone surfaces in the surge zone were several low, flattened gastropods, including the fissurellid limpets *Diodora carolinensis* (Conrad, 1875) and *Hemitoma retiporosa* (Dall, 1903) (Plate 58, E), the acmaeid limpet *Acmaea actina* Woodring, 1928 (Plate 58, J; also found in the Pliocene of Jamaica), the calyptraeids *Dispotaea ramosa* (Conrad, 1842) (Plate 58, C), *Crucibulum costatum* (Say, 1820) (Plate 58, F), and *C. multilineatum* (Conrad, 1841) (Plate 58, I), and the hipponicid *Hipponix ceras* Woodring, 1928 (also found in the Gatunian Province). Large clumps of the sessile gregarious vermetid gastropod *Petaloconchus floridanus* Olsson and Harbison, 1953 (Plate 58, H) also grew among the coral masses. These, and many of the other reef crest gastropods, were often heavily encrusted with the lithothamniid coralline alga *Goniolithon* species (Plate 58, F). Along the very edge of the reef crest, in the zone of the heaviest surf, this coralline alga formed massive ridges.

The Dichocoenia tuberosa Community

Behind the reef crest, there existed a wide shallow area (1-3 m) with high-energy, highly oxygenated water conditions and this comprised the front part of the main reef platform. This reef zone contained the richest scleractinian coral fauna of the entire pseudoatoll system, with twenty-three species occurring together. In response to the high-energy water conditions, all genera on this section of the reef platform formed large, massive, rounded heads. The caryophyllid genus *Dichocoenia* dominated this area, with five species making up at least 30% of the individual coral heads. These included *Dichocoenia tuberosa* Duncan, 1863 (namesake of the community; Plate 59, F), *D. caloosahatcheensis* Weisbord, 1974 (Plate 59, A), *D. eminens* Weisbord, 1974, *D.* species (*D. eminens* complex) (Plate 59, H), and *D.* cf. *stokesi* Edwards and Haime, 1848. Other caryophyllids occurred with the *Dichocoenia* species, including *Barysmilia intermedia* Duncan, 1863 (Plate 59, J) and *Stephanocoenia spongiformis* (Duncan, 1864). Also abundant were the montastreids *Montastrea endothecata* (Duncan, 1863) (Plate 59, C), *M. brevis* (Duncan, 1863) (Plate 59, I), *M. limbata* Duncan, 1863,

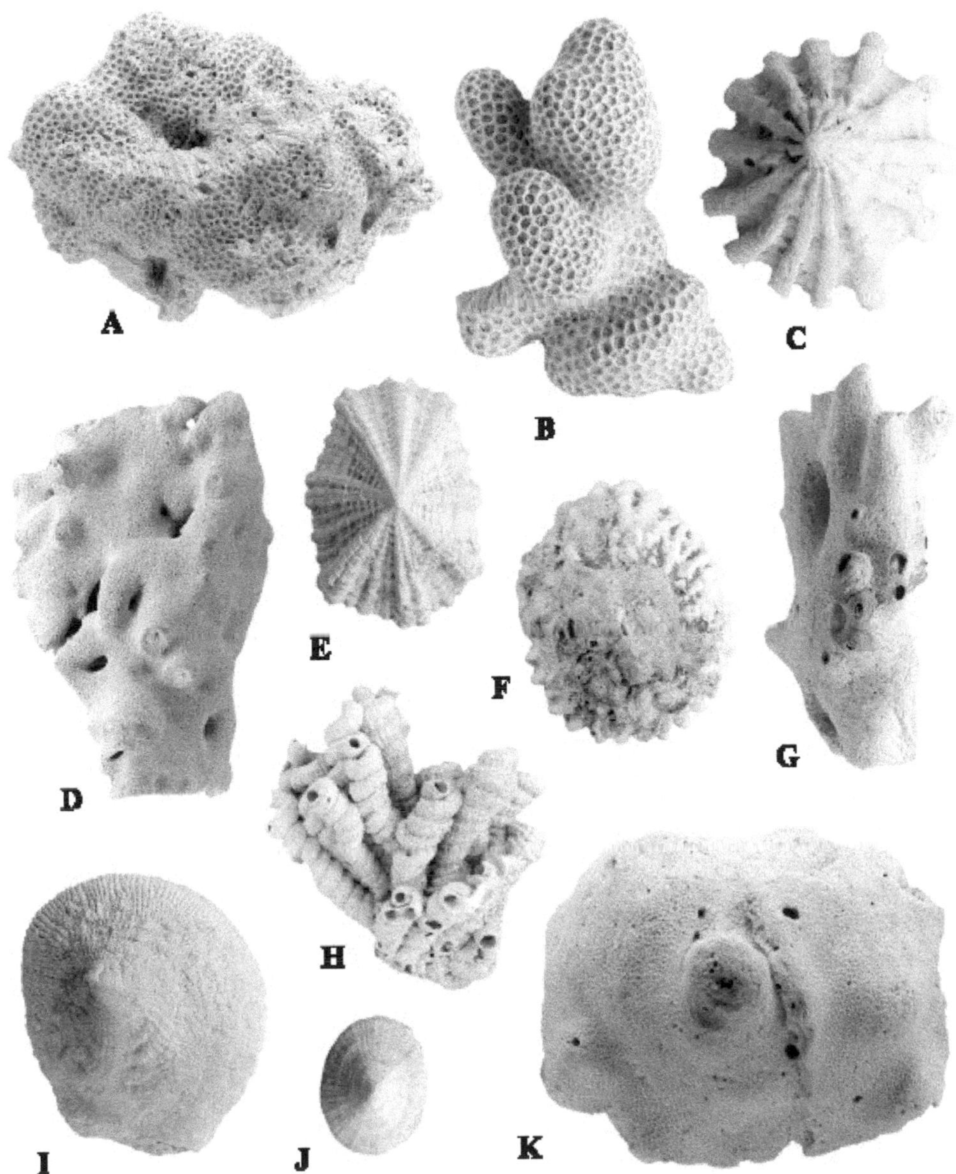

Plate 58. Index fossils and characteristic organisms of the *Pocillopora crassoramosa* Community (reef crest during Golden Gate time) of the Everglades Pseudoatoll. Included are: A= *Pocillopora crassoramosa* Duncan, 1863, length 115 mm; B= *Astrocoenia meinzeri* Vaughan, 1919, length 55 mm; C= *Dispotaea ramosa* (Conrad, 1842), length 43 mm; D= *Millepora alcicornis* Linnaeus, 1758, variety *ramosa*, length 71 mm; E= *Hemitoma retiporosa* (Dall, 1903), length 15 mm; F= *Crucibulum costatum* (Say, 1820), length 38 mm, encrusted with *Goniolithon* coralline algae; G= *Ceratoconcha prefloridana* (Brooks and Ross, 1960), individuals embedded in *Millepora alcicornis*, length 58 mm; H= *Petaloconchus floridanus* Olsson and Harbison, 1953, clump length 42 mm; I= *Crucibulum multilineatum* (Conrad, 1841), length 42 mm; J= *Acmaea actina* Woodring, 1928, length 11 mm; K= *Goniopora jacobiana* Vaughan, 1919, length 124 mm. all specimens were collected in the Golden Gate Member of the Tamiami Formation in the Florida Rock Naples Quarry (old Mule Pen Quarry), Naples, Collier County, Florida and at the Bird Road excavation, Miami, Dade County, Florida (Petuch, 1986).

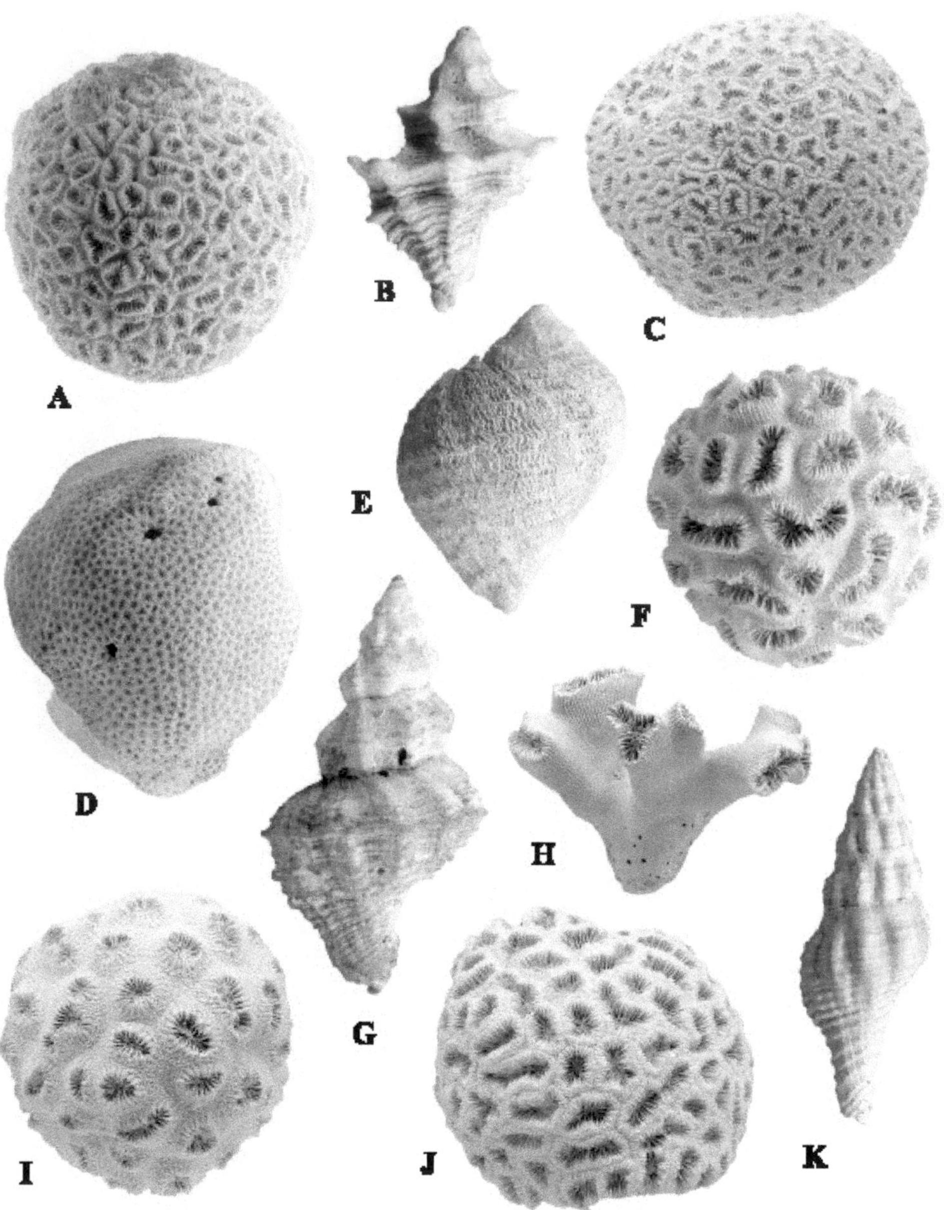

Plate 59. Index fossils and characteristic organisms of the *Dichocoenia tuberosa* Community (front part of the reef platform during Golden Gate time) of the Everglades Pseudoatoll. Included are: A= *Dichocoenia caloosahatcheensis* Weisbord, 1974, length 107 mm; B= *Babelomurex lindae* Petuch, 1988, length 15 mm; C= *Montastrea endothecata* (Duncan, 1863), length 128 mm; D= *Siderastrea pliocenica* Vaughan, 1919, length 119 mm, with apertural tube openings of *Magilus streami*; E= *Magilus streami* Petuch, 1994, length 21 mm; F= *Dichocoenia tuberosa* Duncan, 1863, length 63 mm; G= *Babelomurex kissimmeensis* Petuch, 1991, length 36 mm; H= *Dichocoenia* species, length 65 mm; I= *Montastrea brevis* (Duncan, 1863), length 58 mm; J= *Barysmilia intermedia* Duncan, 1863, length 61 mm; K= *Latirus miamiensis* Petuch, length 45 mm. All specimens were collected in the Golden Gate Member of the Tamiami Formation in the Florida Rock Naples Quarry (old Mule Pen Quarry), Naples, Collier County, Florida.

M. cylindrica (Duncan, 1863), and *M.* species (*M. annularis* complex), *Solenastrea globosa* (Duncan, 1863), *S. distans* (Duncan, 1863), and *S.* species (*S. hyades* complex), and the siderastreids *Siderastrea dalli* Vaughan, 1919, *S. pliocenica* Vaughan, 1919 (Plate 59, D), *S. grandis* Duncan, 1863, *S.* species (*S. pourtalesi* complex), and *S.* species (*S. hyades* complex). Faviid corals were also abundant in this zone and included *Favia* species (*F. fragum* complex), *F.* species, and *Colpophyllia* cf. *natans* (Mueller, 1775). Several coral-feeding gastropods lived on and within the massive heads, and some of these included the coralliophilids *Babelomurex lindae* Petuch, 1988 (Plate 59, B), *B. kissimmeensis* Petuch, 1991 (Plate 59, G), and *Coralliophila miocenica* (Guppy, 1873) (also found in the Gatunian Province), and *Magilus streami* Petuch, 1994 (Plate 59, E) (living inside coral heads; Plate 59, D). Also living in this high-energy zone were several algivorous species, including *Astraea (Lithopoma) tectariaeformis* Petuch, 1986, *A. (Lithopoma) aora* Woodring, 1928 (also found in Jamaica), and the cerithiids "*Clypeomorus*" *obesum* (Gabb, 1873) (an undescribed genus that has morphologically converged on the Indo-Pacific *Clypeomorus*), *Cerithium turriculum* Gabb, 1873, *C. dominicensis* Gabb, 1873 (all also found in the Dominican Republic), *C. leonensis* Mansfield, 1930, and *Ochetoclava* species. Several cryptic carnivorous gastropods also occurred in this zone, and some of these included the fasciolariids *Latirus miamiensis* Petuch, 1986 (Plate 59, K) and *L. stephensae* Lyons, 1991, the buccinids *Hesperisternia miamiensis* (Petuch, 1994), *H. joelshugari* (Petuch, 1994), and *H. dadeensis* (Petuch, 1994), and the columbellids *Parametaria lindae* Petuch, 1986, *P. hertweckorum* Petuch, 1991, and *Columbella submercatoria* Olsson, 1922 (also widespread in the Gatunian region).

The Isophyllia desotoensis Community

Farther back on the shallow reef platform quieter water conditions existed, and these allowed for the growth of a large fauna of delicate "flower" and "brain" corals. This zone contained a rich coral assemblage, with seventeen species occurring together. Here, the mussid corals dominated and included species such as *Isophyllia desotoensis* Weisbord, 1974 (namesake of the community; Plate 60, B), *I.* cf. *sinuosa* (Ellis and Solander, 1786) (Plate 60, I), *I.* cf. *sinuosa* form *multiflora*, *Mussa affinis* (Duncan, 1863), *Syzygophyllia dentata* (Duncan, 1863) (Plate 60, A), *Asterosmilia exarata* Duncan, 1867, *A. abnormalis* Duncan, 1863, *Mycetophyllia* species (*M. lamarckiana* complex), and *Scolymia* species (*S. lacera* complex). Faviid corals were also common in this zone and included large peduncular species such as *Diploria sarasotana* Weisbord, 1974 (Plate 60, J), *D.* cf. *strigosa* (Dana, 1846), and *D.* species (*D. clivosa* complex), and *Thysanus excentricus* Duncan, 1863 (Plate 60, E). Occurring with the mussids and faviids were the meandriniids *Meandrina* cf. *meandrites* (Linnaeus, 1758) (forming large, mushroom-shaped peduncular corallites) and *Dendrogyra* species (Plate 60, F), the caryophyllid *Eusmilia* species (*E. fastigiata* complex), and the agariciid *Agaricia dominicensis* Vaughan, 1919 (Plate 60, D). Gorgonian octocorals must have been abundant on this reef platform area, as attested by the presence of an exceptionally large fauna of gorgonian feeding ovulid gastropods. Some of these included *Cyphoma viaavensis* Petuch, 1986 (Plate 60, H), *C. carolae* Petuch, 1986 (Plate 60, G), *C. miamiensis* Petuch, 1986, *C. finkli* Petuch, 1986, and *Simnialena diegelae* Petuch, 1991. A large fauna of coral-boring bivalves also occurred in this zone and included the petricolids *Petricoxenica concoralla* (H. Vokes, 1976) (Plate 60, K) and *Rupellaria pectarosa* (Conrad, 1834), the gastrochaenid *Gastrochaena ligula* H. Lea, 1843, the pholadid *Martesia ovalis* (Say, 1820), and the lithophagid mussels *Lithophaga yorkensis* Olsson, 1914 and *L.* species. Living among the corals and under coral rubble were several reef-associated carnivorous gastropods, some of which included vermivores such as the conid *Virgiconus miamiensis* (Petuch, 1986) (Plate 60, L) and the mitrid *Scabricola lindae* (Petuch, 1986) (Plate 60, C),

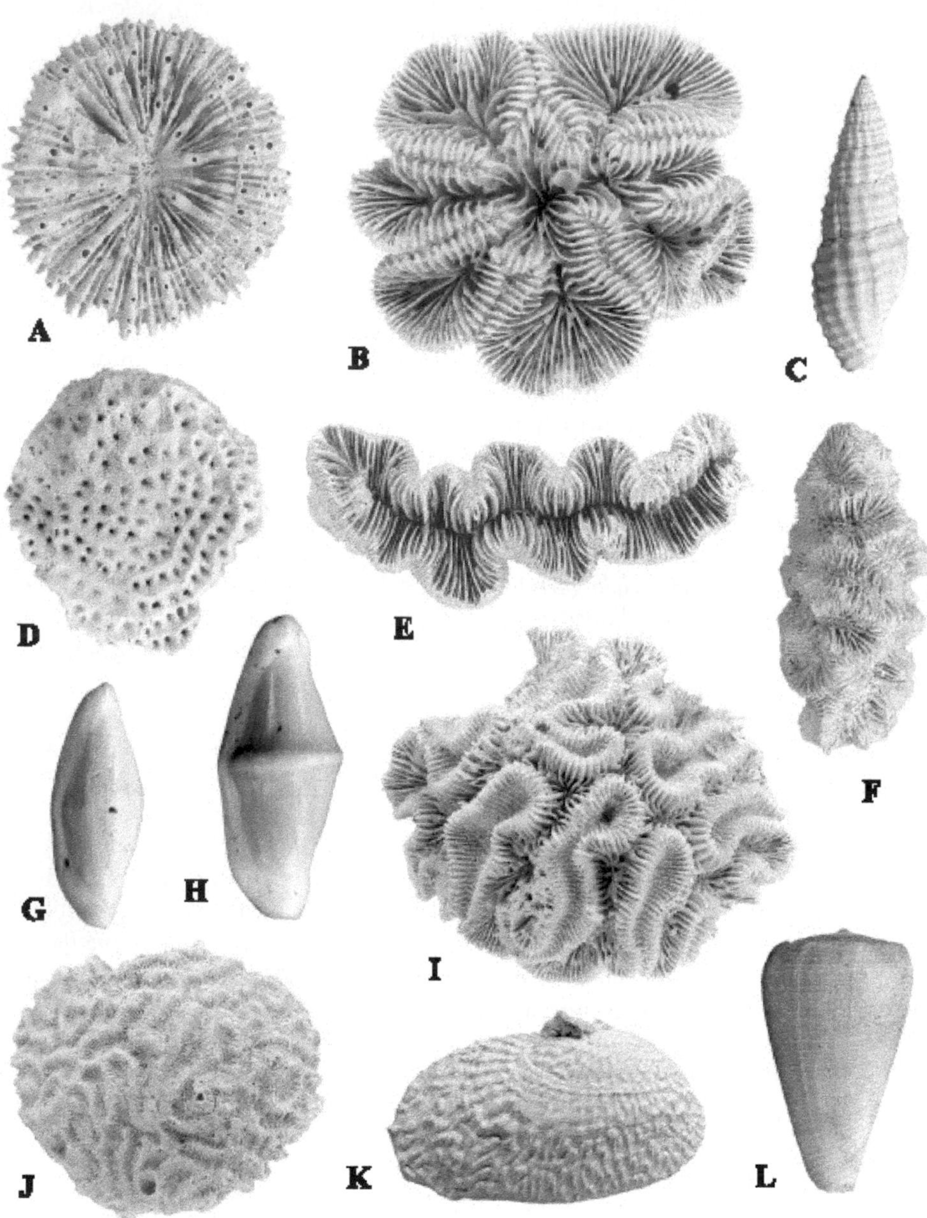

Plate 60. Index fossils and characteristic organisms of the *Isophyllia desotoensis* Community (back part of the reef platform during Golden Gate time) of the Everglades Pseudoatoll. Included are: A= *Syzygophyllia dentata* (Duncan, 1863), length 43 mm; B= *Isophyllia desotoensis* Weisbord, 1974, length 89 mm; C= *Scabricola lindae* (Petuch, 1986), length 26 mm; D= *Agaricia dominicensis* Vaughan, 1919, length 54 mm; E= *Thysanus excentricus* Duncan, 1863, length 114 mm; F= *Dendrogyra* species, length 43 mm; G= *Cyphoma carolae* Petuch, 1986, length 20 mm; H= *Cyphoma viaavensis* Petuch, 1986, length 26 mm; I= *Isophyllia* cf. *sinuosa* (Ellis and Solander, 1786), length 87 mm; J= *Diploria sarasotana* Weisbord, 1974, length 104 mm; K= *Petricoxenica concoralla* (H. Vokes, 1976), length 44 mm; L= *Virgiconus miamiensis* (Petuch, 1986), length 35 mm. All specimens were collected in the Golden Gate Member of the Tamiami Formation in the Florida Rock Naples Quarry (old Mule Pen Quarry), Naples, Collier County, Florida.

drilling molluscivores such as the muricids *Trossulasalpinx curtus* (Dall, 1890), *T. vokesae* Petuch, 1986, *Subpterynotus miamiensis* Petuch, 1994, *Muricopsis lyonsi* Petuch, 1986, *Vitularia linguabison* E. Vokes, 1967, and *Chicoreus miamiensis* Petuch, 1994, and mollusk and echinoderm-feeders such as the ranellid *Cymatium (Septa) henicum* Woodring, 1959, the bursids *Bursa (Marsupina) proavus* Pilsbry, 1922 and *Bursa (Lampasopsis) amphitrites* Maury, 1917, and the tonnid *Malea densicostata* (Rutsch, 1934) (all also found in the Gatunian Province), and the general carnivore/scavenger buccinid *Solenosteira mulepenensis* Petuch, 1994.

The Stylophora affinis Community

In the deeper (3-10 m), quiet water, low energy areas of the back reef, delicate branching corals and fragile "mushroom" corals grew in dense thickets, containing at least seventeen species. Here, these coral forests were dominated by seriatoporids, including *Stylophora affinis* Duncan, 1863 (namesake of the community; Plate 61, A), *S. granulata* Duncan, 1864 (Plate 61, H), *S. minor* Duncan, 1863, and *Madracis* species (*M. mirabilis* complex). Living together with the seriatoporids were other large, delicate branching species such as the acroporids *Acropora* cf. *panamensis* Vaughan, 1919 (Plate 61, E) and *A.* species (Plate 61, G), and *Astreopora* species, the poritids *Porites barracoaensis* Vaughan, 1919, *P. matanzasensis* Vaughan, 1919, and *P.* species (*P. astreoides* complex), and the ahermatype oculinid *Oculina sarasotaensis* Weisbord, 1974. Several faviid "mushroom" corals also occurred in these coral forests and included *Thysanus corbicula* Duncan, 1863 (Plate 61, I) and *T.* cf. *floridanus* Weisbord, 1974. Attached to coral rubble were several encrusting species, including the astrangiid *Astrangia floridana* (Gane, 1895), the caryophyllid *Cladocora* species, and the rhizangiid *Septastrea matsoni* Vaughan, 1919. A large and highly endemic gastropod fauna lived in this pseudoatoll back reef environment, some of which included the large herbivorous cerithiids *Cerithioclava dalli* (Olsson and Petit, 1964) and *C. turriculus* Petuch, 1994, vermivores such as the conids *Eugeniconus paranobilis* (Petuch, 1991) (Plate 61, C), *E. irisae* Petuch, new species (Plate 61, L; see Systematic Appendix; genus now confined to the Indo-Pacific Region), and *Dauciconus bassi* (Petuch, 1991) (Plate 61, B), and the turbinellids *Hystrivasum chilesi* Petuch, 1994 (Plate 61, F) and *H. squamosum* (Hollister, 1971), the coral-feeding ovulid *Jenneria violetae* Petuch, 1991 (Plate 61, J), the urochordate-feeding triviids *Pusula miamiensis* Petuch, 1991 and *Decoriatrivia miccosukee* Petuch, 1991, the hydroid and sponge-feeding trochid *Calliostoma sincerum* Olsson and Harbison, 1953, and the molluscivore fasciolariid *Pleuroploca lindae* Petuch, new species (Plate 61, D and K; see Systematic Appendix; first record of the genus *Pleuroploca* s.s. in North America).

The Antillia bilobata Community

The reef lagoon area adjacent to the back reef housed a rich assemblage of solitary "cup" and "rose" corals, containing at least eleven species. Half-buried in the carbonate sand and mud, these solitary corals formed dense beds that interfingered with small patches of open sand and Turtle Grass (*Thalassia*). Dominating this unique lagoonal community were several species of fungiids, including *Antillia bilobata* Duncan, 1863 (namesake of the community; Plate 62, E), *A. dubia* (Duncan, 1863) (Plate 62, I), *A. walli* Duncan, 1863, and *Antillophyllia* species (Plate 62, H). Faviids were also abundant here, and included the "rose" corals *Manicina pliocenica* Gane, 1900 and *M.* species (*M. areolata* complex), and *Placocyathus variabilis* Duncan, 1863, *P. barretti* Duncan, 1863, and *P. alveolus* (Duncan, 1863). Ahermatypic flabellid corals were also common in these fungiid and faviid beds and included two species, the smaller, oval *Flabellum dubium* Duncan, 1863 and the larger,

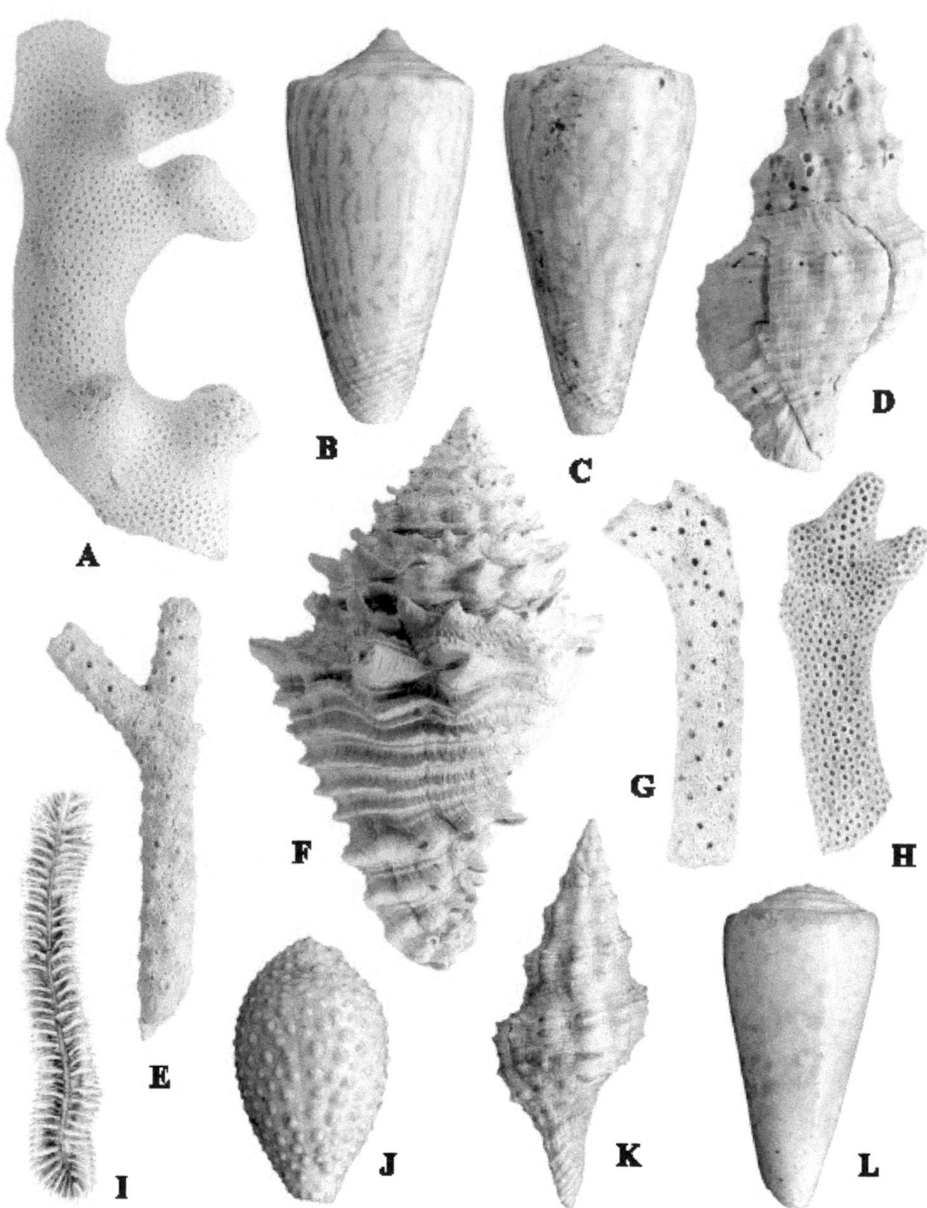

Plate 61. Index fossils and characteristic organisms of the *Stylophora affinis* Community (back reef during Golden Gate time) of the Everglades Pseudoatoll. Included are: A= *Stylophora affinis* Duncan, 1863, length 121 mm; B= *Dauciconus bassi* (Petuch, 1991), length 35 mm; C= *Eugeniconus paranobilis* (Petuch, 1991), length 36 mm; D= *Pleuroploca lindae* Petuch, new species, paratype, length 83 mm; E= *Acropora* cf. *panamensis* Vaughan, 1919, length 74 mm; F= *Hystrivasum chilesi* Petuch, 1994, length 126 mm; G= *Acropora* species, length 63 mm; H= *Stylophora granulata* Duncan, 1863, length 87 mm; I= *Thysanus corbicula* Duncan, 1863, length 66 mm; J= *Jenneria violetae* Petuch, 1991, length 23 mm; K= *Pleuroploca lindae* Petuch, new species, holotype, length 48 mm; L= *Eugeniconus irisae* Petuch, new species, holotype, length 31 mm. All specimens were collected in the Golden Gate Member of the Tamiami Formation in the Florida Rock Naples Quarry (old Mule Pen Quarry), Naples, Collier County, Florida.

elongated *F. exaratum* Duncan, 1863. Underscoring the unique nature of this pseudoatoll back reef lagoon fauna was a rich and highly endemic assemblage of gastropod mollusks. Some of these included molluscivores such as the busyconids *Busycoarctum superbus* (Petuch, 1994) (Plate 62, C), *Lindafulgur miamiensis* (Petuch, 1991) (Plate 62, J), and *Fulguropsis radula* Petuch, 1994, the melongenid *Melongena sarasotaensis* Petuch, 1994, the scaphelline volutid *Scaphella danielleae* Petuch, 1994, the pleioptygmatids *Pleioptygma ronaldsmithi* Petuch, 1994 and *P. debrae* Petuch, 1994, the fasciolariids *Triplofusus harveyensis* (Mansfield, 1930) (Plate 62, K), *Fasciolaria (Cinctura) sarasotaensis* Petuch, 1994, and *Terebraspira okeechobeensis* Petuch, 1994, specialized algivores such as the ampullospirid *Pachycrommium guppyi* (Gabb, 1873) (Plate 62, D; also found in the Gatunian Province), and vermivores such as the conids *Seminoleconus trippae* (Petuch, 1991, *Spuriconus jeremyi* (Petuch, 1994), *S. martinshugari* (Petuch, 1994), *Conasprella hertwecki* (Petuch, 1988), and *Contraconus schmidti* Petuch, 1991, and the giant terebrids *Myurellina miamica* (Olsson, 1967), *M. hunterae* (Olsson, 1967), and *M. aclinica* (Olsson, 1967). Living on the Turtle Grass patches scattered between the coral beds were several cypraeids, some of which included *Pseudadusta lindae* (Petuch, 1986) (Plate 62, A and B), *Siphocypraea (Pahayokea ?) grovesi* Petuch, 1999 (Plate 62, F and G), and *S. dimasi* Petuch, 1999, the modulid *Modulus basileus* (Guppy, 1873), the small muricid *Calotrophon emilyae* Petuch, 1988, and the strombids *Eustrombus dominator* (Pilsbry and Johnson, 1917) (ancestor of the modern Queen Conch, *E. gigas*, and also found in the Gatunian Province), *Macrostrombus mulepenensis* (Petuch, 1994), and *Strombus sarasotaensis* Petuch, 1994.

The Pyrazisinus scalinus Community

During the highest sea level stand of late Piacenzian (Sarasota Unit 4 time), the Myakka Lagoon System was flooded to its maximum size, and contained the largest and best-developed mangrove jungles of the entire Tamiami Subsea. These extensive estuarine forests and intertidal mud flats, which probably resembled the mangrove jungles of southern New Guinea, housed the richest estuarine molluscan fauna known from the American Pliocene (Petuch, 1994, 1997). Although having a high degree of endemism, the Myakka Lagoon fauna also shared several species with the contemporaneous Satilla Lagoon System of the Charleston Sea (Chapter 6) and the Texas Pliocene coastal estuaries. On the organic-rich mud flats adjacent to the mangroves (probably *Rhizophora*), immense aggregations of potamidid gastropods covered the surface, feeding on algal and bacterial films. These included *Pyrazisinus scalinus* (Olsson, 1967) (namesake of the community; Plate 63, I), *P. lindae* Petuch, 1994 (Plate 63, H), *P. sarasotaensis* Petuch, 1994 (Plate 63, F), *Potamides matsoni* Dall, 1913 (Plate 63, G), *P. gracilior* Dall, 1913 (Plate 63, E), *Pachycheilus cancelloides* (Aldrich, 1911), and *P. anagrammatus* Dall, 1913. Living along with the potamidids on the mud flats, and also feeding on algal films, were large aggregations of the neritids *Neritina sphaerica* Olsson and Harbison, 1953 (Plate 63, K) and *Smaragdia merida* (Dall, 1903) (Plate 63, J). The roots of the mangroves were heavily encrusted with clumps of the small, heavy oyster *Crassostrea* cf. *westi* (Mincher, 1941), and these were the principal prey of the drilling muricoideans *Thais (Stramonita) sarasotana* Petuch, 1994 (Plate 63, C) and *Calotrophon myakka* Petuch, 1994 (Plate 63, A). Also living on the mangrove roots and tree oysters were the air-breathing ellobiids *Melampus metae* Petuch, 1991 and *M. jonesae* Petuch, 1991, the high intertidal potamidid *Cerithidea lindae* Petuch, 1994 (Plate 63, B), and the littorinid *Littorinopsis carolinensis* (Conrad, 1836). Several predatory gastropods also occurred in abundance on the potamidid beds, and some of these included the bullid *Bulla sarasotaensis* Petuch, 1994 (Plate 63, D) (omnivorous, also feeding on plants and detritus) and the melongenid *Melongena draperi* Petuch, 1994 (feeding on the potamidids).

Plate 62. Index fossils and characteristic organisms of the *Antillia bilobata* Community (reef lagoon during Golden Gate time) of the Everglades Pseudoatoll. Included are: A,B= *Pseudadusta lindae* (Petuch, 1986), length 37 mm; C= *Busycoarctum superbus* (Petuch, 1994), length 147 mm; D= *Pachycrommium guppyi* (Gabb, 1873), length 23 mm; E= *Antillia bilobata* Duncan, 1863, length 75 mm; F,G= *Siphocypraea grovesi* Petuch, 1999, length 58 mm; H= *Antillophyllia* species, length 47 mm; I= *Antillia dubia* (Duncan, 1863), length 45 mm; J= *Lindafulgur miamiensis* (Petuch, 1991). Length 152 mm; K= *Triplofusus harveyensis* (Mansfield, 1930), length 287 mm. All specimens were collected in the Golden Gate Member of the Tamiami Formation in the Florida Rock Naples Quarry (old Mule Pen Quarry), Naples, Collier County, Florida.

The Mulinia sapotilla Community

In the deeper (1-3 m) parts of the Myakka Lagoon during Unit 4 time, the mactrid bivalve *Mulinia sapotilla* Dall, 1898 (namesake of the community; Plate 64, C) formed extensive monoculture beds on the organic-rich mud bottoms and in tidal channels. These dense mactrid pavements served as the principal food source for several large molluscivorous gastropods, some of which included the melongenids *Echinofulgur jonesae* Petuch, 1994 (Plate 64, A) and *Tropochasca metae* Petuch, 1994 (Plate 64, B), the busyconids *Pyruella laevis* Petuch, 1982 (Plate 64, F) and *P. schmidti* Petuch, 1994, and the fasciolariid *Fasciolaria (Cinctura) sarasotaensis* Petuch, 1994. Along the edges of the mud banks and tidal channels, patches of sea grasses (probably *Halodule* and *Syringodium*) grew on open areas between the *Mulinia* beds. These supported an unusual fauna of cypraeid gastropods, some of which included the estuarine-restricted *Calusacypraea tequesta* (Petuch, 1996) (Plate 64, H and I) and *C. (Myakkacypraea) myakka* Petuch, new species (Plate 64, D and E; see Systematic Appendix), and *Pseudadusta metae* (Petuch, 1994) (Plate 64, J and K). A large polychaete and enteropneust biomass was also present in the weed beds and this supported several vermivorous gastropods, including the abundant turbinellid *Hystrivasum lindae* Petuch, 1994 (Plate 64, G), the turrid *Cymatosyrinx aepytuberculata* Mansfield, 1930, and the conids *Jaspidiconus sarasotaensis* Petuch, 1994 and *Contraconus berryi* Petuch, 1994.

The Perna conradiana Community

During the late Piacenzian (Unit 3 time), after the maximum development of the Myakka estuaries, sea levels again began to drop and large intertidal shoals formed within the Myakka Lagoon System and Everglades Pseudoatoll lagoon. On the late Tamiami shoals, immense beds of the mytilid mussel *Perna conradiana* (d'Orbigny, 1852) (namesake of the community; Plate 65, D) grew in the intertidal areas and formed low, biohermal structures. These large nacreous mussels (which often formed pearls; Plate 65, C) lived half-buried in the sediment and provided the substrate for several suspension-feeding and algivorous calyptraeid gastropods, including *Trochita floridana* Olsson and Petit, 1964 (Plate 65, B), the triangular *Crepidula cannoni* Petuch, 1994 (Plate 65, K), *C. (Bostrycapulus) spinosa* Conrad, 1843, and *Calyptraea concentrica* (H. Lea, 1843). The mussels, with their immense biomass, were the principal prey items of a large number of molluscivorous gastropods, some of which included the drilling muricids *Hexaplex jameshoubricki* Petuch, 1994 (Plate 65, A), *Pterorhytis squamulosa* Petuch, 1994 (Plate 65, E and F), *P. roxaneae* Petuch, 1994 (Plate 65, H and I), *Chicoreus miccosukee* Petuch, 1991 (Plate 65, G), and *Trossulasalpinx lindae* Petuch, 1991, the melongenids *Melongena cannoni* Petuch, 1994 and *Echinofulgur helenae* Olsson, 1967, the busyconids *Busycoarctum tropicalis* (Petuch, 1994), *Pyruella federicoae* Petuch, 1994, and *P. turbinalis* Petuch, 1982, and the fasciolariids *Terebraspira calusa* Petuch, 1994 and *Heilprinia hasta* Petuch, 1994. On patches of sea grasses (probably *Halodule* or *Syringodium*) scattered between the mussel beds, one of the last large Tamiami cypraeid radiations occurred, and included *Siphocypraea cannoni* Petuch, 1994 (Plate 66, G and H), *S. parodizi* Petuch, 1994 (Plate 66, A and B), *Akleistostoma rilkoi* (Petuch, 1999) (Plate 66, C and D), *A.* species, *Calusacypraea sarasotaensis* (Petuch, 1994) (Plate 66, E and F), *C. (Myakkacypraea) kelleyi* Petuch, 1999, *Pseudadusta kalafuti* (Petuch, 1994) (Plate 66, I and J), and *P. ketteri* (Petuch, 1994). Living with the cowries was a large fauna of carnivorous gastropods, some of which included vermivores such as the conids *Spuriconus streami* (Petuch, 1994) (Plate 66, K), *Jaspidiconus susanae* (Petuch, 1994), *Contraconus berryi* subspecies, and *Ximeniconus marylandicus* subspecies, and the turbinellids *Turbinella streami* Petuch, 1994, *Hystrivasum chilesi* Petuch, 1994 and *H. jacksonense* (E.

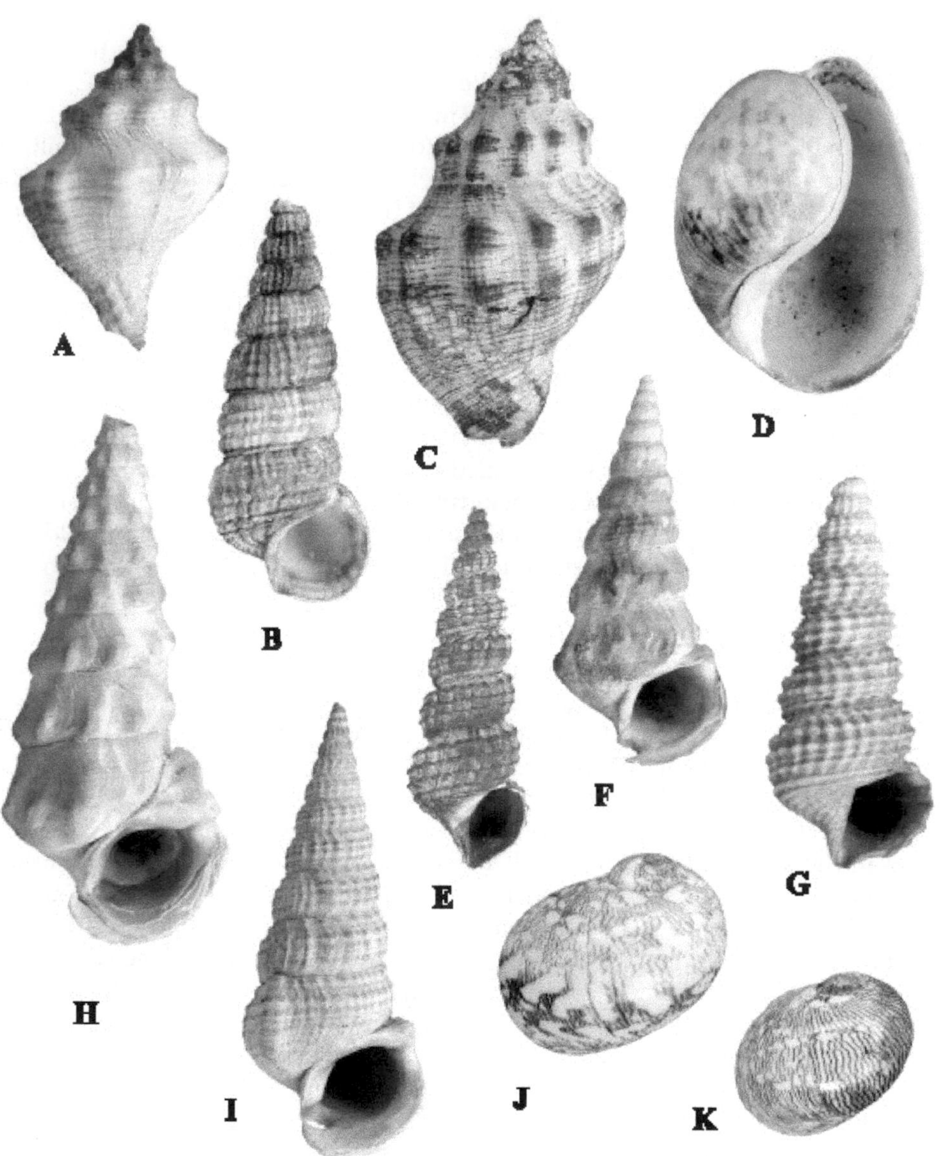

Plate 63. Index fossils and characteristic organisms of the *Pyrazisinus scalinus* Community (mangrove forests and mud flats during Fruitville (Unit 4) time) of the Myakka Lagoon System. Included are: A= *Calotrophon myakka* Petuch, 1994, length 20 mm; B= *Cerithidea lindae* Petuch, length 29 mm; C= *Thais (Stramonita) sarasotana* Petuch, 1994, length 30 mm; D= *Bulla sarasotaensis* Petuch, 1994, length 20 mm; E= *Potamides gracilior* Dall, 1913, length 16 mm; F= *Pyrazisinus sarasotaensis* Petuch, 1994, length 38 mm; G= *Potamides matsoni* Dall, 1913, length 12 mm; H= *Pyrazisinus lindae* Petuch, 1994, length 62 mm; I= *Pyrazisinus scalinus* (Olsson, 1967), length 48 mm; J= *Smaragdia merida* (Dall, 1903), length 5 mm; K= *Neritina sphaerica* Olsson and Harbison, 1953, length 4 mm. All specimens were collected in Unit 4 ("Black Layer;" see Petuch, 1982) of the Fruitville Member of the Tamiami Formation in the APAC quarry, Sarasota, Sarasota County, Florida.

Plate 64. Index fossils and characteristic organisms of the *Mulinia sapotilla* Community (shallow muddy lagoon bivalve beds during Fruitville (Unit 4) time) of the Myakka Lagoon System. Included are: A= *Echinofulgur jonesae* Petuch, 1994, length 75 mm; B= *Tropochasca metae* Petuch, 1994, length 94 mm; C= *Mulinia sapotilla* Dall, 1898, length 28 mm; D,E= *Calusacypraea (Myakkacypraea) myakka* Petuch, new species, holotype, length 59 mm; F= *Pyruella laevis* Petuch, 1982, length 64 mm; G= *Hystrivasum lindae* Petuch, 1994, length 84 mm; H,I= *Calusacypraea tequesta* (Petuch, 1996), length 75 mm; J,K= (Petuch, 1994), length 63 mm. All specimens were collected in Unit 4 ("Black Layer;" see Petuch, 1982) of the Fruitville Member of the Tamiami Formation in the APAC quarry, Sarasota, Sarasota County, Florida.

Vokes, 1966), the suctorial-feeding cancellariid *Cancellaria cannoni* Petuch, 1994, and general carnivores/scavengers such as the volutids *Scaphella gravesae* Petuch, 1994 (Plate 65, L) and *S. maureenae* Petuch, 1994, and the buccinid *Calophos nanus* Petuch, 1994 (Plate 66, L). Large numbers of sea bird bones, particularly of the cormorant *Phalacrocorax* (Plate 65, J), are often encountered in the *Perna* beds in Unit 3 of the Fruitville Member. These bone beds may represent mass bird kills that resulted from the consumption of dead, poisoned fish during "red tide" plankton bloom events.

The Chama emmonsii Community

At the end of Tamiami time, during the deposition of the last bed of the Fruitville Member (Unit 2), sea levels continued to drop and many of the deeper water areas of the pseudoatoll became shallow lagoons (averaging only 2-10 m). Here, the chamid bivalve *Chama emmonsii* Nicol, 1953 (namesake of the community; Plate 67, E), together with the delicate branching coral *Stylophora minor* Duncan, 1863, formed large biohermal structures. The chamids were the main prey items of several molluscivorous gastropods, including the muricids *Hexaplex trippae* Petuch, 1994, *Phyllonotus globosus* (Emmons, 1858), *Trossulasalpinx maryae* Petuch, 1991, *Murexiella petuchi* E. Vokes, 1994 (photograph on frontispiece), *Pterorhytis conradi* subspecies, *Aspella petuchi* E. Vokes, 1995, and *Chicoreus judeae* Petuch, 1994 (Plate 67, F), the fasciolariid *Latirus duerri* Petuch, new species (Plate 67, D; see Systematic Appendix) and the thaidid *Thais (Stramonita) caloosana* (Tucker and Wilson, 1933). Small carnivorous gastropods, such as the buccinid *Monostiolum petiti* Olsson, 1967 (Plate 67, K) and the fasciolariid *Dolicholatirus metae* Petuch, 1994, also occurred on the *Chama* and *Stylophora* reefs. On sand patches scattered among the *Chama* bioherms, several species of pectinids lived together in large colonies. Some of these scallops included *Argopecten evergladensis* subspecies, *Nodipecten vaccamavensis* (Olsson, 1914) (Plate 67, C), *Stralopecten ernestsmithi* (Tucker, 1931) (Plate 67, G), and *Carolinapecten senescens* (Dall, 1898). Feeding on these scallops were several large molluscivorous gastropods, some of which included the fasciolariid *Terebraspira acuta* (Emmons, 1858), the melongenids *Echinofulgur* species and *Melongena penningtonorum* Petuch, 1994, and the busyconids *Busycoarctum tropicalis* subspecies, and *Sinistrofulgur robesonense* (Gardner, 1948). Small patches of sea grasses (probably *Thalassia* and *Halodule*) also grew between the chamid bioherms, and these supported several cypraeids, including *Siphocypraea alligator* Petuch, 1994, *Pseudadusta marilynae* (Petuch, 1994) (Plate 67, A and B), and *Calusacypraea (Myakkacypraea) schnireli* Petuch, new species (Plate 67, I and J; see Systematic Appendix) (the last-living species of the genera *Pseudadusta* and *Calusacypraea*). Also occurring in the sea grass beds were several species of large carnivorous gastropods, some of which included vermivores such as the conids *Spuriconus jonesorum* (Petuch, 1994) and *S. (?) yaquensis* (Gabb, 1873) (Plate 67, H), and the turbinellids *Hystrivasum hyshugari* Petuch, 1994 and *H. hertweckorum* Petuch, 1994, the suctorial-feeding cancellariids *Cancellaria miamiensis* Petuch, 1994 and *C. floridana* Olsson and Petit, 1964, and the echinoderm-feeding cassid *Cassis ketteri* Parodiz and Tripp, 1992.

The Pyrazisinus kissimmeensis Community

During Unit 3 time, the Kissimmee Embayment was filled with extensive mangrove forests and mud flats, much like the Myakka Lagoon System. These mangrove jungles may have contained different, and more sparsely-distributed, species of trees than did the Myakka area. This compositional variance is reflected in the molluscan faunas, which differ greatly between the two estuaries. Like the Myakka mangrove areas, the mud flats of the Kissimmee Embayment supported large aggregations of potamidids, but in this case

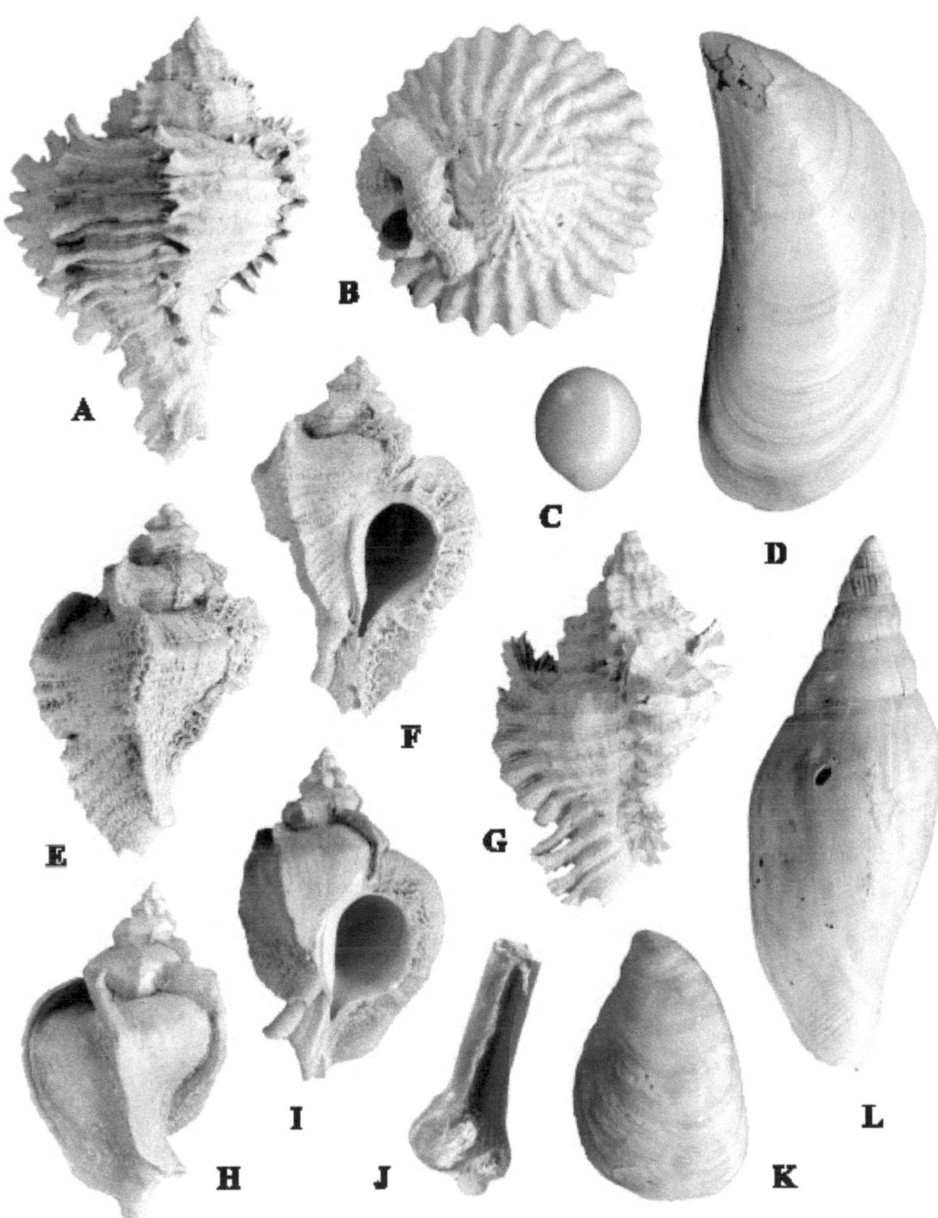

Plate 65. Index fossils and characteristic organisms of the *Perna conradiana* Community (shallow lagoon mussel beds during Fruitville (Unit 3) time) of the Myakka Lagoon System. Included are: A= *Hexaplex jameshoubricki* Petuch, 1994, length 113 mm; B= *Trochita floridana* Olsson and Petit, 1964, length 42 mm, with the vermetid *Serpulorbis granifera* (Say, 1824) attached; C= fossil pearl from *Perna conradiana*, diameter 9 mm; D= *Perna conradiana* (d'Orbigny, 1852), length 109 mm; E,F= *Pterorhytis squamulosa* Petuch, 1994, length 36 mm; G= *Chicoreus miccosukee* Petuch, 1991, length 42 mm; H,I= *Pterorhytis roxaneae* Petuch, 1994, length 38 mm; J= proximal end of Cormorant (*Phalacrocorax*) femur, length (broken) 29 mm; K= *Crepidula cannoni* Petuch, 1994, length 33 mm; L= *Scaphella gravesae* Petuch, 1994, length 153 mm. All specimens were collected in Unit 3 of the Fruitville Member of the Tamiami Formation in Quality Aggregates Phase 6 pit, Sarasota, Sarasota County, Florida.

Plate 66. Index fossils and characteristic organisms of the *Perna conradiana* Community (shallow lagoon mussel beds during Fruitville (Unit 3) time) of the Myakka Lagoon System. Included are: A,B= *Siphocypraea parodizi* Petuch, 1994, length 57 mm; C,D= *Akleistostoma rilkoi* (Petuch, 1999), length 66 mm; E,F= *Calusacypraea sarasotaensis* (Petuch, 1994), length 41 mm; G,H= *Siphocypraea cannoni* Petuch, 1994, length 74 mm; I,J= *Pseudadusta kalafuti* (Petuch, 1994), length 82 mm, K= *Spuriconus streami* (Petuch, 1994), length 53 mm; L= *Calophos nanus* Petuch, 1994, length 27 mm. All specimens were collected in Unit 3 of the Fruitville Member of the Tamiami Formation in Quality Aggregates Phase 6 pit, Sarasota, Sarasota County, Florida.

Plate 67. Index fossils and characteristic organisms of the *Chama emmonsi* Community (shallow lagoon chamid bivalve bioherms during late Fruitville (Unit 2) time) of the Myakka Lagoon System. Included are: A,B= *Pseudadusta marilynae* (Petuch, 1994), length 42 mm; C= *Nodipecten vaccamavensis* (Olsson, 1914), length 74 mm; D= *Latirus duerri* Petuch, new species, holotype, length 52 mm; E= *Chama emmonsi* Nicol, 1953, length 43 mm; F= *Chicoreus judeae* Petuch, 1994, length 59 mm; G= *Stralopecten ernestsmithi* (Tucker, 1931), length 63 mm; H= *Spuriconus (?) yaquensis* (Gabb, 1873), length 44 mm; I,J= *Calusacypraea (Myakkacypraea) schnireli* Petuch, new species, holotype, length 41 mm; K= *Monostiolum petiti* Olsson, 1967, length 24 mm. All specimens were collected in Unit 2 of the Fruitville Member of the Tamiami Formation in Quality Aggregates Phase 9 pit, Sarasota, Sarasota County, Florida.

only a single species, *Pyrazisinus kissimmeensis* (Olsson, 1967) (namesake of the community; Plate 68, H). Unlike the Myakka area, the Kissimmee mud flats housed an unusual and distinctive fauna of large scavenger/carrion-feeding nassariid gastropods, some of which included *Globinassa* cf. *gastrophila* (Olsson, 1916) (Plate 68, B), *Scalanassa scalaspira* (Conrad, 1868) (Plate 68, C), *Paranassa arata* (Say, 1824) (Plate 68, E), and *Ilyanassa marthae* Olsson, 1967 (Plate 68, F). On these organic-rich mud flats, the melongenid *Tropochasca kissimmeensis* Petuch, 1994 (Plate 68, A) (last-living member of the genus *Tropochasca*) was the primary predator of the *Pyrazisinus* snails. Living on the mangrove roots and trunks (Plate 68, J) were two species of high intertidal potamidids, the small *Cerithidea diegelae* Petuch, 1994 (Plate 68, D) and the larger *C. briani* Petuch, 1994 (Plate 68, G). As in Recent mangrove forests, specialized littorinids lived on the high branches and leaves; in this case a single species, *Littorinopsis sheaferi* (Olsson, 1967) (Plate 68, I). Interestingly, the estuarine-restricted cypraeid genus *Calusacypraea* (and its subgenus *Myakkacypraea*), which underwent such a large species radiation in the Myakka Lagoon System, is absent from the similar-appearing Kissimmee estuarine environment.

The Siphocypraea kissimmeensis Community

During the high sea level stand of Unit 4 time, the Kissimmee Embayment reached its maximum development. At this time, much of the central area was filled with wide, shallow (1-5 m) siliciclastic banks and tidal channels and these supported extensive Turtle Grass growth (Figure 28). Associated with these immense *Thalassia* beds was a highly-endemic and characteristic molluscan fauna, with its own unique species radiations. Primary among these was an endemic radiation of the gastropod family Cypraeidae, with at least thirteen gregarious grass bed species having evolved during Units 4-2 time. The oldest member of this rapidly-evolving radiation, *Akleistostoma transitoria* (Olsson and Petit, 1964) (Plate 69, C and D), lived on smaller, less-developed Turtle Grass beds in the southern part of the Kissimmee Embayment during Pinecrest-Unit 7 time. By Unit 4 time, the expanded Thalassia beds housed a descendant fauna composed of *Akleistostoma hughesi* (Olsson and Petit, 1964) (Plate 69, G and H) and *Siphocypraea (Pahayokea) kissimmeensis* Petuch, 1994 (namesake of the community; Plate 69, A and B) (progenitor of the new *subgenus Pahayokea* Petuch; see Systematic Appendix). Occurring with the cowrie aggregations on the Kissimmee Turtle Grass beds, and on open sand patches, was a large fauna of shallowly-infaunal bivalves, some of which included the carditids *Cardita seminolensis* Olsson, 1967, *Pleuromeris pitysia* (Olsson, 1967), and *Carditamera dasytes* Olsson, 1967, the arcids *Arca williamsi* Olsson, 1967 and *Larkinia sellardsi* (Mansfield, 1932), and the cardiid *Trachycardium evergladeensis* (Mansfield, 1931). These grass bed bivalves were the principal prey items for a number of molluscivorous gastropods, including the muricids *Pterorhytis lindae* Petuch, 1994 (Plate 69, I and J) and *Trossulasalpinx kissimmeensis* Petuch, 1994, the fasciolariids *Terebraspira lindae* Petuch, 1994 (Plate 69, E), *T. osceolai* Petuch, 1994, and *T. kissimmeensis* Petuch, 1994, the busyconids *Busycon titan* Petuch, 1994 and *Pyruella carraheri* Petuch, 1994, the buccinid *Ptychosalpinx kissimmeensis* Petuch, 1994, and the scaphelline volutid *Volutifusus obtusus* (Emmons, 1858). A large vermivorous gastropod fauna also occurred on the sand patches between the grass beds and some of these included the conids *Gradiconus duerri* (Petuch, 1994) (Plate 69, F), *Jaspidiconus laurenae* (Petuch, 1994), and *J. jaclynae* (Petuch, 1994), the turbinellids *Hystrivasum vokesae* (Hollister, 1971) (Plate 69, K), *H. olssoni* (E. Vokes, 1966), *H. kissimmense* (Hollister, 1971), and *H. squamosum* (Hollister, 1971), and the terebrid *Strioterebrum kissimmeensis* (Mansfield, 1924). The large olivid *Oliva (Strephona) carolinensis* (Conrad, 1863) was also common on sand patches near the grass beds and often occurred in large swarms. By Unit 3 time, *Siphocypraea (Pahayokea) kissimmeensis* had

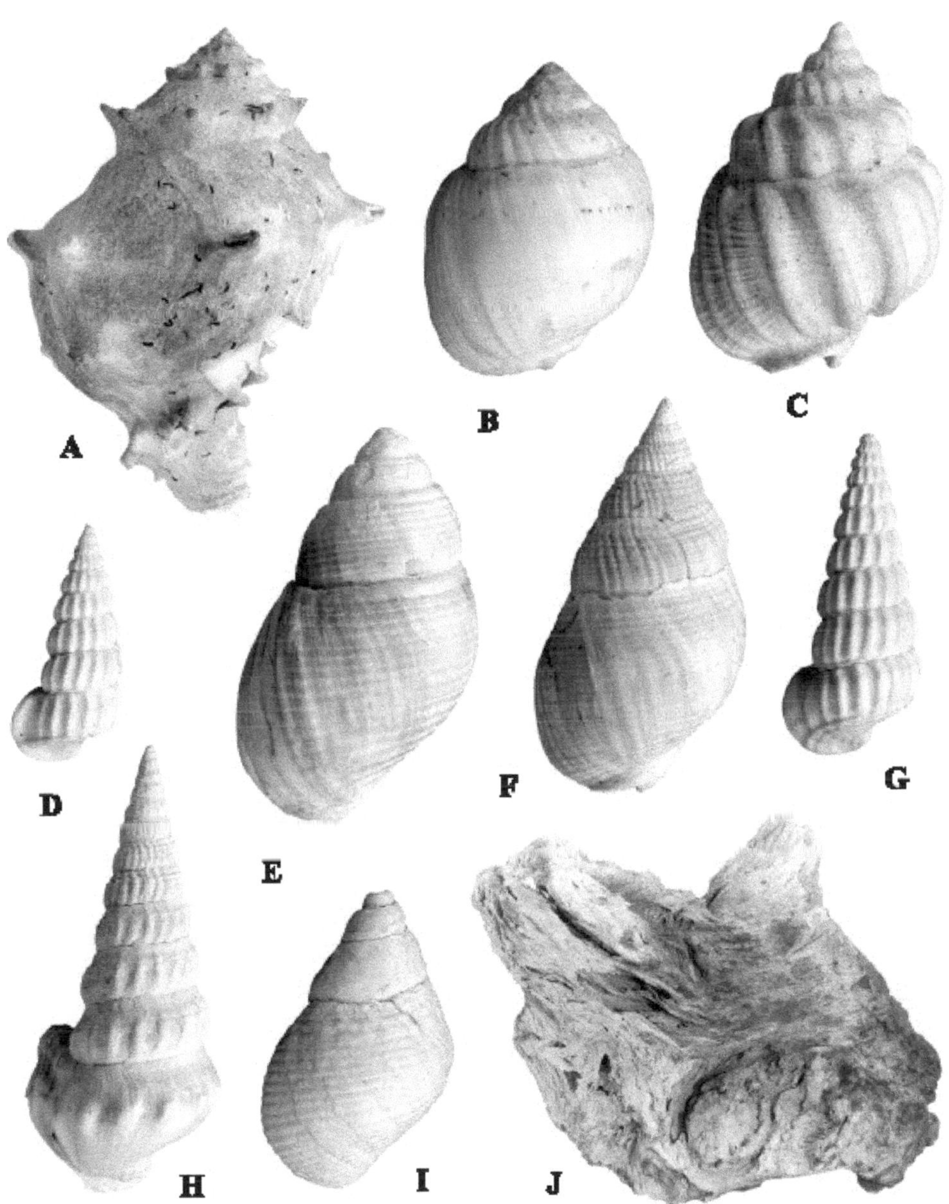

Plate 68. Index fossils and characteristic organisms of the *Pyrazisinus kissimmeensis* Community (mangrove forests and mud flats during Unit 3 time) of the Kissimmee Embayment. Included are: A= *Tropochasca kissimmeensis* Petuch, 1994, length 73 mm; B= *Globinassa* cf. *gastrophila* (Olsson, 1916), length 15 mm; C= *Scalanassa scalaspira* (Conrad, 1868), length 18 mm; D= *Cerithidea diegelae* Petuch, 1994, length 11 mm; E= *Paranassa arata* (Say, 1824), length 27 mm; F= *Ilyanassa marthae* Olsson, 1967, length 26 mm; G= *Cerithidea briani* Petuch, 1994, length 17 mm; H= *Pyrazisinus kissimmeensis* (Olsson, 1967), length 82 mm; I= *Littorinopsis sheaferi* (Olsson, 1967), length 11 mm; J= fossil mangrove wood (possibly *Rhizophora*), length 184 mm. All specimens were collected in the Tamiami Formation (Unit 3 equivalent) at a Kissimmee River dredging near Fort Basinger, Highlands County, Florida.

Plate 69. Index fossils and characteristic organisms of the *Siphocypraea kissimmeensis* Community (sand flats and *Thalassia* beds during Unit 4 time) of the Kissimmee Embayment. Included are: A,B= *Siphocypraea (Pahayokea) kissimmeensis* Petuch, 1994, length 53 mm; C,D= *Akleistostoma transitoria* (Olsson and Petit, 1964), length 70 mm (from a Unit 7 equivalent); E= *Terebraspira lindae* Petuch, 1994, length 142 mm; F= *Gradiconus duerri* (Petuch, 1994) length 62 mm; G,H= *Akleistostoma hughesi* (Olsson and Petit, 1964), length 48 mm; I,J= *Pterorhytis lindae* Petuch, 1994, length 47 mm; K= *Hystrivasum vokesae* (Hollister, 1971), length 75 mm. All specimens were collected in the Tamiami Formation (Units 4 and 7 equivalents) at a Kissimmee River dredging near Fort Basinger, Highlands County, Florida.

evolved into its descendant, *S. (Pahayokea) basingerensis* Petuch, new species (Plate 71, C and D; see Systematic Appendix) and, likewise, the sympatric *Akleistostoma hughesi* was replaced by its descendant, *A. bairdi* Petuch, new species (Plate 71, E and F). These two newly-evolved cowries dominated the *Thalassia* beds, forming the same large aggregations as did their Unit 4 ancestors.

The Siphocypraea penningtonorum Community

The lowered sea levels during Unit 2 time caused widespread intertidal shoals to form within the Kissimmee Embayment, particularly along the western side. Extensive Turtle Grass beds formed along the deeper eastern side of the embayment and these housed the last endemic Kissimmee cowrie fauna. Here, the subgenus *Pahayokea* underwent its final radiation, with three species evolving during Unit 2 time. These included *Siphocypraea (Pahayokea) penningtonorum* Petuch, 1994 (namesake of the community; Plate 70, F and G), the flattened *S. (Pahayokea) gabrielleae* Petuch, new species (Plate 71, G, H, and I; see Systematic Appendix), and *S. (Pahayokea) rucksorum* Petuch, new species (Plate 71, J K, and L). Large aggregations of these three sibling species occurred together with the descendant of *Akleiststoma bairdi, A. diegelae* (Petuch, 1994) (Plate 71, A and B). Some niche partitioning occurred between the *Pahayokea* species, with the flattened *S. gabrielleae* preferring intertidal mud flats and with *S. penningtonorum* preferring Turtle Grass beds. The large, inflated *S. rucksorum* preferred slightly deeper water areas and may have lived on shell rubble bottoms. Living in large beds within the Turtle Grass beds was the noetiid *Noetia carolinensis* (Conrad, 1863) (Plate 70, I). These, and a host of smaller mactrid and tellinid bivalves, were the principal food resource for several molluscivorous macrogastropods, including the fasciolariids *Terebraspira seminole* Petuch, 1994 (Plate 70, J), *T. okeechobeensis* Petuch, 1994 (Plate 70, J), *T. maryae* Petuch, 1994, and *Heilprinia florida* (Olsson and Harbison, 1953), the busyconids *Brachysycon kissimmeensis* Petuch, 1994 (Plate 70, K), *B. canaliferum* (Conrad, 1862), *Busycon auroraensis* Petuch, 1994, *Busycotypus concinnum* (Conrad, 1873), *Sinistrofulgur adversarius* (Conrad, 1863), and *Pyruella basingerensis* Petuch, 1994, the pleioptygmatid *Pleioptygma kissimmeensis* Petuch, 1994 (Plate 70, A), the volutid *Volutifusus typus* Conrad, 1863, and the muricid *Chicoreus shirleyae* E. Vokes, 1974 (Plate 70, E). Polychaete and enteropneust worms were abundant on the organic-rich shallow mud flats and grass beds and were the principal food resource for several large vermivorous gastropods, including the conids *Seminoleconus trippae* (Petuch, 1991) (Plate 70, C) and *Contraconus petiti* Petuch, new species (see Systematic Appendix), and the turbinellids *Hystrivasum shrinerae* (Hollister, 1971) (Plate 70, D) and *H. palmerae* (Hollister, 1971) (Plate 70, H). The suctorial-feeding cancellariid *Ventrilia kissimmeensis* Petuch, 1994 also occurred along the edges of the Turtle Grass beds. Interestingly, the strombid genus *Macrostrombus*, which is so common in Turtle Grass ecosystems elsewhere in the Tamiami Subsea, is absent from the Kissimmee Embayment. This uppermost Tamiami Subsea grass-associated fauna is noteworthy in that it shared many prominent large gastropods with the late Yorktown Subsea (Chowan River Formation; with *Busycon auroraensis, Busycotypus concinnum, Brachysycon canaliferum, Sinistrofulgur adversarius, Volutifusus typus*, and *Contrconus petiti*). On large open sand patches between the grass beds, the scallop *Carolinapecten bertiensis* (Mansfield, 1936) (photograph at end of chapter) formed dense monoculture beds. Like the previously-mentioned gastropods, this large scallop, and the noetiid *Noetia carolinensis*, also occurred in the late Yorktown Subsea (Chowan River Formation; see Chapter 8).

Plate 70. Index fossils and characteristic organisms of the *Siphocypraea penningtonorum* Community (sand flats and *Thalassia* beds during Unit 2 time) of the Kissimmee Embayment. Included are: A= *Pleioptygma kissimmeensis* Petuch, 1994, length 95 mm; B= *Terebraspira seminole* Petuch, 1994, length 130 mm; C= *Seminoleconus trippae* (Petuch, 1991), length 65 mm; D= *Hystrivasum shrinerae* (Hollister, 1971), length 102 mm; E= *Chicoreus shirleyae* E. Vokes, 1974, length 80 mm; F,G= *Siphocypraea (Pahayokea) penningtonorum* Petuch, 1994, length 54 mm; H= *Hystrivasum palmerae* (Hollister, 1971), length 76 mm; I= *Noetia carolinensis* (Conrad, 1863), length 57 mm; J= *Terebraspira okeechobeensis* Petuch, 1994, length 94 mm; K= *Brachysycon kissimmeensis* Petuch, 1994, length 96 mm. All specimens were collected in the Tamiami Formation (Unit 2 equivalent), lowest beds in the Rucks Pit, Fort Drum, Okeechobee County, Florida.

Plate 71. Endemic Cowries of the Kissimmee Embayment, Tamiami Subsea. A,B= *Akleistostoma diegelae* (Petuch, 1994), length 29 mm, Unit 2 equivalent, Kissimmee River dredging at Fort Basinger, Highlands County, Florida; C,D= *Siphocypraea (Pahayokea) basingerensis* Petuch, new species, holotype, length 58 mm, Unit 3 equivalent, Kissimmee River dredging at Fort Basinger, Highlands County, Florida; E,F= *Akleistostoma bairdi* Petuch, new species, holotype, length 51 mm, Unit 3 equivalent, Kissimmee River dredging at Fort Basinger, Highlands County, Florida; G,H,I= *Siphocypraea (Pahayokea) gabrielleae* Petuch, new species, holotype, length 65 mm, Unit 2 equivalent, lowest beds of the Rucks Pit, Fort Drum, Okeechobee County, Florida; J,K,L= *Siphocypraea (Pahayokea) rucksorum* Petuch, new species, holotype, length 63 mm, Unit 2 equivalent, lowest beds of the Rucks Pit, Fort Drum, Okeechobee County, Florida.

Carolinapecten bertiensis (Mansfield,1936), late Tamiami Subsea, Okeechobean Sea and late Yorktown Subsea, Abemarle Sea, length 93 mm. Specimen from the uppermost bed of the Tamiami Formation in Rucks Pit, Fort Drum. Okeechobee County, Florida.

Chapter 8. Latest Pliocene and Earliest Pleistocene Seas

After a brief period of cooling and an accompanying sea level drop, the marine climate warmed again in the late Piacenzian, just prior to the Plio-Pleistocene boundary. Although the intervening cold time was short, it was severe enough to have caused the extinction of many key Caloosahatchian taxa (see Chapter 10). Based on the relatively low peak on the sea level curves shown by Krantz (1991), this latest Piacenzian warming event was also cooler than those that occurred previously during mid-Piacenzian time. Cooler water conditions such as these, and the spatial reduction of the paleosea basins by deltaic infilling, further enhanced the impoverishment of the Caloosahatchian Province faunas in the Albemarle Sea (both late Yorktown and Croatan Subseas) and Charleston Sea (Waccamaw Subsea). This pattern of faunal impoverishment is particularly obvious when comparing the late Yorktown Subsea fauna (contained in the Chowan River Formation) and the Croatan Subsea fauna (contained in the James City Formation; see the following section) with earlier Pliocene faunas from the same area (in the Yorktown Formation). Similarly, the Choctaw Sea was almost completely infilled by this time and the endemic Jackson Subsea faunas became extinct. Only the Okeechobean Sea (Caloosahatchee Subsea), with its deeper interior basin, warmer enclosed oceanographic conditions, and remoteness from muddy deltas, remained structurally intact and acted as a refugium for many eutropical taxa (contained in the Caloosahatchee Formation). Also at this time, the southern part of the Charleston Sea basin was infilled and formed a long chain of sand barrier islands and sheltered, shallow coastal lagoons. This barrier island-lagoon system fused with a similar and larger system along Georgia and Florida to produce a new latest Pliocene-early Pleistocene geomorphological feature, the Nashua Lagoon System (Figures 29 and 30).

At the Plio-Pleistocene boundary, another climatic degeneration took place, resulting in further extinctions of marine taxa and the general impoverishment of the northern faunas. This second cooling event was followed by the Calabrian Age warm time at the beginning of the Pleistocene, when the marine climate briefly returned to oceanographic conditions similar to those in the latest Piacenzian. By this time, the Albemarle Sea had infilled to the extent that it was only a shallow, shoal-filled embayment and was nearing its final stages of deposition. Likewise, the Charleston Sea was nearing finality and was composed mostly of shallow coastal lagoons and sand barrier island chains. The low species-richness, last remnants of the Albemarle and Charleston faunas are contained in the James City and upper Waccamaw Formations. By the end of the Calabrian, the Charleston coastal lagoons had fused with the Nashua Lagoon System, producing a continuous lagoonal environment that extended from South Carolina to southern Florida. This expanded Nashua Lagoon System presaged the barrier island chains of the later Pleistocene and Recent Carolinas, Georgia, and Florida. Of the late Calabrian eastern North American paleoseas, only the landlocked Okeechobean Sea (Caloosahatchee Subsea) contained tropical marine environments. Even with this mild marine climate, the Okeechobean faunas were still affected by the Piacenzian-Calabrian boundary extinction event, as exemplified by the abrupt disappearance of the prominent endemic cypraeid genera *Akleistostoma*, *Calusacypraea*, *Myakkacypraea*, and *Pseudadusta*. Only the genera *Siphocypraea* and *Pahayokea* survived into Calabrian time.

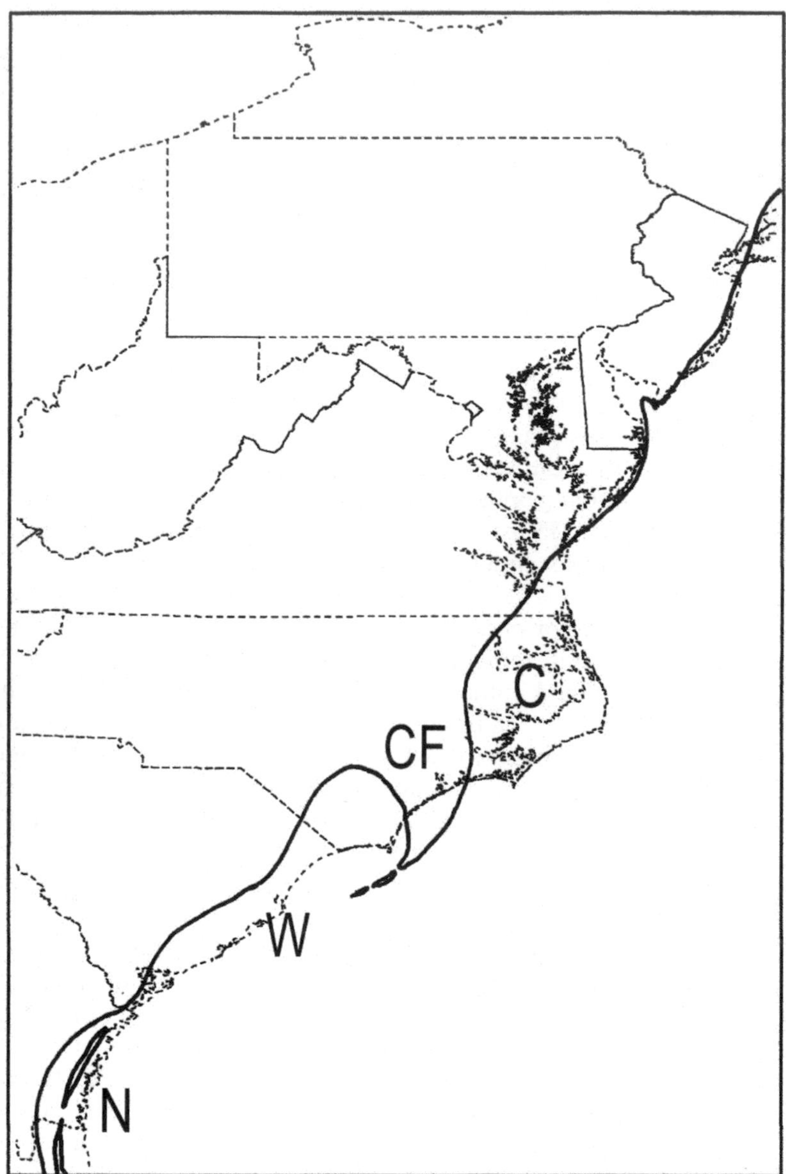

Figure 29. Approximate configuration of the coastline of eastern North America during latest Piacenzian Pliocene and Calabrian Pleistocene times, superimposed upon the outline of the mid-Atlantic United States. Prominent features included: C= Croatan Subsea of the Albemarle Sea, W= Waccamaw Subsea of the Charleston Sea, N= Nashua Lagoon System, CF= Cape Fear Peninsula.

Communities and Environments of the late Yorktown Subsea

From the fossil beds of the Chowan River Formation, two main marine communities can be distinguished for the late Piacenzian Albemarle Sea (late Yorktown Subsea): the *Glyphis pamlicoensis* Community (shallow lagoon and tidal channel coral bioherms) and the *Glycymeris americana* Community (shallow sandy lagoon and intertidal sand flats). As discussed at the end of Chapter 7, many of the common mollusks of the Chowan River fauna have also been found in the late Tamiami Subsea (in the uppermost beds of the Tamiami Formation).

The Glyphis pamlicoensis Community

In the shallow lagoon and tidal channel areas (5-20 m) of the late Yorktown Subsea during Chowan River time, the nontropical, ahermatypic coral *Septastrea* species (close to the Miocene *Septastrea marylandica*; Plate 72, G) formed large bioherms on the open sand bottoms. These coralline structures supplied the substrate for several algivorous and filter-feeding mollusks, including gastropods such as the encrusting vermetid *Serpulorbis granifera* (Say, 1824) (Plate 72, G), the fissurellids *Glyphis pamlicoensis* (Ward and Blackwelder, 1987) (namesake of the community; Plate 72, H) and *Diodora auroraensis* Ward and Blackwelder, 1987, the liotiid *Arene pergemma* (Gardner, 1948), the calyptraeids *Crepidula (Bostrycapulus) chamnessi* Petuch, 1994 and *Crucibulum lawrencei* Ward and Blackwelder, 1987, and encrusting bivalves such as the chamid *Chama emmonsi* Nicol, 1953 and the anomiids *Anomia ephippium* Conrad, 1845 and *Pododesmus fragosus* (Conrad, 1875). Feeding on these bioherm-dwelling mollusks were several carnivorous gastropods, some of which included the drilling ocenebrine muricids *Pterorhytis conradi* (Dall, 1890) (Plate 72, C and D), *Urosalpinx gilmorei* Petuch, 1994 (Plate 72, E), *U. suffolkensis* Gardner, 1948 (Plate 72, F), and *U. gardnerae* Campbell, 1993. Other specialized carnivorous gastropods that on the coral bioherms included the large trochid *Calliostoma philanthropus* (Conrad, 1834), which fed on encrusting sponges and hydroids and the epitoniids *Epitonium sohli* Ward and Blackwelder, 1987 and *E. leai* Ward and Blackwelder, 1987, which fed on zoantherians. Scattered between the coral structures were open patches of sand bottom, and these housed a highly endemic fauna of bivalves, including the surface-dwelling pectinids *Carolinapecten bertiensis* (Mansfield, 1936) and *Leptopecten auroraensis* Ward and Blackwelder, 1987, and shallowly-infaunal species such as the astartid *Astarte berryi* Gardner, 1944 and the small carditid *Pleuromeris auroraensis* Ward and Blackwelder, 1987. Also occurring on these scallop and bivalve beds were several carnivorous gastropods, including molluscivores such as the busyconid *Brachysycon canaliferum* (Conrad, 1862) (Plate 73, I) and *Fulguropsis* species and the drilling naticid *Neverita emmonsi* subspecies, vermivores such as *Contraconus petiti* Petuch, new species (Plate 72, I and J) and the turrid *Cymatosyrinx lunata* (H. Lea, 1843), and general scavengers/carnivores such as the volutid *Volutifusus auroraensis* Petuch, 1994 (Plate 72, A and B) and the buccinid *Ptychosalpinx chesapeakensis* Campbell, 1993 (Plate 72, K).

The Glycymeris americana Community

Closer to shore, on the intertidal sand flats and in the shallow lagoons (0-5 m) during Chowan River time, mixed assemblages of large bivalves and turritellid gastropods formed dense, pavement-like beds on the sea floor. Some of the more abundant and conspicuous bivalves included the glycymerids *Glycymeris americana* (Defrance, 1826) (namesake of the community; Plate 73, H), *G. sloani* Ward and Blackwelder, 1987, and *Costaglycymeris hummi* Ward and Blackwelder, 1987, the carditid *Cyclocardia* species (*C. granulata* complex), and the venerid *Mercenaria permagna* subspecies. Only two small tur-

Plate 72. Index fossils and characteristic organisms of the *Glyphis pamlicoensis* Community (shallow lagoon and tidal channel coral bioherms during late Piacenzian time) of the late Yorktown Subsea. Included are: A,B= *Volutifusus auroraensis* Petuch, 1994 , length 110 mm; C,D= *Pterorhytis conradi* (Dall, 1890), 55 mm; E= *Urosalpinx gilmorei* Petuch, 1994, length 18 mm; F= *Urosalpinx suffolkensis* Gardner. 1948, length 32 mm; G= *Septastrea* species, with attached colony of *Serpulorbis granifera* (Say, 1824), length 178 mm; H= *Glyphis pamlicoensis* (Ward and Blackwelder, 1987), length 37 mm; I,J= *Contraconus petiti* Petuch, new species, holotype, length 87 mm; K= *Ptychosalpinx chesapeakensis* Campbell, 1993, length 46 mm. All specimens were collected in the Chowan River Formation, in the Lee Creek Texasgulf Mine, Aurora, Beaufort County, North Carolina.

ritellids occurred on the open sand bottoms of the early Croatan Sea, and these included *Torcula perexilis* (Conrad, 1873) (Plate 73, J) and *Torculoidella beaufortensis* (Ward and Blackwelder, 1987). Both species lived only in small, scattered patches and not in the large beds that typified the older paleoseas. Feeding on the immense bivalve biomass were several large molluscivorous gastropods, some of which included the busyconids *Busycon auroraensis* Petuch, 1994 (Plate 73, F), *Sinistrofulgur adversarius* (Conrad, 1863) (Plate 73, C), and *Busycotypus concinnum* (Conrad, 1873) (Plate 73, G), and the fasciolariid *Fasciolaria (Cinctura) beaufortensis* Ward and Blackwelder, 1987 (Plate 73, D). Several large general carnivores/scavengers also occurred on the bivalve beds, and included the scaphelline volutid *Volutifusus typus* Conrad, 1863 (Plate 73, A and B) and the buccinid *Ptychosalpinx tuomeyi* (H. Lea, 1843) (Plate 73, E).

Communities and Environments of the Croatan Subsea

During the late Piacenzian, the lagoons and sounds behind the Chuckatuck Archipelago had filled with sediments, and only the southern end of the Albemarle Sea contained marine conditions. By the latest Piacenzian, this final Albemarle reflooding (the Croatan Subsea) was only half the size of the older Yorktown Subsea and was filled with extensive systems of sand shoals, shallow tidal channels, and barrier islands. For the Calabrian Pleistocene (Croatan Subsea), and from data contained in the fossil beds of the James City Formation, only one community is recognized: the *Terebraspira cronleyensis* Community (shallow sandy lagoons and tidal channels). Many of the James City taxa are also found in the upper beds of the Waccamaw Formation, in the upper beds of the Nashua Formation, and in the Caloosahatchee Formation.

The Terebraspira cronleyensis Community

During the Calabrian, the Croatan Subsea was essentially a sandy, shoal-filled open embayment. In the shallow lagoon and tidal channel areas (2-20 m), dense beds of large bivalves occurred on the open sand bottoms. These were made up primarily of the venerids *Merceneria permagna* (Conrad, 1838), *Macrocallista greeni* Ward and Blackwelder, 1987, and *Pitar chioneformis* (Gardner, 1944), the arcid *Anadara aequicostata* (Conrad, 1845), the noetiid *Noetia limula* (Conrad, 1832), the crassatellid *Marvacrassatella kauffmani* Ward and Blackwelder, 1987, and the cardiid *Dinocardium hazeli* Ward and Blackwelder, 1987. Feeding on these bivalves were several molluscivorous gastropods, including the distinctive endemic fasciolariids *Terebraspira cronleyensis* (Gardner, 1948) (namesake of the community; Plate 74, A and B) and *Heilprinia malcolmi* (Ward and Blackwelder, 1987) (Plate 74, E), the giant fasciolariid *Triplofusus acmaensis* (B. Smith, 1940), and the busyconids *Busycon gilmorei* Petuch, 1994, *Sinistrofulgur pamlico* Petuch, 1994, and *Pyruella bladenense* (Gardner, 1948). Several other carnivorous gastropods also occurred on these bivalve beds, including vermivores such as the conids *Ximeniconus waccamawensis* (B. Smith, 1930) (Plate 74, K) and *Seminoleconus diegelae* (Petuch, 1994), suctorial-feeders such as the cancellariids *Ventrilia betsiae* (Olsson and Petit, 1964) (Plate 74, H) and *V. elizabethae* (Olsson and Petit, 1964) (Plate 74, J), and general carnivores/scavengers such as the scaphelline volutid *Volutifusus halscotti* Petuch, 1994 (Plate 74, F and G) and the nassariid *Ilyanassa granifera* (Conrad, 1868) (Plate 74, I). Large biohermal clumps of the ostreid *Conradostrea lawrencei* Ward and Blackwelder, 1987 also grew in the shallow lagoons, and were the principal prey item of the ocenebrine muricid *Urosalpinx auroraensis* Petuch, 1994. Colonial tunicates formed large encrusting masses on the *Conradostrea* oyster bars and these supported the urochordate-feeding triviid gastropod *Trivia lindae* Petuch, 1994 (Plate 74, C and D).

Plate 73. Index fossils and characteristic organisms of the *Glyphis pamlicoensis* Community (shallow lagoon and tidal channel coral bioherms during late Piacenzian time) and *Glycymeris americana* Community (shallow sandy lagoon and intertidal sand flats during late Piacenzian time) of the late Yorktown Subsea. Included are: A,B= *Volutifusus typus* Conrad, 1863, length 142 mm; C= *Sinistrofulgur adversarius* (Conrad, 1863), length 205 mm; D= *Fasciolaria (Cinctura) beaufortensis* Ward and Blackwelder, 1987, length 83 mm; E= *Ptychosalpinx tuomeyi* (H. Lea, 1843), length 72 mm; F= *Busycon auroraensis* Petuch, 1994, length 165 mm; I= *Brachysycon canaliferum* (Conrad, 1862), length 136 mm; G= *Busycotypus concinnum* (Conrad, 1873), length 283 mm; J= *Torcula perexilis* (Conrad, 1873), length 51 mm. All specimens were collected in the Chowan River Formation, in the Lee Creek Texasgulf Mine, Aurora, Beaufort County, North Carolina.

Plate 74. Index fossils and characteristic organisms of the *Terebraspira cronleyensis* Community (shallow sandy lagoons and tidal channels during Calabrian time) of the Croatan Subsea. Included are: A,B= *Terebraspira cronleyensis* (Gardner, 1948), length 152 mm; C,D= *Trivia lindae* Petuch, 1994, length 16 mm; E= *Heilprinia malcolmi* (Ward and Blackwelder, 1987), length 58 mm; F,G= *Volutifusus halscotti* Petuch, 1994, length 122 mm; H= *Ventrilia betsiae* (Olsson and Petit, 1964), length 38 mm; I= *Ilyanassa granifera* (Conrad, 1868), length 20 mm; J= *Ventrilia elizabethae* (Olsson and Petit, 1964), length 37 mm; K= *Ximeniconus waccamawensis* (B. Smith, 1930), length 28 mm. All specimens were collected in the James City Formation, in the Lee Creek Texasgulf Mine, Aurora, Beaufort County, North Carolina.

Communities and Environments of the Waccamaw Subsea

During the late Piacenzian and Calabrian, the Charleston Sea basin was almost filled with sediments and its last vestige, the Waccamaw Subsea, like the Croatan Subsea, was reduced to a simple, open embayment. Paleoenvironmental studies of type Waccamaw Formation outcrops along the Waccamaw River (Horry County, South Carolina; Dubar and Howard, 1963) have shown the shallow (10-30 m), sand-bottom offshore areas housed only a single ecosystem, the *Argopecten vicenarius* Community. This simple scallop-based ecosystem closely resembled the *Argopecten gibbus carolinensis* Community of the Recent Carolinas.

The Argopecten vicenarius Community

During Calabrian time (equivalent to the upper beds of the Waccamaw Formation), the Waccamaw Subsea scallop beds were dominated by only two species, *Argopecten vicenarius* (Conrad 1843) (namesake of the community; Plate 75 G) and *Carolinapecten senescens* (Dall, 1898). Scattered open patches between the scallop beds supported several abundant bivalves, including the glycymerids *Glycymeris americana* (Defrance, 1826) and *G. arata* (Conrad, 1841), the lucinids *Parvilucina multilineata* (Tuomey and Homes 1856) and *Bellucina waccamawensis* (Dall, 1903) the venerids *Lirophora latilirata* subspecies (Plate 75, H) and *Puberella marcottae* (Olsson and Petit, 1964), the anomiid *Pododesmus waccamawensis* (Gardner, 1943) (attached to shell rubble), and the mactrid *Spisula similis* (Say, 1822). Feeding on this rich biomass of bivalves was a large fauna of molluscivorous gastropods, including the busyconids *Busycon gilmorei* Petuch, 1994 (Plate 75, C), *Sinistrofulgur pamlico* Petuch, 1994, *Busycotypus rucksorum* Petuch, 1994, *Brachysycon amoenum* (Conrad, 1875) (Plate 75, F), and *Pyruella bladenense* (Gardner, 1948) (Plate 75, I), the ocenebrine muricids *Urosalpinx auroraensis* Petuch, 1994 (Plate 75, A and B), *U.* species, and *Eupleura* species, and the fasciolariids *Heilprinia portelli* Petuch, 1994 (Plate 75, D) and *Triplofusus acmaensis* (B. Smith, 1940), and the naticids *Neverita duplicata* subspecies and *Sinum multiplicatus* (Dall, 1892). Also living on the sand patch areas was a large component of carnivorous gastropods, some of which included vermivores such as conids *Contraconus scotti* Petuch, 1994, *Ximeniconus waccamawensis* (B. Smith, 1930), and *Gradiconus presozoni* (Olsson and Petit, 1964) and the terebrid *Strioterebrum petiti* (Olsson, 1067), the suctorial-feeding cancellariids *Ventrilia betsiae* (Olsson and Petit, 1964) and *V. elizabethae* (Olsson and Petit, 1964), and general carnivores/scavengers such as the scaphelline volutids *Volutifusus halscotti* Petuch, 1994 (Plate 75, E) and *Scaphella brennmortoni* Olsson and Petit, 1964, the olivid *Oliva (Strephona) roseae* Petuch, 1991, and the large nassariid *Globinassa johnsoni* (Dall, 1892) (Plate 75, J).

Communities and Environments of the Nashua Lagoon System

By Calabrian time, the southern part of the Charleston Sea and the entire eastern coastal region of Florida, south to the Okeechobean Sea, had developed a continuous long chain of barrier islands. This archipelago sheltered the single largest series of lagoons and estuaries known from the early Pleistocene, the Nashua Lagoon System (Figure 30). Oceanographically, this coastal lagoon feature was, at best, only warm temperate in climate and water temperatures fluctuated wildly over the course of a year; with warm, subtropical temperatures in the summer and cold temperatures in the winter. The Nashua Lagoon System also ranged into the northern end of the Okeechobean Sea basin, into the old Kissimmee Embayment area, in its southernmost extension. This area, the Rucks Embayment (Figure 30), was connected to the main coastal lagoons by a series of channels between the barrier islands and shared the same oceanographic conditions.

Plate 75. Index fossils and characteristic organisms of the *Argopecten vicenarius* Community (shallow lagoon and open sound pectinid beds) of the Waccamaw Subsea. Included are: A,B= *Urosalpinx auroraensis* Petuch, 1994, length 36 mm; C= *Busycon gilmorei* Petuch, 1994, length 138 mm; D= *Heilprinia portelli* Petuch, 1994, length 40 mm; E= *Volutifusus halscotti* Petuch, 1994, length 117 mm; F= *Brachysycon amoenum* (Conrad, 1875), length 123 mm; G= *Argopecten vicenarius* (Conrad, 1843), length 53 mm; H= *Lirophora latilirata* subspecies, length 39 mm; I= *Pyruella bladenense* (Gardner, 1948), length 112 mm; J= *Globinassa johnsoni* (Dall, 1892). All specimens were collected in the Waccamaw Formation, in the Calabash pit, Shallotte, Brunswick County, North Carolina.

Because of both fluctuating water temperatures and salinities, the fauna of the Nashua Lagoon System was impoverished, with a much lower species-richness than is seen in the contemporaneous Waccamaw and Caloosahatchee Subseas. Although the majority of the Nashua molluscan fauna was shared with the cooler water Waccamaw Subsea, a large number of prominent species were endemic to the coastal lagoons and these occurred together with a small component of warm water Caloosahatchee taxa. From assemblages preserved in the highly-fossiliferous Nashua Formation in the Rucks Embayment area, two main communities can be identified. These include the *Mulinia lateralis* Community (shallow muddy coastal lagoons) and the *Mercenaria permagna* Community (shallow sand flats and intertidal sand bars). The cooler water nature of the Nashua Lagoon System is demonstrated by the absence of several key tropical index groups such as the Melongenidae, Strombidae, Turbinellidae, Potamididae, and Modulidae.

The Mulinia lateralis Community

In the deeper (5-20 m), muddier areas of the Nashua lagoons, immense beds of the mactrid bivalve *Mulinia lateralis* (Say, 1822) (namesake of the community; Plate 76, E) literally carpeted the sea floor in a solid pavement. Also occurring on the dense mactrid beds were several other bivalves, including the shallowly-infaunal noetiid *Noetia limula* (Conrad, 1832) (Plate 76, C), the ostreid *Conradostrea lawrencei* Ward and Blackwelder, 1987 (Plate 76, D) (which occurred in small, scattered clumps), and the pectinids *Carolinapecten solaroides* (Heilprin, 1886) (Plate 76, A) and *C. senescens* (Dall, 1898) (Plate 76, F) (which lived in small, scattered beds). This simple, but high-biomass, bivalve fauna was the principal food resource for several molluscivorous gastropods, some of which included the busyconids *Brachysycon amoenum* (Conrad, 1875) (Plate 77, B), *Pyruella bladenense* (Gardner, 1948), and *Busycotypus scotti* Petuch, 1994 (Plate 77, I), the fasciolariids *Fasciolaria (Cinctura) rucksorum* Petuch, 1994 (Plate 77, F) and *Triplofusus acmaensis* (B. Smith, 1940) (Plate 77, C), the naticid *Sinum* species, and the ocenebrine muricids *Urosalpinx rucksorum* Petuch, 1994 (Plate 76, G) and *Eupleura calusa* Petuch, 1994. On open sand patches between the mactrid beds, and in shallower water areas adjacent to sand banks, several carnivorous gastropods occurred, some of which included vermivores such as the conid *Ximeniconus waccamawensis* (B. Smith, 1930) (Plate 76, I), suctorial-feeders such as the cancellariid *Ventrilia rucksorum* Petuch, 1994 (Plate 76, H), and general carnivores/scavengers such as the olivids *Oliva (Strephona) roseae* Petuch, 1991 (Plate 76, J) and *O. (Strephona) rucksorum* Petuch, 1994 (Plate 76, B) and the scaphelline volutid *Volutifusus halscotti* Petuch, 1994 (Plate 77, E). In deeper areas with cleaner water, stronger tidal currents, and normal salinities, the ahermatypic coral *Oculina* species (*O. sarasotaensis* complex) formed large, ramose colonies. Some of these were large enough to form biohermal structures, and often grew together with small beds of *Conradostrea lawrencei*. These bioherms acted as the main substrate for large stacks of the sessile, filter-feeding calyptraeid gastropod *Crepidula roseae* Petuch, 1991.

The Mercenaria permagna Community

On shallow sand banks and intertidal sand flats (0-5 m) along the western sides of the Nashua barrier islands and along the mainland, the venerid bivalve *Mercenaria permagna* (Conrad, 1838) (namesake of the community; Plate 77, A) formed extensive beds. These large and massive bivalves were so dominant and so closely-packed that only one other large bivalve occurred with them; the giant cardiid *Dinocardium hazeli* Ward and Blackwelder, 1987 (Plate 77, G). In some exposures of the Nashua Formation, such as in the Rucks Pit at Fort Drum, Florida, *Mercenaria permagna* forms solid, thick, monoculture beds, attesting to the species' abundance on these Calabrian sand flats. This simple ecosystem

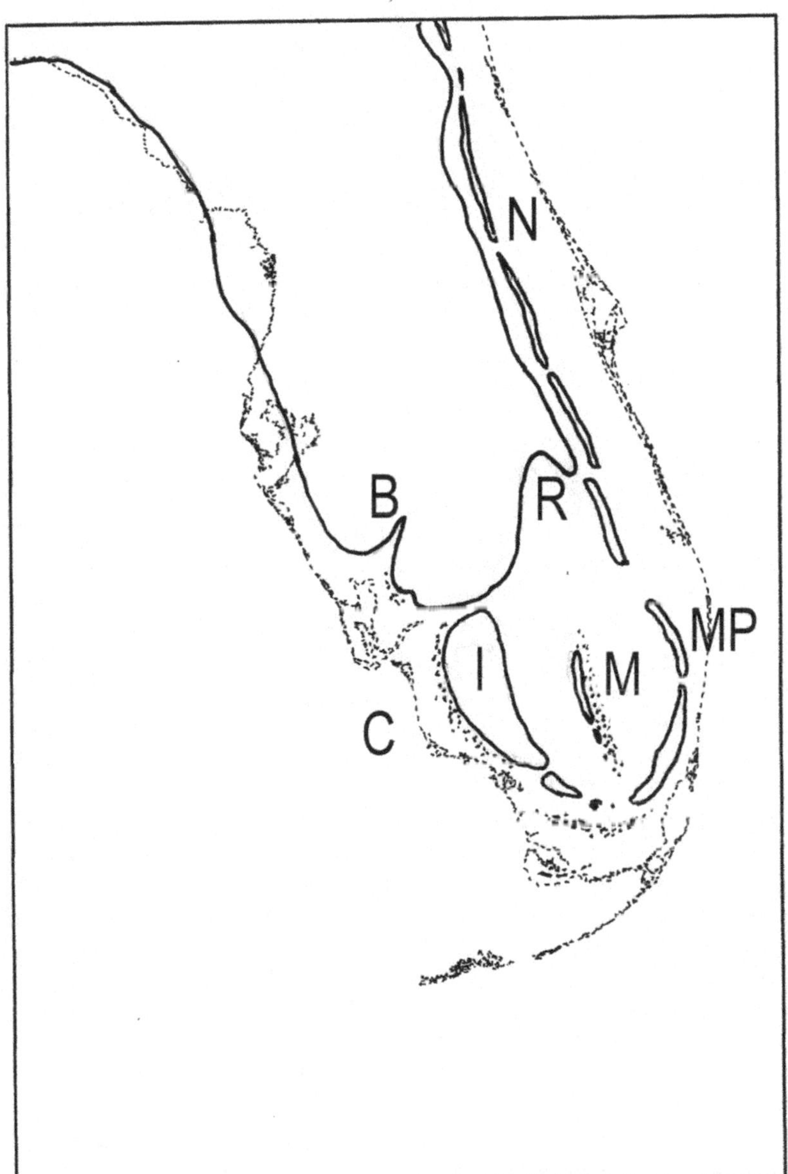

Figure 30. Approximate configuration of the Florida coastline during Calabrian Pleistocene time, superimposed upon the outline of the Recent Floridian Peninsula. Prominent features included: C= Caloosahatchee Subsea of the Okeechobean Sea, N= Nashua Lagoon System, M= Miccosukee Island and Reef Tract, I= Immokalee Island and Archipelago, MP= Miami and Palm Beach Archipelagos, R= Rucks Embayment, B= Brantley Embayment.

contained only two large molluscan predators, the endemic busyconids *Busycon rucksorum* Petuch, 1994 (Plate 77, D) and *Sinistrofulgur yeehaw* Petuch, 1994 (Plate 77, H). Within the southern part of the Rucks Embayment, water temperatures may have been seasonally warm enough to allow some tropical Caloosahatchee taxa to migrate, temporarily, into the lagoon system. Here, the cypraeid *Siphocypraea* species, the fasciolariid *Heilprinia caloosaensis* (Heilprin, 1886), and the conid *Contraconus* cf. *tryoni* (Heilprin, 1886) are known to have occurred, but only rarely and sporadically.

Communities and Environments of the Caloosahatchee Subsea

Following the latest Piacenzian climatic cooling and extinction event, the Okeechobean Sea faunas rebounded during the Calabrian warm time and underwent rapid evolution, producing many new species and species complexes. These survivors inhabited a much altered Okeechobean Sea, where the Everglades Pseudoatoll reef tracts had become buried in sand and had formed a large semicircle of islands. This new paleosea, the Caloosahatchee Subsea (Figure 30), initiated its development in the latest Piacenzian (equivalent to Unit 1 at Sarasota, an un-named lowest member of the Caloosahatchee Formation; see Petuch, 1982) but did not reach its maximum size and complexity until the Calabrian. At that time, the buried pseudoatoll formed the bases for the Miami, Palm Beach, and Immokalee Archipelagos and these developed smaller secondary reef systems along their intralagoonal coasts. A prominent new feature that appeared at this time was Miccosukee Island and Reef Tract (Petuch, 1997), a large mangrove and coralline structure that had developed within the central lagoon, along the eastern edge of the Hendry Platform and bordering the western edge of the deep Loxahatchee Trough. Another prominent feature was the Brantley Embayment, a large shallow, elongated bay that extended inland along the southwestern tip of the Floridian Peninsula. During the Calabrian, the Caloosahatchee fauna and oceanographic conditions extended northward along the western coast of Florida, to at least Recent St. Petersburg. When considering that the entire eastern coast of Florida contained a cool-water, Waccamaw-type fauna at this time, the presence of the tropical Caloosahatchee Subsea fauna along western Florida demonstrates that the Calabrian Gulf of Mexico was much warmer than the Atlantic Ocean.

As in the Tamiami Subsea, the Calabrian Caloosahatchee Subsea underwent three separate depositional events in quick succession. These correspond to the three Pleistocene members of the Caloosahatchee Formation: the lower Fort Denaud Member, the middle Bee Branch Member, and the upper Ayer's Landing Member (DuBar, 1958). From the fossil assemblages preserved within these units, eight distinct communities can be recognized and these include the *Pyrazisinus scalatus* Community (mangrove forests and intertidal mud flats during Fort Denaud time), the *Siphocypraea problematica* Community (*Thalassia* beds during Fort Denaud time), the *Arcoptera wagneriana* Community (shallow sand bottom lagoon during Fort Denaud time), the *Dichocoenia eminens* Community (coral reefs and coral bioherms during Fort Denaud time), the *Anomalocardia caloosana* Community (shallow sand bottom lagoons during Bee Branch time), the *Siphocypraea brantleyi* Community (*Thalassia* beds during Bee Branch time), the *Oliva jenniferae* Community (shallow sand bottom lagoons during Ayer's Landing time), and the *Siphocypraea griffini* Community (*Thalassia* beds during Ayer's Landing time).

The *Pyrazisinus scalatus* Community

During Fort Denaud time (early Calabrian), the eastern sides of Miccosukee and Immokalee Islands and the southern tip of the mainland Floridian Peninsula were heavily forested with dense mangrove (probably *Rhizophora* and *Avicennia*) jungles. These, and the organic-rich intertidal mud flats that formed between mangrove islands, supported the

Plate 76. Index fossils and characteristic organisms of the *Mulinia lateralis* Community (shallow muddy coastal lagoons) of the Nashua Lagoon System. Included are: A= *Carolinapecten solaroides* (Heilprin, 1886), length 175 mm; B= *Oliva (Strephona) rucksorum* Petuch, 1994, length 38 mm; C= *Noetia limula* (Conrad, 1832), length 55 mm; D= *Conradostrea lawrencei* Ward and Blackwelder, 1987, length 80 mm; E= *Mulinia lateralis* (Say, 1822), length 27 mm; F= *Carolinapecten senescens* (Dall, 1898), length 82 mm; G= *Urosalpinx rucksorum* Petuch, 1994, length 21 mm; H= *Ventrilia rucksorum* Petuch, 1994, length 29 mm; I= *Ximeniconus waccamawensis* (B. Smith, 1930), length 35 mm; J= *Oliva (Strephona) roseae* Petuch, 1991, length 84 mm. All specimens were collected in the Nashua Formation, in the Rucks Pit, Fort Drum, Okeechobee County, Florida.

Plate 77. Index fossils and characteristic organisms of the *Mulinia lateralis* Community (shallow muddy coastal lagoons) and the *Mercenaria permagna* Community (shallow sand flats and intertidal sand bars) of the Nashua Lagoon System. Included are: A= *Mercenaria permagna* (Conrad, 1838), length 125 mm; B= *Brachysycon amoenum* (Conrad, 1875), length 112 mm; C= *Triplofusus acmaensis* (B. Smith, 1940), length 288 mm; D= *Busycon rucksorum* Petuch, 1994, length 158 mm; E= *Volutifusus halscotti* Petuch, 1994, length 84 mm; F= *Fasciolaria (Cinctura) rucksorum* Petuch, 1994, length 97 mm; G= *Dinocardium hazeli* Ward and Blackwelder, 1987, length 96 mm; H= *Sinistrofulgur yeehaw* Petuch, 1994, length 218 mm; I= *Busycotypus scotti* Petuch, 1994, length 163 mm. All specimens were collected in the Nashua Formation, in the Rucks Pit, Fort Drum, Okeechobee County, Florida.

richest-known tropical estuarine molluscan fauna found anywhere in early Pleistocene eastern North America. Although not as diverse as the fauna of the older Myakka Lagoon System, the Caloosahatchee mangrove-mud flat community was similar and was dominated by potamidid gastropods. Covering the mud flats at low tide, and feeding on algal and bacterial films, were immense numbers of two species of the genus *Pyrazisinus*; *P. scalatus* (Heilprin, 1886) (namesake of the community; Plate 78, H) and *P. ecarinatus* (Dall, 1892) (Plate 78, B). These served as the principal prey items of the large melongenids *Echinofulgur echinatum* (Dall, 1890) (Plate 78, G) and *Melongena caloosahatcheensis* Petuch, 1994. Scattered among, and growing on, the mangrove roots was the unusual ostreid oyster *Crassostrea labellensis* (Olsson and Harbison, 1953) (Plate 78, E). This ornate, elongated oyster was the principal prey item of the drilling thaidids *Thaisiella trinitatensis brujensis* (M. Smith, 1946) (Plate 78, A; first record of the genus *Thaisiella* from north of Panama) and *Thais (Stramonita)* species (*T. haemostoma* complex). Crawling on the mangrove roots, branches, and leaves were several arboreal gastropods, including the potamidids *Cerithidea jenniferae* Petuch, 1994 (Plate 78, D) and *C. xenos* Petuch, 1991 (Plate 78, K), the littorinids *Littorinopsis caloosahatcheensis* (Petuch, 1991) (Plate 78, F) and *L. seminole* (Petuch, 1991) (Plate 78, I), and the neritid *Nerita (Theliostyla) hertweckorum* Petuch, 1994 (Plate 78, C). On open areas of the mud flats, between the potamidid beds, an unusually large fauna of scavenger/carrion-feeding nassariid gastropods occurred. Some of these nassariids included *Globinassa roseae* (Petuch, 1991) (Plate 78, J), *G. floridana* (M. Smith, 1936), *G. schizopyga* (Dall, 1892), *Paranassa deleonensis* (Tucker, 1931), *Scalanassa corbis* (Olsson, 1967), *S. olssoni* (Petuch, 1994) (Plate 78, L), and *Ilyanassa palmbeachensis* Petuch, 1994.

The Siphocypraea problematica Community

In shallow water (1-5 m) areas adjacent to the mud flats during Fort Denaud time (early Calabrian), extensive beds of Turtle Grass (*Thalassia*) carpeted the sea floor. These sea grass prairies housed a new, early Pleistocene species radiation of the grass bed cowrie genus *Siphocypraea*. Already by the early Calabrian, a new subgenus (*Okeechobea*; see Systematic Appendix) had evolved and this co-occurred with the other two subgenera that had survived the late Piacenzian extinction. On typical Fort Denaud Turtle Grass beds, representatives of these three groups lived in large aggregations and included *Siphocypraea* (s.s.) *problematica* (Heilprin, 1886) (namesake of the community; Plate 79, C and D), *S. (Okeechobea) philemoni* Fehse, 1997 (Plate 79, F and G), and *S. (Pahayokea) josiai* Fehse, 1997 (Plate 79, K and L). Living on the individual grass blades, and feeding on epiphytic algae and diatoms were several small gastropods, including the modulids *Modulus caloosahatcheensis* Petuch, 1994 (Plate 79, I) and *M.* species (*M. basileus* complex), the trochid *Tegula calusa* Petuch, 1994 (Plate 79, J), the turbinids *Astraea (Lithopoma) precursor* (Dall, 1892) and *A. (Lithopoma) scolopax* Olsson and Harbison, 1953, and the cerithiids *Cerithium triticium* Olsson and Harbison, 1953, *C. willcoxi* Olsson and Harbison, 1953, and *C. litharium* Dall, 1892. The large turbinid *Turbo (Taenioturbo) rhectogrammicus* Dall, 1892 (Plate 79, E) also occurred in the grass beds and often served as the attachment substrate for clumps of the ostreid oyster *Crassostrea subdigitalina* (Olsson and Harbison, 1953). Living nestled between the Turtle Grass roots were numerous large bivalves, including the lucinids *Miltha caloosaensis* (Dall, 1898) and *Stewartia intermixta* (Olsson and Harbison, 1953), the cardiids *Laevicardium wagnerianum* Olsson and Harbison, 1953 and *Acrosterigma dalli* (Heilprin, 1886), the arcids *Caloosarca crassicosta* Heilprin, 1886), *Cunearca scalarina* (Heilprin, 1886), *Dallarca petersburgensis* Olsson and Harbison, 1953, and *Anadara lienosa* (Say, 1832), and numerous smaller tellinids and mactrids. Feeding on this large bivalve

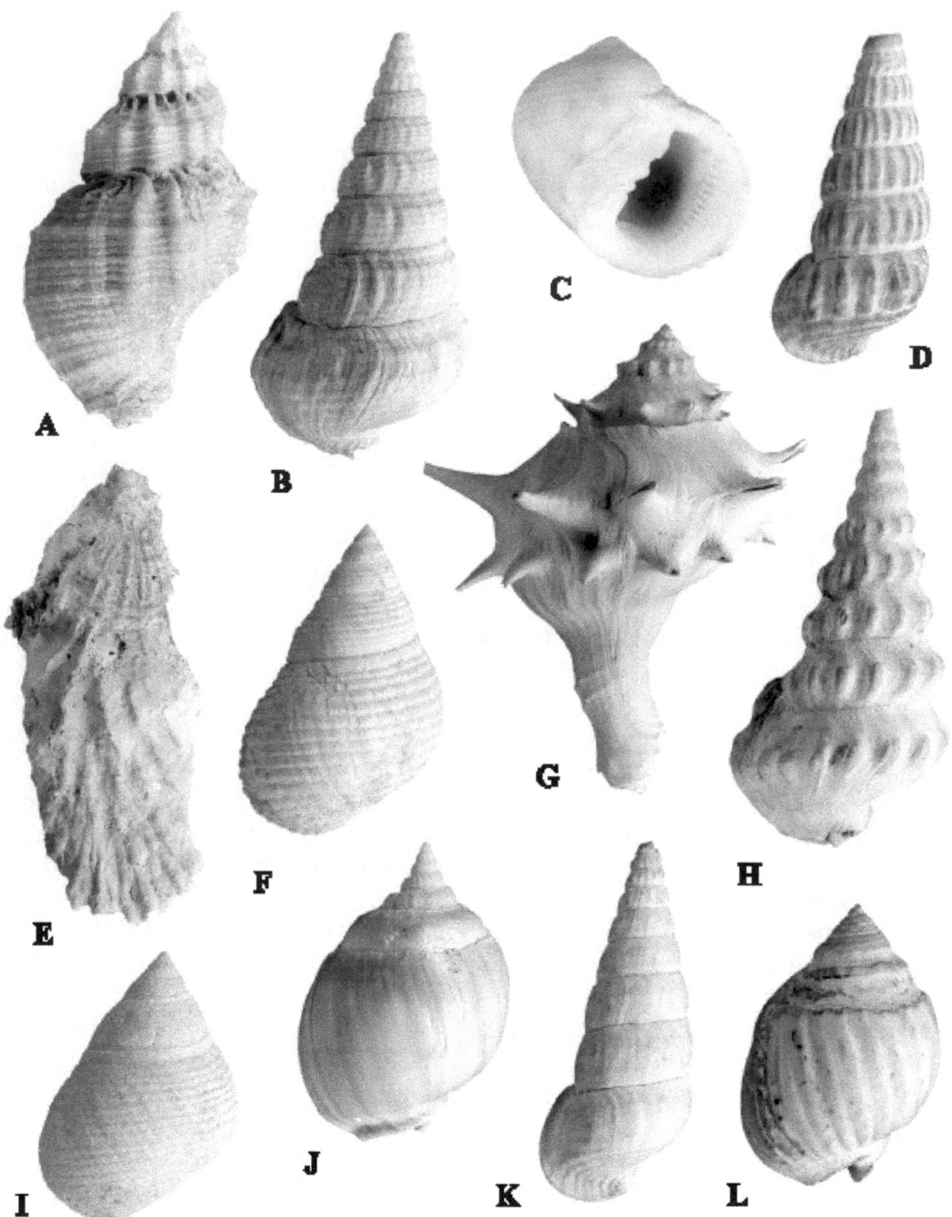

Plate 78. Index fossils and characteristic organisms of the *Pyrazisinus scalatus* Community (mangrove forests and intertidal mud flats during early Calabrian time) of the Caloosahatchee Subsea. Included are: A= *Thaisiella trinitatensis brujensis* (M. Smith, 1946), length 44 mm; B= *Pyrazisinus ecarinatus* (Dall, 1892), length 61 mm; C= *Nerita (Theliostyla) hertweckorum* Petuch, 1994, length 17 mm; D= *Cerithidea jenniferae* Petuch, 1994, length 27 mm; E= *Crassostrea labellensis* (Olsson and Harbison, 1953), length 119 mm; F= *Littorinopsis caloosahatcheensis* (Petuch, 1991), length 18 mm; G= *Echinofulgur echinatum* (Dall, 1890), length 88 mm; H= *Pyrazisinus scalatus* (Heilprin, 1886), length 78 mm; I= *Littorinopsis seminole* (Petuch, 1991), length 12 mm; J= *Globinassa roseae* (Petuch, 1991), length 23 mm; K= *Cerithidea xenos* Petuch, 1991, length 25 mm; L= *Scalanassa olssoni* (Petuch, 1994), length 27 mm. All specimens were collected in the Fort Denaud Member of the Caloosahatchee Formation, along the Miami Canal at the Holey Land Levee, Palm Beach County, Florida.

Plate 79. Index fossils and characteristic organisms of the *Siphocypraea problematica* Community (*Thalassia* beds during the early Calabrian) of the Caloosahatchee Subsea. Included are: A= *Macrostrombus leidyi* (Heilprin, 1886), length 147 mm; B= *Turbinella regina* Heilprin, 1886, length 286 mm; C,D= *Siphocypraea problematica* Heilprin, 1886, length 64 mm; E= *Turbo (Taenioturbo) rhectogrammicus* Dall, 1892, length 43 mm; F,G= *Siphocypraea (Okeechobea) philemoni* Fehse, 1997, length 61 mm; H= *Terebraspira scalarina* (Heilprin, 1886), length 160 mm; I= *Modulus caloosahatcheensis* Petuch, 1994, length 10 mm; J= *Tegula calusa* Petuch, 1994, length 15 mm; K,L= *Siphocypraea (Pahayokea) josiai* Fehse, 1997, length 43 mm. All specimens were collected in the Fort Denaud Member of the Caloosahatchee Formation, along the Miami Canal south of Lake Harbor, Palm Beach County, Florida.

fauna were several molluscivorous gastropods, including the fasciolariids *Terebraspira scalarina* (Heilprin, 1886), *Fasciolaria calusa* Petuch, 1994, and *Triplofusus acmaensis* (B. Smith, 1940), the busyconids *Busycoarctum rapum* (Heilprin, 1886) and *Sinistrofulgur palmbeachensis* Petuch, 1994, and the muricids *Calotrophon attenuatus* (Dall, 1890), *C. perplexus* (Olsson and Harbison, 1953), *Favartia alta* (Dall, 1890), *Chicoreus calusa* Petuch, 1991, *C. susanae* Petuch, 1994, and *Vokesimurex pahayokee* Petuch, 1994. Interstitial enteropneust and polychaete worms were abundant on the Fort Denaud Turtle Grass beds and were the main prey items of the giant grass bed turbinellids, *Turbinella regina* Heilprin, 1886 (Plate 79, B) and *T. scolymoides* Dall, 1890. Along the edges of the grass beds, large aggregations of the strombid *Macrostrombus leidyi* (Heilprin, 1886) (Plate 79, A) grazed on epiphytic algae and these were commonly preyed upon by the giant fasciolariid *Triplofusus acmaensis*. In shallow areas adjacent to open sand bottoms, individual heads of the faviid coral *Manicina pliocenica* subspecies occurred in abundance.

The Arcoptera wagneriana Community

In the shallow lagoons (5-20 m) adjacent to the Turtle Grass beds during Fort Denaud time (early Calabrian), extensive sand bottom substrates supported immense beds of the shallowly infaunal arcid bivalve *Arcoptera wagneriana* (Dall, 1898) (namesake of the community; Plate 80, K). These arcid colonies were particularly dense and well developed within the Brantley Embayment and along Miccosukee Island. Living within the *Arcoptera* beds were large numbers of other smaller bivalves, including the cardiids *Americardia columba* (Heilprin, 1886), *Trigoniocardia willcoxi* (Dall, 1900), and *Trachycardium oedalium* (Dall, 1900), the arcid *Calloarca taeniata* (Dall, 1898), and the tellinids *Tellina (Acorylus) suberis* (Dall, 1900), *T.* species (*T. alternata* complex), and *T.* species (*T. tayloriana* complex). These bivalves, and *Arcoptera*, were the principal food resources for several molluscivorous gastropods, including the busyconids *Pyruella planulatum* (Dall, 1890), *P. soror* Petuch, 1994, *Sinistrofulgur caloosahatcheensis* Petuch, 1994, and *Fulguropsis floridanum* (Olsson and Harbison, 1953), the fasciolariids *Liochlamys bulbosa* (Heilprin, 1886) (Plate 80, G), *Fasciolaria monocingulata* Dall, 1890, *F. (Cinctura) apicina* Dall, 1890, and *Heilprinia caloosaensis* (Heilprin, 1886), and the muricids *Eupleura calusa* Petuch, 1994, *E. intermedia* Dall, 1890, *Phyllonotus evergladesensis* Petuch, 1994 (Plate 80, J), *Subpterynotus textilis* (Gabb, 1873), *Murexiella macgintyi* (M. Smith, 1938), and *Acantholabia floridana* Olsson and Harbison, 1953. Living on open sand patches scattered between the arcid aggregations were small beds of the pectinids *Carolinapecten solaroides* (Heilprin, 1886) (Plate 80, A), *Stralopecten caloosaensis* (Dall, 1898) (Plate 80, I), *Argopecten anteamplicostatus* (Mansfield, 1936), and *Leptopecten irremotis* (Olsson and Harbison, 1953). In some areas (as along the eastern shore of Miccosukee Island), large beds of turritellid gastropods also occurred along with the scallop beds and these were composed primarily of *Bactrospira perattenuata* (Heilprin, 1886), *Apicula apicalis* (Heilprin, 1886), *A. wagneriana* (Olsson and Harbison, 1953), and *Torculoidella subannulata* (Heilprin, 1886). Living with the scallops and turritellids on open sand bottoms in shallower water, and feeding on surficial algal films, were large aggregations of the small strombid *Strombus keatonorum* Petuch, 1994 (Plate 80, F). These, and the pectinids and turritellids, co-occurred with an exceptionally rich and highly diverse fauna of specialized carnivorous gastropods, some of which included vermivores such as the conids *Calusaconus spuroides* (Olsson and Harbison, 1953) (Plate 80, D), *Contraconus osceolai* Petuch, 1991 (Plate 80, B), *C. tryoni* (Heilprin, 1886) (Plate 80, E), *Seminoleconus diegelae* (Petuch, 1994), *Gradiconus joelshugari* (Petuch, 1994), *G. parkeri* (Richards and Harbison, 1947), *Ximeniconus robertsi* (Olsson and Harbison, 1953), and *Jaspidiconus wilsoni* (Petuch, 1994), and the turrids *Knefastia lindae* Petuch, 1994, *Cymatosyrinx acila* Dall, 1890, *Polystira calusa*

Plate 80. Index fossils and characteristic organisms of the *Arcoptera wagneriana* Community (shallow sand bottom lagoon during the early Calabrian) of the Caloosahatchee Subsea. Included are: A= *Carolinapecten solaroides* (Heilprin, 1886), length 172 mm; B= *Contraconus osceolai* Petuch, 1991, length 88 mm; C= *Oliva (Strephona) erici* Petuch, 1994, length 87 mm; D= *Calusaconus spuroides* (Olsson and Harbison, 1953), length 37 mm; E= *Contraconus tryoni* (Heilprin, 1886), length 158 mm; F= *Strombus keatonorum* Petuch, 1994, length 73 mm; G= *Liochlamys bulbosa* (Heilprin, 1886), length 95 mm; H= *Oliva (Porphyria) paraporphyria* Petuch, 1991, length 67 mm; I= *Stralopecten caloosaensis* (Dall, 1898), length 65 mm; J= *Phyllonotus evergladesensis* Petuch, 1994, length 86 mm; K= *Arcoptera wagneriana* (Dall, 1898), length 98 mm. All specimens were collected in the Fort Denaud Member of the Caloosahatchee Formation, along the Miami Canal south of Lake Harbor, Palm Beach County, Florida.

Petuch, 1994, and *Hindsiclava perspirata* (Dall, 1890), suctorial-feeders such as the cancellariid *Cancellaria conradiana* Dall, 1890, echinoderm-feeders such as the cassid *Phalium alligator* Petuch, 1991 and the ficid *Ficus caloosahatchiensis* (B. Smith, 1907), and general carnivores/scavengers such as the olivids *Oliva (Strephona) erici* Petuch, 1994 (Plate 80, C), *O. (Strephona) roseae* Petuch, 1991, and *O.(Porphyria) paraporphyria* Petuch, 1991 (Plate 80, H), the scaphelline volutid *Scaphella floridana* Heilprin, 1886, the pleioptygmatid *Pleioptygma lineolata* (Heilprin, 1886), and the small nassariids *Ilyanassa wilmingtonensis* Gardner, 1948, *I. granifera* (Conrad, 1868), and *Scalanassa evergladesensis* Petuch, 1994.

The Dichocoenia eminens Community

During Fort Denaud time (early Calabrian), large coral bioherms developed in shallow water areas (2-10 m) along the eastern coast of Miccosukee Island, the western and northern coasts of Immokalee Island, and in the Brantley Embayment. In being unzonated and having a low species diversity, these Caloosahatchee reefs differed greatly from the reef tracts of the Tamiami Everglades Pseudoatoll. Typically, the Caloosahatchee biohermal reefs were dominated by only two caryophyllid corals, *Dichocoenia eminens* Weisbord, 1974 (namesake of the community; Plate 81, B; always peduncular) and *D. caloosahatcheensis* Weisbord, 1974 (Plate 81, J). These occurred together with rounded, massive corals such as the montastreids *Montastrea* cf. *annularis* (Ellis and Solander, 1736), *Solenastrea* cf. *bournoni* Edward and Haime, 1849, and *S.* cf. *hyades* (Dana, 1846), and the siderastreid *Siderastrea pliocenica* Vaughan, 1919, and branching, ramose corals such as the oculinid *Oculina* species and the rhizangiid *Septastrea crassa* (Holmes, 1858). In the areas adjacent to open sand bottoms and Turtle Grass beds, the faviid *Placocyathus* species (Plate 81, K) grew in pockets of sand in coral rubble. Here, the rhizangiids *Septastrea matsoni* Vaughan, 1919 (Plate 81, D) and *Coenangia* species (Plate 81, C) grew on living gastropod shells and often were carried around by hermit crabs. The large biohermal structures formed by these corals provided the habitat for a large number of reef-dwelling mollusks. Some of these included bivalves such as the spondylid *Spondylus rotundatus* Heilprin, 1886, the pectinid *Lindapecten harrisii* Dall, 1898), the chamid *Chama macerophylla* subspecies, the coral-nesting limids *Lima (Ctenoides) floridana* Olsson and Harbison, 1953 and *L. (Promantellum) florpacifica* Olsson and Petit, 1964, the plicatulid *Plicatula* species, and the arcids *Arca aquila* Heilprin, 1886 and *Barbatia irregularis* Dall, 1898. These coral-dwelling bivalves were the principal prey items for a large fauna of muricid gastropods, some of which included *Pterorhytis wilsoni* Petuch, 1994 (Plate 81, H and I), *Muricopsis* cf. *lyonsi* Petuch, 1986, *Vokesinotus lepidotus* (Dall, 1890), *Trossulasalpinx subsidus* (Dall, 1890), and *Dermomurex engonatus* (Dall, 1890). An especially rich and diverse fauna of specialized carnivorous gastropods was also present, including coral-feeders such as the ovulid *Jenneria richardsi* Olsson, 1967 (Plate 81, A) and the coralliophilid *Babelomurex mansfieldi* (McGinty, 1940), octocorallian-feeders such as the ovulid *Cyphoma* species, mollusk and echinoderm-feeders such as the tonnid *Malea springi* Petuch, 1989 (feeding on cidarid sea urchins) and the ranellid *Cymatium (Septa) martinianum* subspecies, vermivores such as the conids *Purpuriconus protocardinalis* (Petuch, 1991) (Plate 81, F) and *Cariboconus harbisonae* (Petuch, 1994) and the mitrid *Scabricola caloosahatcheensis* (Petuch, 1991), the suctorial-feeding cancellariid *Bivetopsia pachia* (M. Smith, 1940), and the large vermivore vasinine turbinellid *Hystrivasum horridum* (Heilprin, 1886) (Plate 81, C) (also found in the *Thalassia* beds). A large fauna of small general carnivores was also present and included the buccinids *Solenosteira mengeana* Dall, 1890, *Hesperisternia olssoni* (Petuch, 1994), *H. calusa* (Petuch, 1994), and *Gemophos maxwelli* Olsson and Harbison, 1953, the columbellids *Anachis clewistonensis* (M. Smith, 1938), and *A. amydra* (Dall, 1890), the fasciolariids *Latirus hypsipettus* Dall, 1890 and *L. caloosahatchiensis* Lyons, 1991, and the moru-

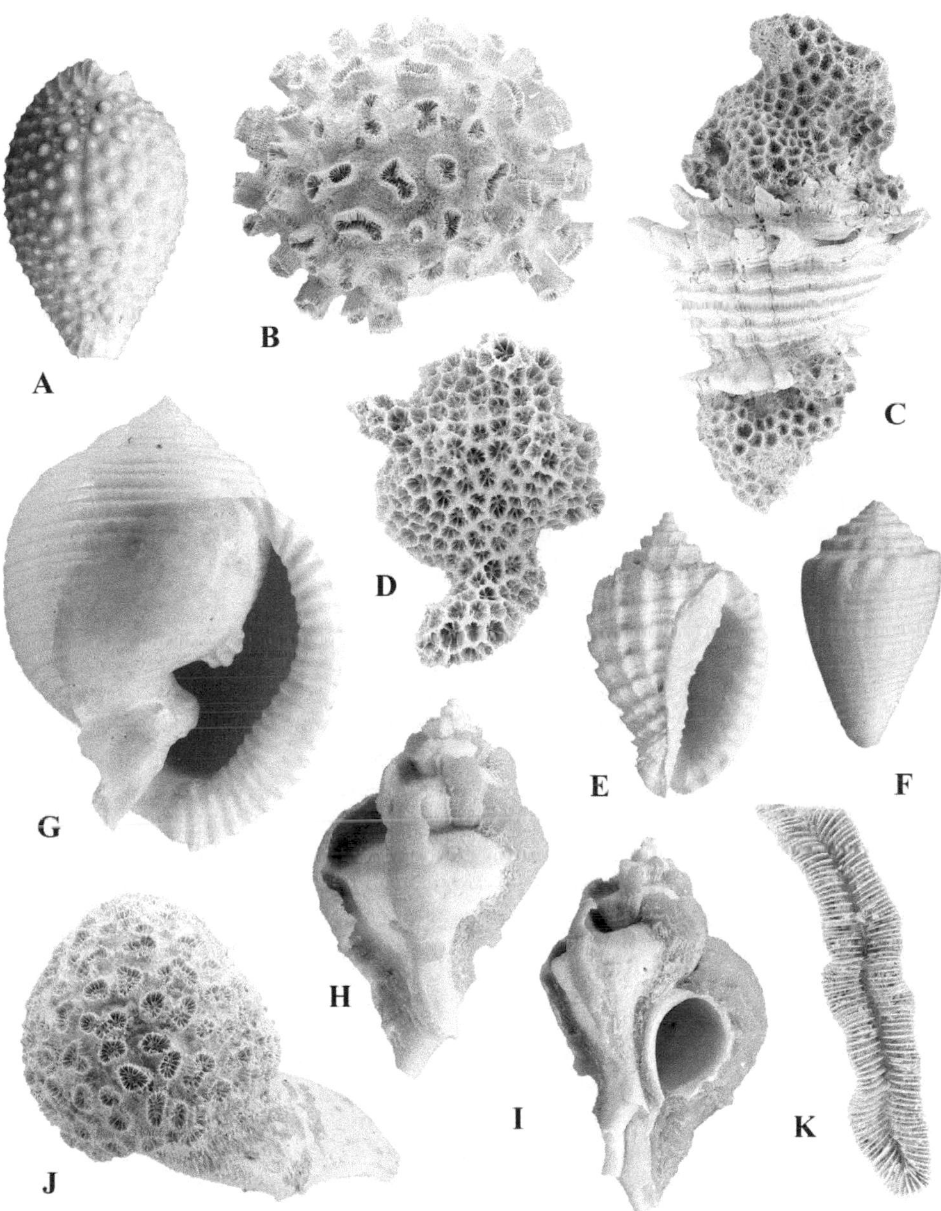

Plate 81. Index fossils and characteristic organisms of the *Dichocoenia eminens* Community (coral reefs and coral bioherms during the early Calabrian) of the Caloosahatchee Subsea. Included are: A= *Jenneria richardsi* Olsson, 1967, length 28 mm; B= *Dichocoenia eminens* Weisbord, 1974, length 146 mm; C= *Hystrivasum horridum* (Heilprin, 1886) covered with *Coenangia* species, length 126 mm; D= *Septastrea matsoni* Vaughan, 1919, length 64 mm; E= *Cancellomorum macgintyi* (M. Smith, 1937), length 20 mm; F= *Purpuriconus protocardinalis* (Petuch, 1991), length 19 mm; G= *Malea springi* Petuch, 1989, length 186 mm; H,I= *Pterorhytis wilsoni* Petuch, 1994, length 40 mm; J= *Dichocoenia caloosahatcheensis* Weisbord, 1974, length 76 mm (growing on *Arcoptera wagneriana* valve); K= *Placocyathus* species, length 141 mm. All specimens were collected in the Fort Denaud Member of the Caloosahatchee Formation, along the Miami Canal at the Broward County Levee, Palm Beach-Broward County line, Florida.

minine harpids *Cancellomorum macgintyi* (M. Smith, 1937) (Plate 81, E) and *Morum floridanum* Tucker and Wilson, 1933.

The Anomalocardia caloosana Community

At the end of Fort Denaud time, a brief, but severe, climatic degeneration took place and the Okeechobean faunas were again stressed by cooling and loss of habitats. This pattern was reversed by Bee Branch time (mid-Calabrian), when sea levels again began to rise and the marine climate warmed within the Okeechobean Sea. In response to this warming event, the Caloosahatchee molluscan fauna underwent a spurt of rapid speciation, producing an entire new fauna. Due to sedimentary infilling during the intervening low water stand, much of the Caloosahatchee Subsea area was made up of sand shoals and banks. In the shallow (0-5 m) sand bottom lagoons, and on sand bars and intertidal sand flats, the venerid bivalve *Anomalocardia caloosana* (Dall, 1900) (namesake of the community; Plate 82, K) formed extensive beds of closely packed individuals. These occurred together with several other bivalves, including the venerids *Mercenaria campechiensis* subspecies and *Chione* species (*C. elevata* complex), the lucinids *Lucinisca caloosana* (Dall, 1903), *Armimiltha disciformis* (Heilprin, 1886), *Lucina pensylvanica* subspecies, and *Dallucina amabilis* (Dall, 1898), the glycymerid *Glycymeris floridana* Olsson and Harbison, 1953, and the arcid *Cunearca megerata* Olsson and Harbison, 1953. Feeding on the dense beds of bivalves were several large molluscivorous gastropods, including the muricids *Phyllonotus labelleensis* Petuch, 1994 (Plate 82, F) and *Chicoreus* species (*C. calusa* complex), the fasciolariids *Fasciolaria seminole* Petuch, 1994, *F. (Cinctura) lindae* Petuch, 1994 (Plate 82, E), *Terebraspira labelleensis* Petuch, 1994, and *Heilprinia evergladesensis* Petuch, 1994, and the busyconids *Sinistrofulgur labelleensis* Petuch, 1994, *Busycoarctum rapum* subspecies, and *Fulguropsis elongatus* (Gill, 1867. Polychaete worms were abundant on these bivalve beds and sand flats and were the prey items for several common vermivorous gastropods, including the conids *Dauciconus gravesae* (Petuch, 1994) (Plate 82, J), *Gradiconus ronaldsmithi* (Petuch, 1994), *Calusaconus spuroides* subspecies, and *Contraconus* species (*C. osceolai* complex). A large fauna of suctorial-feeding cancellariids was also present in the *Anomalocardia* beds, some of which included *Cancellaria clewistonensis* Olsson and Harbison, 1953 (Plate 82, C), *Ventrilia helenae* (Olsson and Petit, 1964), and *Trigonostoma caloosahatchiensis* Tucker and Wilson, 1932.

The Siphocypraea brantleyi Community

In the shallow lagoons and embayments (1-10 m) of the Caloosahatchee Subsea during Bee Branch time (mid-Calabrian), beds of Turtle Grass (*Thalassia*) covered large areas of the sea floor and supported a rich and highly endemic molluscan fauna. While the bivalve fauna was essentially the same as in the older, early Calabrian *Siphocypraea problematica* Community, the Bee Branch gastropod fauna was greatly altered and had evolved into a suite of new species. Primary among these were the grass bed cowries, including *Siphocypraea problematica* subspecies, *S. (Okeechobea) brantleyi* Petuch, new species (namesake of the community; Plate 82, A and B; see Systematic Appendix), and *S. (Pahayokea) aspenae* Petuch, new species (Plate 82, G and H; see Systematic Appendix). These three species formed large aggregations and dominated the macrofauna of the Turtle Grass beds. Co-occurring with the cowries were several classic grass bed algivorous gastropods, including the modulid *Modulus* species (*M. basileus* complex), the turbinid *Astraea (Lithopoma)* species, and the cerithiids *Cerithium vicinia* Olsson and Harbison, 1953, *C. preatratum* Olsson and Harbison, and *Cerithioclava caloosaense* (Dall, 1892). These small gastropods and the interstitial bivalves were the principal prey items of the large and abundant melongenids *Echinofulgur palmbeachensis* Petuch, 1994 (Plate 82, I) and *Melongena*

Plate 82. Index fossils and characteristic organisms of the *Anomalocardia caloosana* Community (shallow sand bottom lagoons during the mid-Calabrian) and the *Siphocypraea brantleyi* Community (*Thalassia* beds during the mid-Calabrian) of the Caloosahatchee Subsea. Included are: A,B= *Siphocypraea (Okeechobea) brantleyi* Petuch, new species, holotype, length 56 mm, see Systematic Appendix; C= *Cancellaria clewistonensis* Olsson and Harbison, 1953, length 51 mm; D= *Macrostrombus brachior* Petuch, 1994, length 138 mm; E= *Fasciolaria (Cinctura) lindae* Petuch, 1994, length 42 mm; F= *Phyllonotus labelleensis* Petuch, 1994, length 86 mm; G,H= *Siphocypraea (Pahayokea) aspenae* Petuch, new species, holotype, length 59 mm, see Systematic Appendix; I= *Echinofulgur palmbeachensis* Petuch, 199477 mm; J= *Dauciconus gravesae* (Petuch, 1994), length 29 mm (with small *Coenangia* coral attached to spire); K= *Anomalocardia caloosana* (Dall, 1900), length 32 mm. All specimens were collected in the Bee Branch Member of the Caloosahatchee Formation, in the Brantley Pit, south of Arcadia, DeSoto County, Florida.

caloosahatcheensis subspecies. The two melongenids also occurred on the adjacent sand flats and were the most predominant and conspicuous molluscivorous gastropods of Bee Branch time. On open sand patches adjacent to the sea grass beds, the strombid *Macrostombus brachior* (Petuch, 1994) (Plate 82, D) grazed on algae and formed large aggregations. As in all Okeechobean Turtle Grass beds, turbinellid gastropods were the principal predators of interstitial enteropneust and polychaete worms. During the mid-Calabrian these included *Turbinella wendyae* Petuch, 1994 and *Hystrivasum horridum* subspecies. Both the *Anomalocardia caloosana* Community and the *Siphocypraea brantleyi* Community were especially well developed within the Brantley Embayment.

The Oliva jenniferae Community

At the end of Bee Branch time, a second brief but severe climatic degeneration occurred. The resultant sea level drop and prolonged cooler climate caused the extinction of many taxa and forced the evolution of a new molluscan fauna. By the beginning of Ayer's Landing time (late Calabrian), these last remnants of the Buckinghamian Subprovince (Caloosahatchian Province) survived only within the confines of the Caloosahatchee Subsea as a relict fauna. Since this last high sea level stand was the lowest of the three Calabrian eustatic highs, much of the Okeechobean Sea was emergent, with only the eastern side (Loxahatchee Trough and Miccosukee Island areas) containing deeper, lagoonal environments. For this reason, the areal extent of the Ayer's Landing Member is much smaller than that of the older Caloosahatchee members. At this time, the most widespread depositional environment was the shallow (1-10 m) sand bottom lagoonal environment and this supported a low diversity but highly endemic molluscan fauna. Here, several bivalves occurred in abundance, including the venerid *Anomalocardia concinna* Olsson and Harbison, 1953, the corbulids *Bothrocorbula willcoxi* (Dall, 1898) and *Varicorbula caloosae* (Dall, 1898), the mactrid *Spisula solidissima* subspecies, and the arcid *Cunearca scalarina* (Heilprin, 1886). These occurred together with the strombid *Strombus keatonorum* subspecies, which lived in large aggregations that grazed on surficial algal films. The bivalves and strombids were the principal prey items of molluscivorous gastropods such as the busyconids *Busycon duerri* Petuch, 1994 (Plate 83, B), *Pyruella eismonti* Petuch, 1991, and *P. ovoidea* Petuch, 1994, the muricid *Phyllonotus martinshugari* Petuch, 1994, the melongenids *Melongena chickee* Petuch, 1994, *M. acutangulata* Petuch, 1994, and *Echinofulgur griffini* Petuch, 1994, the naticid *Neverita* species (*N. fossata* complex), and the fasciolariid *Liochlamys griffini* Petuch, 1994. General carnivore/scavenger gastropods also occurred with the molluscivorous species, and included the pleioptygmatid *Pleioptygma lineolata* subspecies, the scaphelline volutids *Scaphella oleiniki* Petuch, 1994 (Plate 83, F), *S. griffini* Petuch, 1994, and *S. tomscotti* Petuch, 1994, and the olivids *Oliva (Strephona) jenniferae* Petuch, 1994 (namesake of the community; Plate 83, K), *O. (Strephona) briani* Petuch, 1994, and *O. (Strephona) jeremyi* Petuch, 1994, and the suctorial-feeding cancellariid *Cancellaria calusa* Petuch, 1994. These olive shells occurred in large swarms, with *O. jenniferae* often being so abundant that it dominated the sea floor macrofauna. A large polychaete biomass was also present in these shallow lagoonal environments and this supported several abundant vermivorous conids, including *Contraconus scotti* Petuch, 1994 (Plate 83, A), *C. heilprini* Petuch, 1994 (Plate 83, H), *C. mitchellorum* Petuch, 1994 (Plate 83, I), and *Ximeniconus waccamawensis* (B. Smith, 1930). In the shallowest areas adjacent to intertidal sand and mud flats, the potamidid gastropod *Pyrazisinus intermedius* Petuch, 1994 formed small beds but never occurred in the abundance seen in previous times.

Plate 83. Index fossils and characteristic organisms of the *Oliva jenniferae* Community (shallow sand bottom lagoons during the late Calabrian) and the *Siphocypraea griffini* Community (*Thalassia* beds during the late Calabrian) of the Caloosahatchee Subsea. Included are: A= *Contraconus scotti* Petuch, 1994, length 59 mm; B= *Busycon duerri* Petuch, 1994, length 174 mm; C,D= *Siphocypraea griffini* Petuch, 1991, length 39 mm; E= *Turbinella lindae* Petuch, 1994, length 178 mm; F= *Scaphella oleiniki* Petuch, 1994, length 63 mm; G= *Hystrivasum griffini* Petuch, 1994, length 66 mm; H= *Contraconus heilprini* Petuch, 1994, length 32 mm; I= *Contraconus mitchellorum* Petuch, 1994, length 112 mm; J= *Echinofulgur griffini* Petuch, 1994, length 105 mm; K= *Oliva (Strephona) jenniferae* Petuch, 1994, length 48 mm. All specimens were collected in the Ayer's Landing Member of the Caloosahatchee Formation at two locations; the Griffin Brothers Pit, Holey Land Wildlife Conservation Area, southwestern Palm Beach County, Florida, and in the Palm Beach Aggregates quarry, Loxahatchee, Palm Beach County, Florida.

The Siphocypraea griffini Community

During Ayer's Landing time, the Okeechobean Turtle Grass beds were less widespread than in the early and middle Calabrian and were only well developed along Miccosukee Island and the northern coast of the Palm Beach Archipelago. Here, the last species radiation of the endemic grass bed cowries occurred, with large aggregations of *Siphocypraea griffini* Petuch, 1991 (namesake of the community; Plate 83, C and D), *S.* species, and *S. (Pahayokea)* species dominating the ecosystem. A small fauna of interstitial bivalves also occurred on these last Caloosahatchee sea grass beds and included the lucinid *Miltha caloosaensis* subspecies, the arcid *Caloosarca crassicosta* subspecies, and the cardiids *Trachycardium muricatum* subspecies and *Americardia columba* (Heilprin, 1886), and small tellinids. Several algivorous gastropods occurred with these bivalves, some of which included the cerithiids *Cerithium griffini* Petuch, 1994 and *Cerithioclava weeksae* Petuch, 1994, and the turbinids *Turbo (Marmarostoma) tiara* M. Smith, 1936 and *T. (Taenioturbo)* species. Only a few molluscivorous gastropod predators were present on these grass beds, and these included the fasciolariids *Fasciolaria (Cinctura) apicina* subspecies and *Triplofusus acmaensis* (B. Smith, 1940) and the busyconid *Fulguropsis griffini* Petuch, 1994. The enteropneust and polychaete fauna of the *Thalassia* beds also supported the vermivorous turbinellid gastropods *Turbinella lindae* Petuch, 1994 (Plate 83, E) and *Hystrivasum griffini* Petuch, 1994 (Plate 83, G).

Chapter 9. Early and Late Pleistocene Seas

At the end of Calabrian time, a severe and rapid climatic degeneration took place, causing sea level to drop to its lowest in over three million years. This early Pleistocene low sea level stand was even below that which occurred during the cold Messinian-Zanclean boundary (Krantz, 1991), reflecting the extreme severity of this worldwide cooling event. In eastern North America at this time, all the paleosea basins were emergent, leading to both accelerated infilling by terriginous sediments and the extinction of their endemic index species and communities (see Chapter 10). This late Calabrian cooling event was the precursor to the Nebraskan Glacial Stage, which produced over three hundred thousand years of cooler marine climates and emergent coastlines. At the end of Nebraskan time, climates again began to warm and sea levels rose, initiating the Aftonian Interglacial Stage. By this time, the Albemarle and Charleston Sea basins were completely filled and only a single narrow coastal barrier island-lagoon system extended from present-day New York City to the St. Lucie Inlet of southern Florida. This expanded descendant of the older Nashua Lagoon System, the Socastee Lagoon System (named for the late Pleistocene Socastee Formation of the Carolinas), contained an impoverished marine fauna and had only limited numbers of environments. Because of its deeper basinal structure and because of the slow rate of sedimentary infilling, the Okeechobean Sea was the only original paleosea to remain intact after Nebraskan time. Although the warmer Okeechobean area acted as a refugium for eutropical groups, many classic Caloosahatchian Province molluscan index taxa became extinct by the beginning of the Nebraskan Stage. Some of the more important and obvious of these included bivalves such as the arcids *Cunearca* and *Arcoptera*, and gastropods such as the conid *Contraconus*, the cypraeid *Siphocypraea* (and all its subgenera), the turbinellid *Hystrivasum*, the fasciolariids *Terebraspira* and *Liochlamys*, the muricids *Trossulasalpinx* and *Pterorhytis* (survives as one rare Ecuadorian species), and the melongenid *Echinofulgur*.

In response to major glacial-interglacial pulses during the Pleistocene, the Okeechobean Sea flooded three separate times, with two intervening emergent times. These flooding events took place during the Aftonian Interglacial Stage (the Loxahatchee Subsea, regressing during the Kansan Glacial Stage), the Yarmouthian Interglacial Stage (the Belle Glade Subsea, regressing during the Illinoian Glacial Stage), and the Sangamonian Interglacial Stage (the Lake Worth Subsea, regressing during the Wisconsinan Glacial Stage). Being nearly infilled by the end of Wisconsinan time, the Lake Worth Subsea never reflooded during the Holocene sea level rise. This low, filled basin, instead, became the Recent Everglades. Farther north during Wisconsinan time, in the St. Lawrence River, Vermont, and upper New York State areas, massive continental ice sheets had isostatically depressed the entire region to such low levels that the Gulf of St. Lawrence waters actually extended inland all the way to eastern Lake Ontario. This last paleosea, the Champlain Sea, existed only until the Pleistocene-Holocene boundary, when the lessened weight of the melting continental glaciers caused the area to isostatically rebound and drain the sea water back into the Gulf of St. Lawrence. This was the only eastern North American paleosea to completely contain a high-Arctic marine fauna.

Communities and Environments of the Loxahatchee Subsea

During Nebraskan time, the Okeechobean Sea basin, although emergent, was filled by Lake Immokalee, a giant fresh water paleolake roughly four times larger than present-day Lake Okeechobee. The autochthonous carbonate lake sediments (mostly charaphyte calcarenites), together with siliciclastics brought in by the proto-Kissimmee River to the north, filled most of the deep areas of the paleosea basin. When sea levels rose again during Aftonian time, obliterating Lake Immokalee, the Okeechobean Sea basin was much reduced in size and the pseudoatoll island systems had expanded to enclose the central areas. The resultant, land-locked lagoon, the Loxahatchee Subsea (Figure 31), was structurally complex and contained the largest array of tropical environments found anywhere in Aftonian eastern North America. By this time, the Immokalee, Miami, and Palm Beach Archipelagos had fused into a single, low, sandy, wide landmass, separated only by narrow tidal channels. The sheltered western edge of this fused island chain, along the deep Loxahatchee Trough, was fringed with a narrow, but complex, reef system (the Capeletti Reef Tract). Likewise, the central Miccosukee Island had expanded to form a low, sandy landmass and was fringed by a large coral reef system along its eastern side (the Miccosukee Reef Tract). Two new major geomorpholgical features also appeared at this time, including the Tomeu Islands, a large and complex chain of mangrove islands at the northern end of the Palm Beach Archipelago, and the Monroe Reef Tract, a series of coral reef systems along the outer edge of the Miami Archipelago and the precursor of the Recent Florida Keys.

All the neritic environments of the Loxahatchee Subsea are preserved in the richly-fossiliferous beds of the lower part of the Bermont Formation (DuBar, 1974) of southern Florida. From these units, the indurated Holey Land Member (Petuch, 1990) and the unconsolidated equivalent Loxahatchee units (Petuch, 1994; 1997), four distinct tropical marine communities can be recognized. These include the *Pyrazisinus roseae* Community (mangrove forest and intertidal mud flats), the *Titanostrombus williamsi* Community (Thalassia beds), the *Carolinapecten jamieae* Community (shallow sand bottom lagoon pectinid beds), and the *Arcohelia limonensis* Community (coral reefs and coral bioherms). The Loxahatchee Subsea is particularly important in that it housed a number of Caloosahatchee Subsea relictual genera that had survived the Nebraskan cold time. These relicts, which had speciated into new Loxahatchee taxa, included bivalves such as the pectinid *Carolinapecten* and the lucinid *Miltha*, and gastropods such as the busyconid *Pyruella*, the ovulid *Jenneria*, the potamidid *Pyrazisinus*, the olivid *Oliva (Porphyria)*, the conid *Ximeniconus*, and the triviid *Pusula*. Several new endemic molluscan groups also appeared during the Aftonian, probably in response to the ecological pressures and stresses during Nebraskan time. Found no where else but inside the warm-water, protected Loxahatchee Subsea were gastropods such as the elongated melongenid *Melongena (Miccosukea)*, the spectacular giant olivid *Lindoliva*, and the giant strombid *Titanostrombus* (extant in the Recent only in the Panamic Province and in Brazil). These Loxahatchee genera co-occurred with large endemic species radiations of the gastropod genera *Strombus* and *Macrostrombus* (Strombidae), *Melongena* (Melongenidae), and *Oliva (Strephona)* (Olividae).

The Pyrazisinus roseae Community

During Aftonian time, the northern end of the Palm Beach Archipelago had elongated northwestward to form a long, curved arc of small mangrove islands. This islet chain, the Tomeu Islands (Figure 31), was composed of Red Mangrove (*Rhizophora*) (Plate 84, H) and Black Mangrove (*Avicennia*) (Plate 84, J) forests, and closely resembled the extensive man-

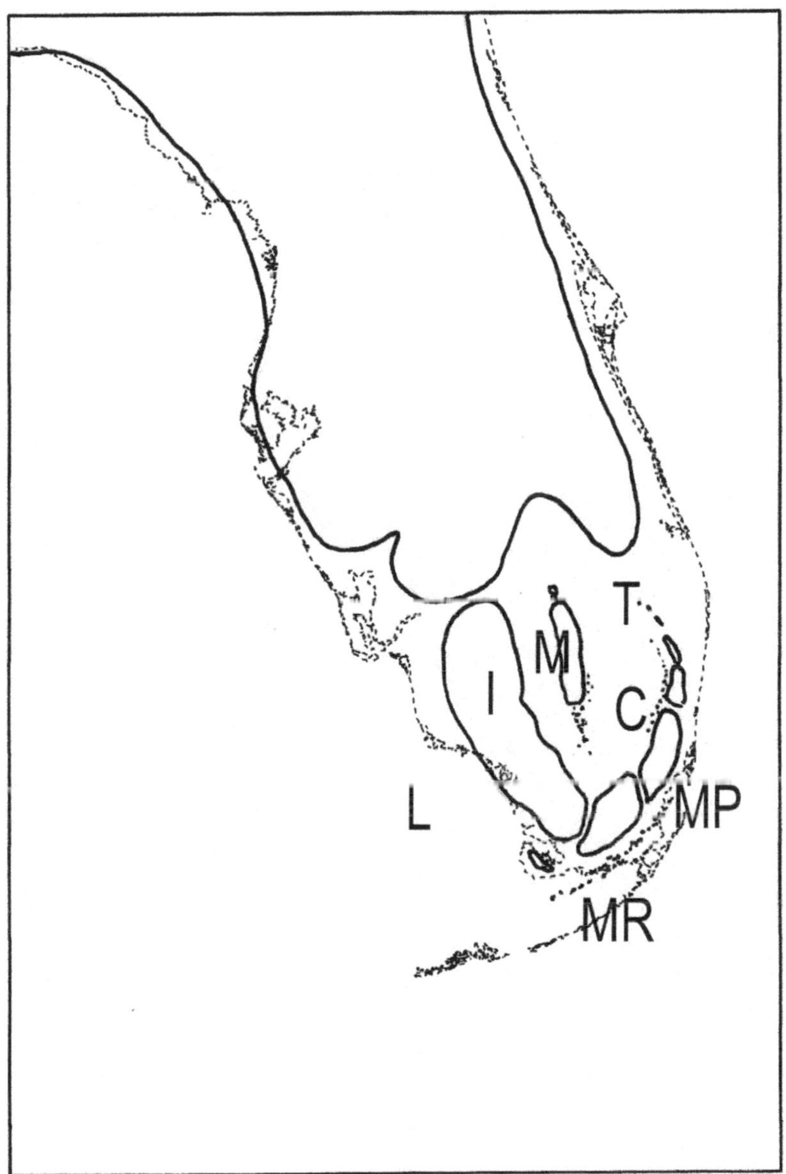

Figure 31. Approximate configuration of the Florida coastline during Aftonian Pleistocene time, superimposed upon the outline of the Recent Floridian Peninsula. Prominent features included: L= Loxahatchee Subsea, I= Immokalee Island, M= Miccosukee Island, T= Tomeu Islands, C= Capeletti Reef Tract, MP= Miami and Palm Beach Archipelagos, MR= Monroe Reef Tract.

grove jungles and islands of Recent Florida Bay. Similar mangrove forests extended along the inner coast of Immokalee Island, but these were not as well developed or as extensive as the Tomeu mangrove environments. In all these areas, the roots of the Red Mangroves were heavily encrusted with massive growths of the mangrove oyster, *Crassostrea rhizophorae* (Guilding, 1828) (Plate 84, A) and this species was the principal prey item of the drilling thaidid gastropod *Thais (Stramonita)* species (*T. haemostoma* complex). As in Recent western Atlantic mangrove forests, littorinid gastropods and mussels were abundant in the Loxahatchee forests and lived in close association with the roots and leaves of the mangrove trees; in this case with the mangrove periwinkle *Littorinopsis* species (*L. angulifera* complex) (Plate 84, C) living on the Red Mangrove leaves and branches, and the mytilid mussel *Brachidontes venustus* Olsson and Harbison, 1953 (Plate 84, B) living in clumps attached to the erect breather roots of the Black Mangrove. The high intertidal potamidid gastropod *Cerithidea duerri* Petuch, 1994 (Plate 84, G) also lived on both the trunk and breather boots of the Black Mangroves. As seen in all the older Okeechobean subseas, extensive organic-rich intertidal mud flats interfingered with the mangrove forests. Here, the potamidid gastropods *Pyrazisinus roseae* Petuch, 1991 (namesake of the community; Plate 84, D), *P. palmbeachensis* Petuch, 1994 (Plate 84, E), and *P. turriculus* Petuch, 1994 (Plate 84, F) formed huge aggregations that literally carpeted the mud flats and fed on surficial algal and bacterial films. These were the principal prey items of the long-siphoned mangrove-dwelling melongenid *Melongena (Miccosukea) cynthiae* Petuch, 1990 (Plate 84, I). Also on the mud flats adjacent to the mangrove forests, the ornate, elongated oyster *Crassostrea* species (*C. labellensis* complex) formed large biohermal clumps. These oyster bars supported an interesting and still unstudied drilling ocenebrine muricid fauna, some of which included *Urosalpinx* species (elongated, *U. cinereus* complex), *Eupleura* species (*E. tampaensis* complex) and two other undescribed *Eupleura* species (one in the *E. leonensis* complex and one in the *E. sulcidentata* complex), and *Vokesinotus* cf. *griffini* Petuch, 1991. Large masses of colonial tunicates also occurred on these oyster bars, and these supported a large fauna of urochordate-feeding triviid gastropods, some of which included *Pusula dadeensis* Petuch, 1994 (originally thought to have come from the Tamiami Formation, but now known to be from the Bermont Formation), *Trivia* species (*T. quadripunctata* complex), *T.* species (*T. maltbiana* complex), and *Niveria* species (*N. pediculus* complex).

The *Titanostrombus williamsi* Community

Of the shallow water (1-5 m) environments of the Loxahatchee Subsea, the Turtle Grass (*Thalassia*) beds were the most widespread and frequently encountered. The greatest sea grass bed development was seen in the wide, shallow areas behind the Tomeu Islands and along the western coast of Miccosukee Island. Here, several large, algivorous grass bed-dwelling strombid gastropods occurred in large, prominent aggregations, and included the giant *Titanostrombus williamsi* (Olsson and Petit, 1964) (namesake of the community; Plate 85, A), *Eustrombus gigas pahayokee* (Petuch, 1994) (Plate 85, I), and the high-spired *Macrostrombus scotti* Petuch, 1994 (Plate 85, K). Living on the individual sea grass blades were a number of small algivorous gastropods, some of which included the modulid *Modulus bermontianus* Petuch, 1994 (Plate 85, G), the turbinids *Turbo (Marmarostoma)* species (*T. castanea* complex), *Astraea (Lithopoma) lindae* Petuch, 1994, and *Astraea (Astralium) phoebia* subspecies, the cerithiids *Cerithium floridanum* Moerch, 1876 and *C. algicola* C. Adams, 1845, the sea grass acmaeid limpet *Acmaea pustulata* subspecies, and the trochid *Tegula fasciata* (Born, 1778). The large relictual cerithiid *Cerithioclava garciai* Houbrick, 1985 also lived with its smaller relatives on these grass beds and on adjacent sand patches, feeding on epiphytic and dasycladacean algae. Occurring with these algivores on and among the grass

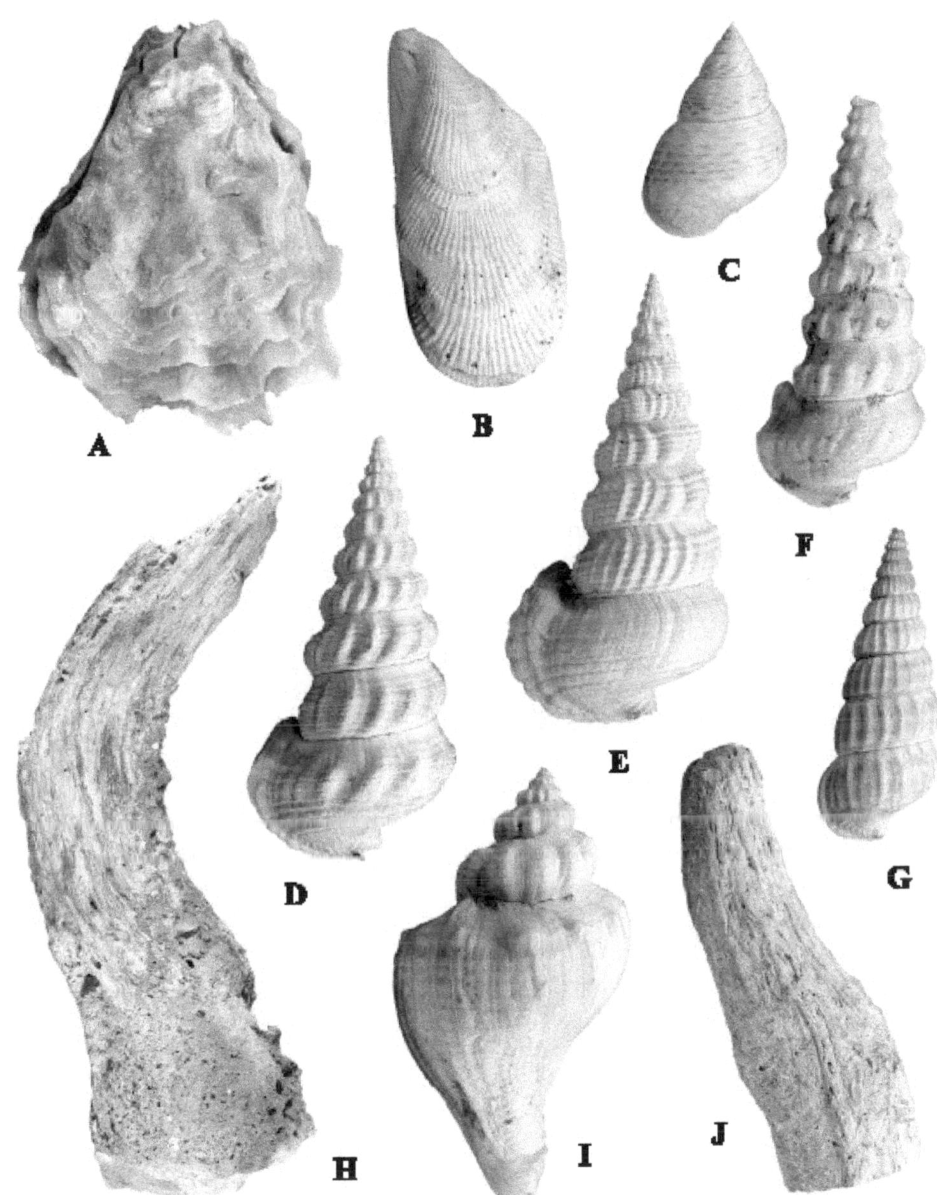

Plate 84. Index fossils and characteristic organisms of the *Pyrazisinus roseae* Community (mangrove forests and intertidal mud flats during Aftonian time) of the Loxahatchee Subsea. Included are: A= *Crassostrea rhizophorae* (Guilding, 1828), length 63 mm; B= *Brachidontes venustus* Olsson and Harbison, 1953, length 32 mm; C= *Littorinopsis* species, length 17 mm; D= *Pyrazisinus roseae* Petuch, 1991, 58 mm; E= *Pyrazisinus palmbeachensis* Petuch, 1994, length 62 mm; F= *Pyrazisinus turriculus* Petuch, 1994, length 58 mm; G= *Cerithidea duerri* Petuch. 1994, length 26 mm; H= Red Mangrove (*Rhizophora*) petrified wood, length 320 mm; I= *Melongena (Miccosukea) cynthiae* Petuch, 1990, length 61 mm; J= Black Mangrove (*Avicennia*) petrified breather root, length 82 mm. All specimens were collected in the "Loxahatchee Member" (unconsolidated Holey Land Member equivalent) of the Bermont Formation, in the Palm Beach Aggregates, Inc. quarry, Loxahatchee, Palm Beach County, Florida.

blades were several species of hydroid and sponge-eating trochids, including *Calliostoma lindae* Petuch, 1994, *C. tampaense* subspecies, and *C.* species (*C. adelae* complex). The solitary faviid corals *Manicina areolata* (Linnaeus, 1786) (Plate 85, J) and *M. gyrosa* (Ellis and Solander, 1786) and the irregular echinoid *Clypeaster rosaceus* subspecies (Plate 85, D) were abundant in the Loxahatchee Turtle Grass beds and often formed large swarms near open sandy areas and on sand patches between the grass beds. Also found on these interspersed open sand patches were the large echinoid-feeding cassid gastropods *Cassis schnireli* Petuch, 1994 (Plate 85, E and F) and *C. jameshoubricki* Petuch, new species (Plate 85, B and C; see Systematic Appendix). These peripheral open sand patches also housed a large polychaete fauna and this, in turn, supported a number of vermivorous gastropods, including the conids *Calusaconus tomeui* Petuch, new species (Plate 85, H; see Systematic Appendix), *Jaspidiconus hyshugari* (Petuch, 1994), and *Ximeniconus palmbeachensis* subspecies (typically with a pustulated shell sculpture; *X. perplexus* complex), and the terebrid *Strioterebrum* species (*S. dislocata* complex). Bivalves such as the lucinids *Miltha carmenae* H. Vokes, 1969, *Lucina pectinata* (Gmelin, 1791), *Codakia orbicularis* (Linnaeus, 1758), and *Ctena orbiculata* (Montagu, 1808), the venerid *Chione elevata* (Say, 1822), and the arcid *Caloosarca catasarca* (Dall, 1898) were especially common on these grass beds, where they lived buried interstitially in the sea grass root mat. These, in turn, were the main food resource for several molluscivorous gastropods, including the muricids *Chicoreus dilectus* (A. Adams, 1855), *Phyllonotus* species (*P. labelleensis* complex), *P.* species (*P. martinshugari* complex), *Vokesimurex diegelae* Petuch, 1994, *Vokesinotus griffini* Petuch, 1991, *Eupleura longior* Petuch, 1994, *E.* species (*E. tampaensis* complex), and *Calotrophon ostrearum* (Conrad, 1846), the fasciolariids *Fasciolaria okeechobeensis* Tucker and Wilson, 1932, *F. (Cinctura) holeylandica* Petuch, 1994, and *Triplofusus giganteus* (Kiener, 1840), and the busyconid *Sinistrofulgur holeylandicum* Petuch, 1994.

The Carolinapecten jamieae Community

In deeper areas (5-20 m) of the Loxahatchee Trough during Aftonian time, open sand bottom environments dominated most of the sea floor. Here, two species of pectinid bivalves formed large, scattered beds, including *Carolinapecten jamieae* Petuch, new species (namesake of the community; Plate 86, C; see Systematic Appendix; the last of its genus) and *Argopecten irradians* subspecies. Living with the scallops was a large fauna of bivalves, some of which included the venerids *Mercenaria campechiensis* subspecies, *Puberella intapurpurea* (Conrad, 1849, *Macrocallista nimbosa* (Lightfoot, 1786), *M. maculata* (Linnaeus, 1758), and *Lirophora latilirata* (Conrad, 1841), the chamid *Arcinella cornuta* Conrad, 1866, the tellinids *Tellina magna* Spengler, 1798, *T. alternata* Say, 1822, and *Tellidora cristata* (Recluz, 1842), the cardiids *Dinocardium robustum* subspecies and *Trachycardium muricatum* subspecies, and numerous small mactrids and corbulids. The scallops and infaunal bivalves were the principal prey items for several molluscivorous gastropods, some of which included the busyconids *Fulguropsis feldmanni* Petuch, 1991 and *Pyruella tomeui* Petuch, new species (Plate 86, G; see Systematic Appendix), the fasciolariid *Fasciolaria (Cinctura) capelettii* Petuch, 1994, and the muricids *Murexiella graceae* (MacGinty, 1940) and *Favartia lindae* Petuch, 1987. Large general carnivores such as the scaphelline volutids *Scaphella capelettii* Petuch, 1994 and *S. seminole* Petuch, 1990 and the giant spotted olivids *Lindoliva spengleri* Petuch, 1988 (Plate 86, F) and *L. griffini* Petuch, 1988 were also probably feeding on the smaller bivalves. The open sand areas between the scallop beds must have supported a rich polychaete worm fauna, as attested to by the abundant presence of several species of vermivorous gastropods, including the conids *Gradiconus capelettii* (Petuch, 1990) (Plate 86, D), *G. loxahatcheensis* (Petuch, 1994) (Plate 86, H), *G. delessertii* subspecies, *Spuriconus spengleri*

Plate 85. Index fossils and characteristic organisms of the *Titanostromus williamsi* Community (*Thalassia* beds during Aftonian time) of the Loxahatchee Subsea. Included are: A= *Titanostrombus williamsi* (Olsson and Petit, 1964), length 265 mm; B,C= *Cassis jameshoubricki* Petuch, new species, holotype, length 183 mm (see Systematic Appendix); D= *Clypeaster roseaceus* subspecies, length 91 mm; E,F= *Cassis schnireli* Petuch, 1994, length 221 mm; G= *Modulus bermontianus* Petuch, 1994, length 14 mm; H= *Calusaconus tomeui* Petuch, new species, holotype, length 34 mm (see Systematic Appendix); I= *Eustrombus gigas pahayokee* Petuch, 1994, length 179 mm; J= *Manicina areolata* (Linnaeus, 1786), length 131 mm; K= *Macrostrombus scotti* (Petuch, 1994), length 183 mm. All specimens were collected in the "Loxahatchee Member" (unconsolidated Holey Land Member equivalent) of the Bermont Formation, in the Palm Beach Aggregates, Inc. quarry, Loxahatchee, Palm Beach County, Florida.

Plate 86. Index fossils and characteristic organisms of the *Carolinapecten jamieae* Community (shallow sand bottom lagoon pectinid beds during Aftonian time) of the Loxahatchee Subsea. Included are: A= *Macrostrombus mayacensis* (Tucker and Wilson, 1933), length 148 mm; B= *Oliva (Porphyria) gravesae* Petuch, 1994, length 57 mm; C= *Carolinapecten jamieae* Petuch, new species, holotype, length 94 mm (see Systematic Appendix); D= *Gradiconus capelettii* (Petuch, 1990), length 42 mm; E= *Spuriconus spengleri* (Petuch, 1991), length 63 mm; F= *Lindoliva spengleri* Petuch, 1988, length 110 mm; G= *Pyruella tomeui* Petuch, new species, holotype length 89 mm (see Systematic Appendix); H= *Gradiconus loxahatcheensis* (Petuch, 1994), length 51 mm; I= *Strombus erici* Petuch, 1994, length 95 mm; J= *Rhyncholampas evergladesensis* Mansfield, 1932, length 63 mm. The specimens were collected in either the "Loxahatchee Member" (unconsolidated Holey Land Member equivalent) of the Bermont Formation, in the Palm Beach Aggregates, Inc. quarry, Loxahatchee, Palm Beach County, Florida (Figures A, B, C, G, H, and I), or in the Holey Land Member of the Bermont Formation, in the Griffin Brothers quarry, Holey Land Wildlife Management Area, Palm Beach County, Florida (Figures D, E, and J), and in the Capeletti Brothers pit #11, Miami, Dade County, Florida (Figure F).

(Petuch, 1991) (Plate 86, E), *S.* species (*S. lemoni* complex), *Dauciconus amphiurgus* (Dall, 1889), and *Jaspidiconus maureenae* (Petuch, 1994). Other carnivorous gastropods included the suctorial-feeding cancellariids *Cancellaria* species and *Ventrilia lindae* Petuch, 1994 and smaller olivid gastropods such as *Oliva (Porphyria) gravesae* Petuch, 1994 (Plate 86, B), *O. (Strephona) adami* Petuch, 1994, *O. (Strephona) ryani* Petuch, 1994, and *O. (Strephona) wendyae* Petuch, 1994. These olivids were particularly abundant and prominent on these sand patches and formed large aggregations. In the shallowest areas closest to the Turtle Grass beds, the strombids *Strombus erici* Petuch, 1994 (Plate 86, I), *S. capelettii* Petuch, 1994, *Macrostrombus mayacensis* (Tucker and Wilson, 1933) (Plate 86, A), and *M. diegelae* (Petuch, 1991) also formed large colonies and lived together with the irregular echinoid *Rhyncholampas evergladesensis* Mansfield, 1932 (Plate 86, J). This small echinoderm was the main prey item of the high-spired cassid *Phalium loxahatcheensis* Petuch, 1994.

The Arcohelia limonensis Community

During Aftonian time, the Capeletti Reef Tract and the Miccosukee Island reefs contained large numbers of coral biohermal complexes, some of which grew in a linear fashion paralleling the inner coast of the island systems. Forming in quiet, protected waters, these coralline structures were unzonated and resembled the intralagoonal microatolls found on Recent South Pacific atolls. North of Miccosukee Island, some of these microatoll-like bioherms formed large coral knolls that rose steeply from the Loxahatchee sea floor. Unlike the reefs and bioherms of the Everglades Pseudoatoll and the Recent South Pacific, these Loxahatchee coralline structures were low-diversity systems, with only a few species of corals being present. Large coral heads made up the main framework of these bioherms and included massive corals such as the montastreids *Montastrea annularis* (Ellis and Solander, 1786), *Solenastrea hyades* (Dana, 1846) form *hispida* (Plate 87, K), and *S. bournoni* Edwards and Haime, 1849, the siderastreids *Siderastrea siderea* (Ellis and Solander, 1786) and *S. radians* (Pallas, 1766), the mussids *Mussa angulosa* (Pallas, 1766) (Plate 87, B) and *Scolymia lacera* (Pallas, 1766) (Plate 87, E), and the faviids *Diploria clivosa* (Ellis and Solander, 1786) and *D. labyrinthiformis* (Linnaeus, 1758). Growing in open areas among these massive hermatypic corals were large, dense thickets of delicate branching corals, including the poritids *Porites furcata* Lamarck, 1816 and *P. divaricata* Lesueur, 1820, and the ahermatypic oculinids *Arcohelia limonensis* Vaughan, 1919 (namesake of the community; Plate 87, J) and *Oculina diffusa* Lamarck, 1816 (Plate 87, C). Small encrusting colonies of the caryophyllid *Cladocora arbuscula* (Lesueuer, 1820), the astrangiid *Astrangia* species (*A. floridana* complex), and the solitary faviid *Thysanus* species (Plate 87, I; an unnamed Caloosahatchee relict) grew between the massive coral heads and the branching coral thickets. The presence of five undescribed species of octocorallian-feeding ovulid gastropods in the lower beds of the Bermont Formation (three *Cyphoma* species, one *Pseudocyphoma* species, and one *Simnialena* species) indicates that these bioherms also housed a rich and diverse gorgonian fauna, comparable to that of the Everglades Pseudoatoll. The Loxahatchee coral bioherms also supported an extremely rich fauna of specialized carnivorous gastropods, including coral-feeders such as the coralliophilids *Babelomurex scalariformis* (Lamarck, 1822) (Plate 87, H) and *Coralliophila caribaea* Abbott, 1958 and the eocypraeinine ovulids *Jenneria loxahatchiensis* M. Smith, 1936 (Plate 88, G) and *J. hepleri* Olsson, 1967 (Plate 88, H; only in the lowest bed of the Holey Land Member), vermivores such as the conids *Cariboconus griffini* (Petuch, 1990) (see Systematic Appendix) and *Purpuriconus* species and the vasinine turbinellid *Vasum floridanum* McGinty, 1940 (Plate 87, F), molluscivores such as the muricids *Acanthotrophon striatoides* E. Vokes, 1981, *Trachypollia didyma* subspecies, *Muricopsis oxytatus* (M. Smith, 1938), *Murexiella glypta* (M.

Smith, 1938), and *Tripterotyphis triangularis* (A. Adams, 1856) (Plate 87, G) (all feeding on small encrusting bivalves such as chamids and plicatulids), general carnivores such as the columbellid *Microcythara caloosahatcheensis* Petuch, 1991 (originally thought to have come from the Caloosahatchee Formation, but now known to be from the Bermont Formation) and the buccinid *Monostiolum thomasi* Olsson, 1967, colonial tunicate-feeders such as the triviids *Pusula lindajoyceae* Petuch, 1994 (Plate 88, E), *Niveria bermontiana* Petuch, 1994 (Plate 88, F), and *Trivia* species (*T. quadripunctata* complex), hydroid and sponge-feeders such as the cypraeids *Luria voleki* Petuch, new species (Plate 88, A and B; see Systematic Appendix), *Pseudozonaria portelli* (Petuch, 1990) (Plate 88, I and J), *Macrocypraea spengleri* (Petuch, 1990) (Plate 88, C and D), and *M. joannae* Petuch, new species (Plate 88, K and L; see Systematic Appendix), echinoderm-feeders such as the tonnid *Malea petiti* Petuch, 1989 (Plate 87, A) (feeding on cidarid echinoids), and the molluscivorous ranellids *Cymatium (Septa) krebsii* Moerch, 1877 and *C. (Septa) nicobaricum* (Roeding, 1798).

Communities and Environments of the Belle Glade Subsea

At the end of Aftonian time, the marine climate again cooled and sea levels began to drop. This climatic degeneration culminated in the Kansan Glacial Stage, when the entire Floridian Peninsula was emergent for over two hundred thousand years. During this time, many of the Loxahatchee Caloosahatchian relict taxa became extinct, in particular the gastropods *Pyruella, Oliva (Porphyria), Microcythara,* and *Malea* and the pectinid bivalve *Carolinapecten*. Also during this cool, emergent time, the Okeechobean Sea basin was again filled with a giant fresh water paleolake, Lake Okeelanta. Within this immense fresh water "sea," which was at least four times larger than present-day Lake Okeechobee and was probably the largest lake in Kansan North America, several species radiations of fresh water gastropods took place. The most dramatic of these was a radiation of the planordid genus *Seminolina*, which resulted in the evolution of at least thirty species, including elongated, scalariform types and discoidal forms (Petuch, 1997). By the end of Kansan time, climates warmed and sea levels again began to rise, flooding Lake Okeelanta with sea water and destroying the endemic molluscan species radiations. This climatic warming led to the Yarmouthian Interglacial Stage, which dawned on an altered, shallower Belle Glade Subsea (Figure 32) and a newly evolved endemic marine fauna. At this time, the circumbasinal island systems had enlarged and widened, essentially enclosing the Loxahatchee Trough area in a continuous landmass. Besides small transecting tidal channels, the Belle Glade Subsea was connected to the Atlantic Ocean and the Gulf of Mexico only by two narrow straits along the northern part of the basin. A new feature, the Bermont Lagoon System, appeared at this time along the southwestern tip of Florida, outside of the main enclosed Belle Glade basin. Being oceanographically disconnected from the warmer, solar-heated waters of the central Belle Glade basin, the Bermont Lagoon System contained a less-diverse, impoverished version of the rich eutropical faunas of the Loxahatchee Trough area.

From the well-preserved fossil assemblages of the upper beds of the Bermont Formation in the Everglades region (in the Everglades Basin equivalent of the type member), three distinct marine communities can be recognized for the Belle Glade Subsea. These included the *Strombus lindae* Community (shallow carbonate banks during Yarmouthian time), the *Caloosarca aequilitas* Community (*Thalassia* beds during Yarmouthian time), and the *Nodipecten pernodosus* Community (shallow sand bottom lagoons during Yarmouthian time). Of particular interest within the Belle Glade Subsea were species radiations of the gastropod families Olividae, Strombidae, and Melongenidae. The end of Yarmouthian time saw the return of cooler climates and lowered sea levels, in

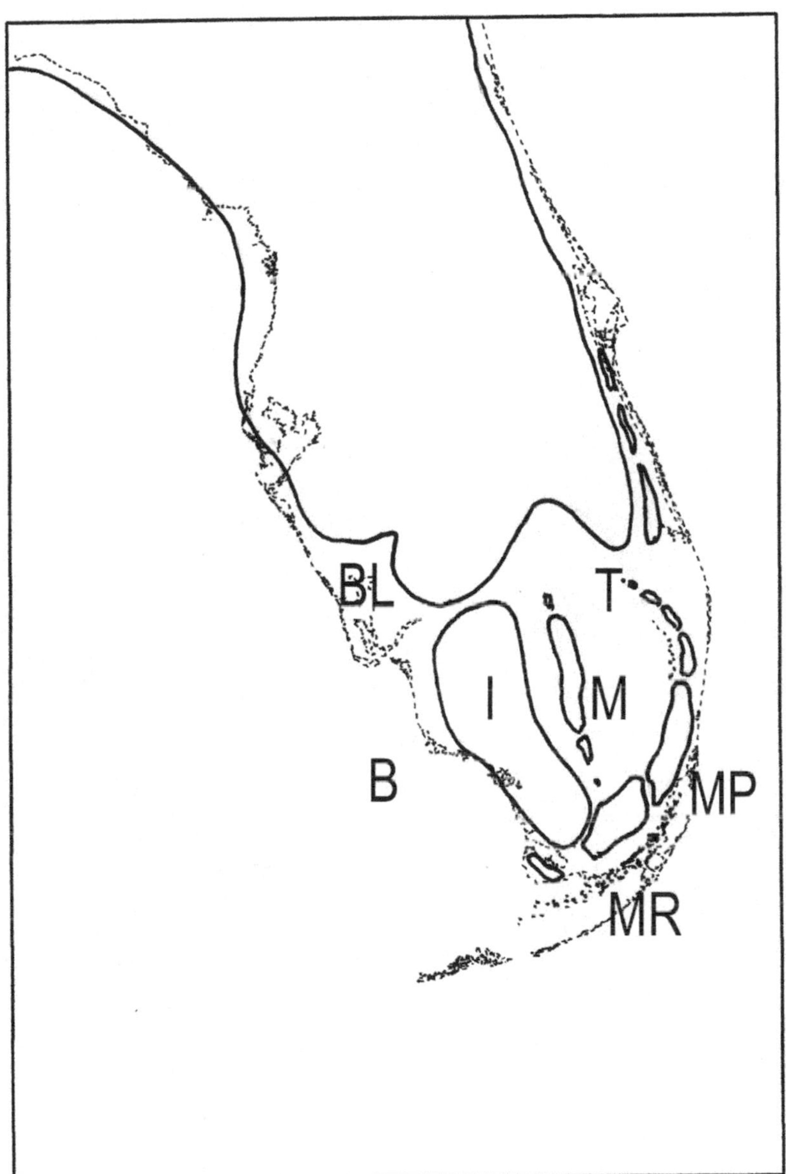

Figure 32. Approximate configuration of the Florida coastline during Yarmouthian Pleistocene time, superimposed upon the outline of the Recent Floridian Peninsula. Prominent features included: B= Belle Glade Subsea, BL= Bermont Lagoon System, I= Immokalee Island, M= Miccosukee Island, T= Tomeu Islands, MP= Miami and Palm Beach Archipelagos, MR= Monroe Reef Tract.

Plate 87. Index fossils and characteristic organisms of the *Arcohelia limonensis* Community (coral reefs and coral bioherms during Aftonian time) of the Loxahatchee Subsea. Included are: A= *Malea petiti* Petuch, 1989, length 122 mm; B= *Mussa angulosa* (Pallas, 1766), length 116 mm; C= *Oculina diffusa* Lamarck, 1816, length 33 mm; D= *Cariboconus griffini* (Petuch, 1990), length 18 mm; E= *Scolymia lacera* (Pallas, 1766), length 72 mm; F= *Vasum floridanum* McGinty, 1940, length 117 mm; G= *Tripterotyphis triangularis* (A. Adams, 1856), length 17 mm; H= *Babelomurex scalariformis* (Lamarck, 1822), length 24 mm; I= *Thysanus* species, length 68 mm; J= *Arcohelia limonensis* Vaughan, 1919, length 41 mm; K= *Solenastrea hyades* (Dana, 1846) form *hispida*, length 131 mm. All specimens were collected in the "Loxahatchee Member" (unconsolidated Holey Land Member equivalent) of the Bermont Formation, in the Palm Beach Aggregates, Inc. quarry, Loxahatchee, Palm Beach County, Florida.

Plate 88. Cypraeoidean gastropods of the *Arcohelia limonensis* Community, Loxahatchee Subsea. A,B= *Luria voleki* Petuch, new species, holotype, length 38 mm (see Systematic Appendix); C,D= *Macrocypraea spengleri* (Petuch, 1990, length 121 mm; E= *Pusula lindajoyceae* Petuch, 1994, length 22 mm; F= *Niveria bermontiana* Petuch, 1994, length 21 mm; G= *Jenneria loxahatchiensis* M. Smith, 1936, length 29 mm; H= *Jenneria hepleri* Olsson, 1967, length 22 mm; I,J= *Pseudozonaria portelli* (Petuch, 1990), length 24 mm; K,L= *Macrocypraea joanneae* Petuch, new species, holotype, length 54 mm (see Systematic Appendix). The specimens were collected in the "Loxahatchee Member" (unconsolidated Holey Land Member equivalent) of the Bermont Formation, in either the Palm Beach Aggregates, Inc. quarry, Loxahatchee, Palm Beach County, Florida (Figures A, B, C, D, G, and H), the Capeletti Brothers pit #11, Miami, Dade County, Florida (Figures E and F), or in the Mecca quarry, North Palm Beach, Palm Beach County, Florida (Figures I, J, K, and L).

this case the most severe of the entire Pleistocene (Petuch, 1997). During this extreme cold time, the Illinoian Glacial Stage, most of the Belle Glade endemic species became extinct and the post-Yarmouthian faunas were noticeably impoverished. During the Illinoian emergent time, some elements of the Belle Glade communities managed to survive by moving into refugia in offshore areas, along the Monroe Reef Tract, or into the Caribbean Basin (see Chapter 10). Several of these are extant in the Recent.

The Strombus lindae Community

During Yarmouthian time, the northern and eastern coasts of Miccosukee Island (Figure 32) contained extensive, shallow (0-2 m) carbonate sand banks and housed a distinctive, geographically restricted, and highly endemic molluscan fauna. Here, as in the Recent Exuma Sound area of the Bahamas, which this area must have resembled, large thickets of *Porites* corals (primarily *Porites furcata*) interfingered with patches of open sand and dense growths of dasycladacean calcareous algae. Feeding on these algal forests and surficial algal films were large aggregations of strombid gastropods, including the small *Strombus lindae* Petuch, 1991 (namesake of the community; Plate 89, H) and *Macrostrombus holeylandicus* (Petuch, 1994) (Plate 89, C), the high-spired turbinid *Turbo (Marmarostoma) duerri* Petuch, 1994, and the dwarf potamidid *Pyrazisinus miamiensis* Petuch, 1994. The venerid bivalve *Chione elevata* subspecies was especially abundant on these shallow flats and formed dense, closely-packed beds. These, several small tellinid bivalves, and the small strombids, were the principal food resources for several molluscivorous gastropods, including the busyconids *Sinistrofulgur roseae* Petuch, 1991 (Plate 89, A), and *Fulguropsis evergladesensis* Petuch, 1994 (Plate 89, I), the melongenid *Melongena (Rexmela) diegelae* Petuch, 1994 (Plate, 89, F), the muricids *Chicoreus duerri* Petuch, 1994, *C. gravesae* Petuch, 1994, and *Vokesinotus griffini* Petuch, 1991, the naticid *Naticarius canrena* subspecies, and the fasciolariid *Fasciolaria (Cinctura) evergladesensis* Petuch, 1991 (Plate 89, K). The open sand patches, as seen in similar biotopes in the Recent Philippines, supported a large and prominent fauna of general carnivore/scavenger olivid gastropods. Some of these endemic Miccosukee Island olive shells included *Oliva (Strephona) lindae* Petuch, 1991 (Plate 89, J), *O. (Strephona) cokyae* Petuch, 1991 (Plate 89, D), *O. (Strephona) edwardsae* subspecies, *O. (Strephona) southbayensis* Petuch, 1994, and *Lindoliva diegelae* Petuch, 1988. Living with the olivids and feeding on infaunal polychaete worms were the endemic Miccosukee Island conids *Spuriconus micanopy* (Petuch, 1994) (Plate 89, B), *Calusaconus evergladesensis* (Petuch, 1991) (Plate 89, E), and *Ximeniconus palmbeachensis* (Petuch, 1994) (*X. perplexus* complex). The unusual columbellid *Eurypyrene miccosukee* Petuch, 1991 also occurred in the algae beds and *Porites* thickets, where it fed on small molluscan prey.

The Caloosarca aequilitas Community

In deeper areas (2-5 m) adjacent to the carbonate sand banks during Yarmouthian time, Turtle Grass (*Thalassia*) formed extensive beds that circled the Loxahatchee Trough region. These were particularly well developed along Miccosukee and Tomeu Islands, were they represented the predominant ecosystem. In these sea grass beds, the arcid bivalve *Caloosarca aequilitas* (Tucker and Wilson, 1932) (namesake of the community; Plate 90, F) formed large beds that nestled, shallowly buried, within the root mats. This endemic arcid co-occurred with several other grass bed bivalves, including the lucinids *Codakia orbicularis* (Linnaeus, 1758), *Lucina pectinata* (Gmelin, 1791), *Anodontia alba* Link, 1817, and *Ctena orbiculata* (Montagu, 1808), the arcid *Anadara lienosa* (Say, 1831), and the venerids *Chione elevata* (Say, 1822), *Mercenaria campechiensis* subspecies, and *Periglypta listeri* (Gray, 1838). These were the principal prey items for several molluscivorous gastropods, including the busy-

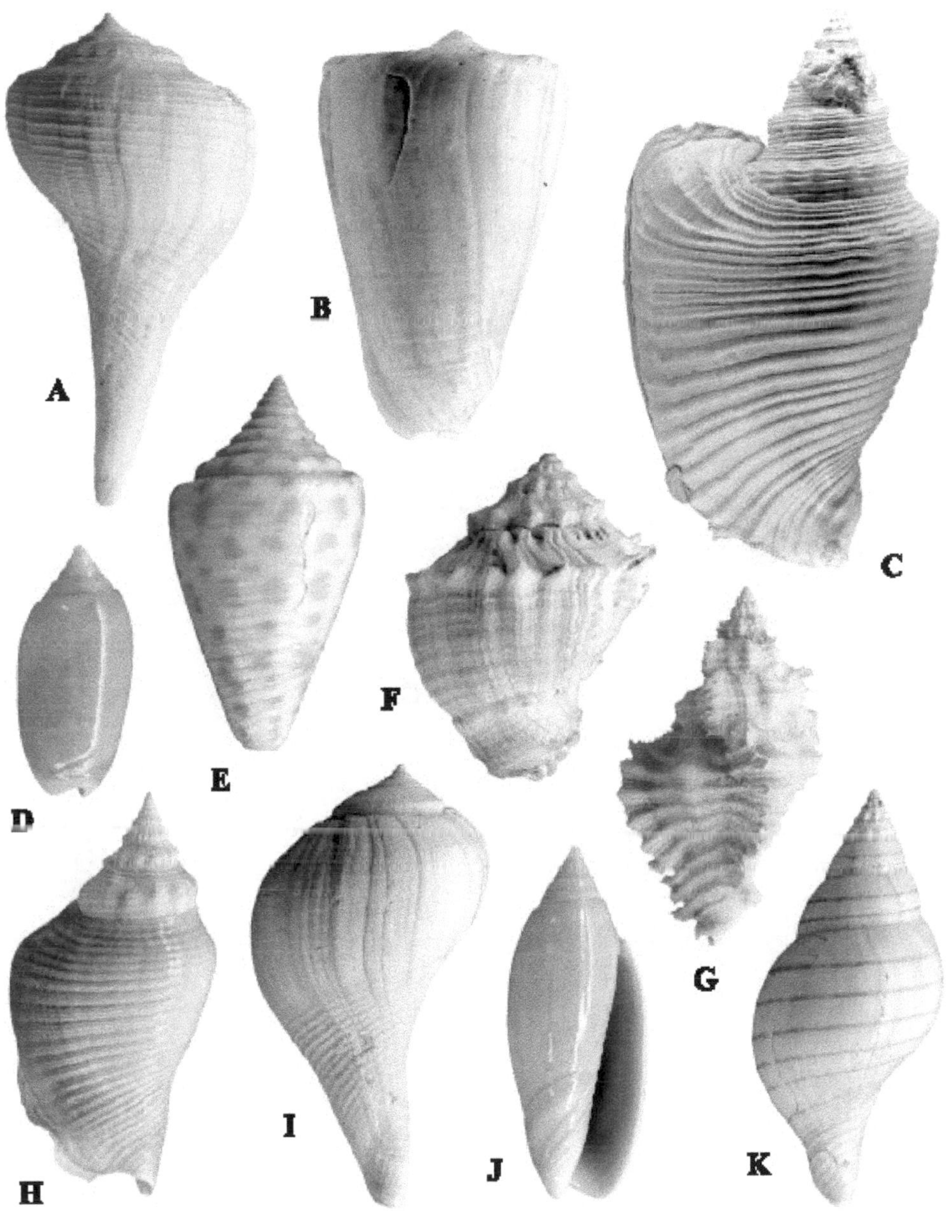

Plate 89. Index fossils and characteristic organisms of the *Strombus lindae* Community (shallow carbonate banks during Yarmouthian time) of the Belle Glade Subsea. Included are: A= *Sinistrofulgur roseae* Petuch, 1991, length 109 mm; B= *Spuriconus micanopy* (Petuch, 1994), length 90 mm; C= *Macrostrombus holeylandicus* (Petuch, 1994), length 194 mm; D= *Oliva (Strephona) cokeyae* Petuch, 1991, length 16 mm; E= *Calusaconus evergladesensis* (Petuch, 1991), length 29 mm; F= *Melongena (Rexmela) diegelae* Petuch, 1994, length 27 mm; G= *Chicoreus duerri* Petuch, 1994, length 32 mm; H= *Strombus lindae* Petuch, 1991, length 44 mm; I= *Fulguropsis evergladesensis* Petuch, 1994, length 51 mm; J= *Oliva (Strephona) lindae* Petuch, 1991, length 28 mm; K= *Fasciolaria (Cinctura) evergladesensis* Petuch, 1991, length 47 mm. All specimens were collected in the carbonate bank beds of the "Belle Glade Member" (equivalent to stratotype) of the Bermont Formation, from dredgings of the North New River Canal, South Bay, Palm Beach County, Florida.

Plate 90. Index fossils and characteristic organisms of the *Caloosarca aequilitas* Community (*Thalassia* beds during Yarmouthian time) and the *Nodipecten pernodosus* Community (shallow sand bottom lagoons during Yarmouthian time) of the Belle Glade Subsea. Included are: A= *Strombus evergladesensis* Petuch, 1991, length 71 mm; B= *Spuriconus lemoni* (Petuch, 1990), length 57 mm; C= *Nodipecten pernodosus* (Heilprin, 1886), length 98 mm; D= *Melongena (Rexmela) bispinosa* (Philippi, 1844), length 34 mm; E= *Vokesimurex bellegladeensis* (E. Vokes, 1963), length 47 mm; F= *Caloosarca aequilitas* (Tucker and Wilson, 1932), length 56 mm; G= *Oliva (Strephona) edwardsae* Olsson, 1967, length 30 mm; H= *Turbinella hoerlei* E. Vokes, 1966, length 170 mm; I= *Fulguropsis feldmanni* Petuch, 1991, length 83 mm; J= *Melongena lindae* Petuch, 1994, length 64 mm. All specimens were collected in the "Belle Glade Member" (equivalent to stratotype) of the Bermont Formation, from dredgings of the North New River Canal, South Bay, Palm Beach County, Florida.

conids *Sinistrofulgur roseae* subspecies and *Fulguropsis capelettii* Petuch, 1994, the fasciolariids *Fasciolaria okeechobeensis* Tucker and Wilson, 1932, *F. (Cinctura) evergladesensis* subspecies, and *Triplofusus giganteus* (Kiener, 1840), the muricids *Phyllonotus pomum* (Gmelin, 1791) and *Chicoreus dilectus* (A. Adams, 1855), and the melongenids *Melongena lindae* Petuch, 1994 (Plate 90, J) and *M. (Rexmela) bispinosa* (Philippi, 1844) (Plate 90, D). Living on the individual grass blades were the small carnivorous columbellid *Zaphrona taylorae* Petuch, 1987 and several small algivorous gastropods, some of which included the modulids *Modulus bermontianus* Petuch, 1994 and *M. calusa* subspecies, the cerithiids *Cerithium algicola* C. Adams, 1845, *C. lutosum* Menke, 1828, and *C. floridanum* Moerch, 1876, the trochid *Tegula fasciata* (Born, 1778), and the turbinids *Astraea (Astralium) phoebia* Roeding, 1798 and *A. (Lithopoma) southbayensis* Petuch, 1994. The turbinellid *Turbinella hoerlei* E. Vokes, 1966 (Plate 90, H) was also a common resident of the Turtle Grass beds, where it fed on polychaete and enteropneust worms. On large, open sand patches scattered between the grass beds, the algivorous strombid gastropod *Strombus evergladesensis* Petuch, 1991 formed immense aggregations and co-occurred with the less common strombids *Macrostrombus* cf. *mayacensis* (Tucker and Wilson, 1933) and *Lobatus wilsonorum* (Petuch, 1994). These sand patches also supported a large fauna of carnivorous gastropods, some of which included the general carnivore/scavenger olivids *Oliva (Strephona) edwardsae* Olsson, 1967 (Plate 90, G), *O. (Strephona) smithorum* Petuch, 1994, and *O. (Strephona) sayana* subspecies and vermivores such as the conids *Spuriconus lemoni* (Petuch, 1990) (Plate 90, B), *Gradiconus anabathrum tranthami* (Petuch, 1995), and *Jaspidiconus jaspideus* subspecies, the turrids *Neodrillia blacki* Petuch, new species (see Systematic Appendix and Chapter 10), *Syntomodrillia moseri* subspecies, and *Crassispira tampaensis* Bartsch and Rehder, 1939, and the terebrid *Strioterebrum dislocata* subspecies. The cassids *Cassis* new species (*C. spinella* complex), *Phalium granulatum* (Born, 1778) and *P. cicatricosum* (Gmelin, 1791) were also common on these sand patches, where they were the principal predators of grass bed-associated echinoids such as *Clypeaster*, *Meoma*, and *Plagiobrissus*.

The Nodipecten pernodosus Community

In the deepest areas (10-100 m) of the Loxahatchee Trough during Yarmouthian time, large beds of pectinid bivalves covered the open carbonate sand bottoms. These beds were composed of only four species, which included *Argopecten amplicostatus* subspecies, "*Aequipecten*" species (smooth and unsculptured, in the *A. lineolaris* complex), *Nodipecten pernodosus* (Heilprin, 1886) (namesake of the community; Plate 90, C), and *Lindapecten* species (*L. muscosus* complex). Living along with the scallops on the open sand bottoms were several other bivalves, some of which included the glycymerids *Glycymeris americana* (Defrance, 1829), *G. decussata* (Linnaeus, 1758) and *G. pectinata* (Gmelin, 1791), the venerids *Puberella intapurpurea* (Conrad, 1849), *Macrocallista maculata* (Linnaeus, 1758), and *Lirophora latilirata* subspecies, and numerous small mactrids and tellinids. Feeding on the scallops and infaunal bivalves were several molluscivorous gastropods, including the busyconid *Fulguropsis feldmanni* Petuch, 1991 (Plate 90, I), the fasciolariid *Fasciolaria (Cinctura) capelettii* subspecies, and the muricids *Vokesimurex bellegladeensis* (E. Vokes, 1963) (Plate 90, E), *V. anniae* (M. Smith, 1940), and *Chicoreus rachelcarsonae* Petuch, 1987. This deeper water environment also housed several other carnivorous gastropods, including vermivores such as the conids *Gradiconus capelettii* subspecies, *G. sennottorum* Rehder and Abbott, 1951, and *G. delessertii* (Recluz, 1843), suctorial-feeders such as the cancellariids *Cancellaria* species (*C. richardpetiti* complex) and *Ventrilia tenera* (Philippi, 1848), and general carnivores/scavengers such as the olivid *Oliva (Strephona) murielae* Olsson, 1967 and the scaphelline volutid *Scaphella capelettii* Petuch, 1994.

Plate 91. Index fossils and characteristic organisms of the *Busycon carica* Community (intertidal sand bars and shallow muddy-sand lagoons) and the *Crassostrea virginica* Community (intertidal and shallow lagoon oyster bars) of the Socastee Lagoon System. Included are: A,B= *Neverita fossata* (Gould, 1847), length 27 mm; C= *Mercenaria campechiensis* (Gmelin, 1791), length 124 mm; D= *Busycon carica* (Gmelin, 1791), length 130 mm; E= *Crassostrea virginica* (Gmelin, 1791), length 122 mm; F= *Oliva (Strephona) sayana* subspecies, length 78 mm; G= *Crepidula fornicata* (Linnaeus, 1758), length 47 mm; H= *Thais (Stramonita) floridana* (Conrad, 1837), length 63 mm; I= *Sinistrofulgur laeostomum* (Kent, 1987), length 135 mm; J= *Callinectes sapidus* Rathbun, 1896), length 64 mm; K= *Oxypode* species, length 141 mm. The specimens were collected in either the Socastee Formation, Intracoastal Waterway, South Myrtle Beach, Horry County, South Carolina (Figures A, B, C, D, E, F, G, and I), in the Sunderland Formation, south of Little Cove Point, Calvert County, Maryland (Figures H and J), or in the Pamlico Formation, in eroding beach dunes, Vero Beach, Indian River County, Florida (Figure K).

Communities and Environments of the Socastee Lagoon System

During both the Yarmouthian and Sangamonian Interglacial Stages, the Socastee Lagoon System contained essentially the same, highly impoverished molluscan fauna. These cold-tolerant and highly eurythermal survivors of the older Nashua fauna changed little during the middle and late Pleistocene and retained the same basic community structure and species compositions. The intervening Illinoian Glacial Stage did little to alter the Socastee assemblages, simply displacing them either seaward or landward in response to eustatic patterns. The greatest development of the Socastee faunas occurred during the warm Sangamonian Stage, when several southern species migrated northward as far as New Jersey (Richards, 1962). Some of these warmer-water migrants included the thaidid *Thais (Stramonita) floridana* (Conrad, 1837), the terebrids *Strioterebrum dislocata* (Say, 1822) and *S. concava* (Say, 1827), and the fissurellid *Diodora cayenensis* (Lamarck, 1822). Today, these are all confined to areas south of Cape Hatteras. In fossiliferous beds from Sangamonian time, in the Cape May, Sunderland, Socastee, Flanner's Beach, and Norfolk Formations, only two marine communities are recognized: the *Busycon carica* Community (intertidal sand bars and shallow muddy-sand lagoons) and the *Crassostrea virginica* Community (intertidal and shallow lagoon oyster bars). With the onset of the severe Wisconsinan Glacial Stage, these communities were further impoverished, leaving behind the modern coastal lagoon and shallow neritic assemblages seen from North Carolina to Cape Cod.

The Busycon carica Community

On muddy-sand bottoms, within tidal channels and shallow barrier island lagoons (1-5 m) and on intertidal sand flats (0-1 m), a few species of bivalves formed extensive, closely packed beds. These included the mactrid *Mulinia lateralis* (Say, 1822), the venerids *Mercenaria mercenaria* (Linnaeus, 1758) and *M. campechiensis* (Gmelin, 1791) (Plate 91, C), the noetiid *Noetia (Eontia) ponderosa* (Say, 1822), the arcid *Larkinia transversa* (Say, 1822), the myid *Mya arenaria* Linnaeus, 1758, the corbulid *Caryocorbula contracta* (Say, 1822), the solenoid *Ensis directus* Conrad, 1843, and the solecurtid *Tagelus plebeius* (Lightfoot, 1786). These bivalves were the principal prey items for molluscivorous gastropods such as the busyconids *Busycon carica* (Gmelin, 1791) (namesake of the community; Plate 91, D), *Sinistrofulgur laeostomum* (Kent, 1987) (Plate 91, I), and *Busycotypus canaliculatum* (Linnaeus, 1758), and the drilling naticids *Neverita fossata* (Gould, 1847) (*N. duplicata* variant?) (Plate 91, A and B) and *Euspira heros* (Say, 1822). In the northern part of the Socastee Lagoon System, the endemic buccinoidean *Atractodon stonei* (Pilsbry, 1893) also occurred on these sand flats and was a principal predator of smaller bivalves. By late Wisconsinan time, however, this unusual neptune-like buccinoidean was extinct. Both the bivalves and small specimens of the molluscivorous gastropods were prey items for the portunid crab *Callinectes sapidus* Rathbun, 1896 (Plate 91, J), which was abundant in this community and also on the oyster bars. Several other types of carnivorous gastropods were also present on the sand flats and included general carnivores such as the olivid *Oliva (Strephona) sayana* subspecies (Plate 91, F), scavengers/carrion-feeders such as the nassariids *Ilyanassa obsoleta* (Say, 1822) and *I. trivittata* (Say, 1822) and the columbellid *Astyris lunata* (Say, 1826), and vermivores such as the terebrids *Strioterebrum dislocata* (Say, 1822) and *S. concava* (Say, 1827), and the turrid *Kurtziella cerina* (Kurtz and Stimpson, 1851).

The Crassostrea virginica Community

In the intertidal areas near river mouths and in muddy estuarine areas (0-2 m),

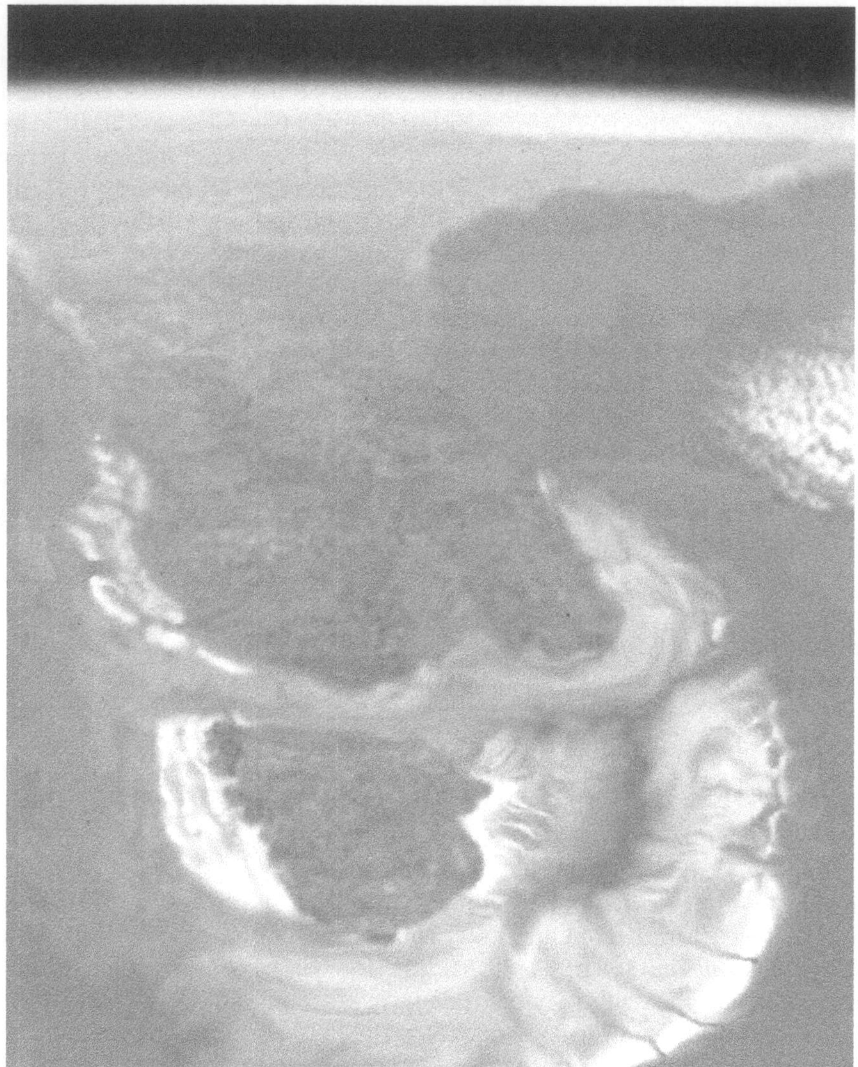

Figure 33. Simulated satellite image of the Floridian Peninsula during the Sangamonian Pleistocene, showing the possible appearance of the Lake Worth Subsea (looking north). Prominent geomorphological features include Immokalee Island in the west, the remnants of the Loxahatchee Trough in the central region, the mangrove keys of the Palm Beach and Tomeu Islands along the east, and the extensive Miami oolite banks along the south. The GIS imagery was done by Charles Roberts, Ph.D., Department of Geography and Geology, Florida Atlantic University, using paleoenvironmental and paleobathymetric data supplied by the author, enhancing a NASA space shuttle photograph of Florida.

immense beds of the ostreid oyster *Crassostrea virginica* (Gmelin, 1791) (namesake of the community; Plate 91, E) formed biohermal structures that dominated the lagoon floors. These served as the substrate for several low-motility or sessile mollusks, including the mytilid bivalve *Ischadium recurvum* (Rafinesque, 1820), and gastropods such as the fissurellid *Diodora cayenensis* (Lamarck, 1822) and the calyptraeids *Crepidula fornicata* (Linnaeus, 1758) (Plate 91, G) and *C. (Ianacus) plana* Say, 1822. Feeding on the oysters were three drilling gastropods, including the thaidid *Thais (Stramonita) floridana* (Conrad, 1837) (Plate 91, H) and the ocenebrine muricids *Urosalpinx cinereus* (Say, 1822) and *Eupleura caudata* (Say, 1822) and several suctorial pyramidellids, including *Odostomia seminuda* (A. Adams, 1839), *O. impressa* (Say, 1822), and *O. acutidens* (Dall, 1883). On open sandy shores adjacent to the oyster bars, ghost crabs of the genus *Oxypode* (Plate 91, K) were abundant and fed on carrion along the high tide line. Farther upsteam in the river mouths, on low salinity mud banks, the mactrid bivalve *Rangia cuneata* (Sowerby, 1831) formed immense monoculture beds.

Communities and Environments of the Lake Worth Subsea

Following the extreme climatic cooling and low sea levels during the Illinoian Glacial Stage, the Okeechobean Sea basin reflooded for the last time during the Sangamonian Interglacial Stage. This final paleosea, the Lake Worth Subsea, had two main pulses of flooding; one at the beginning of Sangamonian time and another, higher sea level stand during the late Sangamonian, with an intervening short, small low stand. During these times, the two members of the Fort Thompson Formation were deposited all across southern Florida, with the lower Okaloacoochee Member being deposited in latest Illinoian-earliest Sangamonian lower sea level stand and the upper Coffee Mill Hammock Member being deposited during the late Sangamonian high sea level stand. During the emergent time of the previous Illinoian Glacial Stage, the Loxahatchee Trough depression of the Okeechobean Sea basin was again filled by another giant fresh water lake, Lake Pahayokee ("River of Grass" in Seminole). Here, a second species radiation of planorbid gastropods took place, along with radiations of viviparid and amnicolid gastropods and corbiculid and fresh water mactrid bivalves. The reflooding of the Okeechobean Sea basin at the beginning of Sangamonian time destroyed these endemic oligohaline lacustrine faunas and repopulated the area with a newly evolved eutropical marine fauna. During the Sangamonian, even in the highest sea level stands, the Lake Worth Subsea was almost completely infilled and was composed mostly of shallow banks and mangrove islands, much like Recent Florida Bay (Figure 33). At this time, only the eastern side of the paleosea was submerged and was composed of three main depositional regions; a northern chain of mangrove islands and extensive Turtle Grass beds (depositional center for the Okaloacoochee and Coffee Mill Hammock Members of the Fort Thompson Formation), a southern set of oolite banks that accumulated behind a reef tract (depositional center for the Miami Formation and the foundation of the Miami ridges, Florida Bay, and the southern Florida Keys), and an extreme southern narrow reef tract that had built upon the older Monroe Reef Tract (the depositional center for the Key Largo Formation and the foundation of the Recent northern Florida Keys).

From the Okaloacoochee fossil beds of the first Lake Worth flooding event (early Sangamonian), two main marine communities are recognized: the *Pyrazisinus gravesae* Community (mangrove forests and intertidal mud flats) and the *Strombus alatus kendrewi* Community (shallow lagoon carbonate sand flats and *Thalassia* beds). From the Coffee Mill Hammock fossil beds of the second Lake Worth flooding event (late Sangamonian), two other marine communities are recognized; the *Chione elevata* Community (shallow

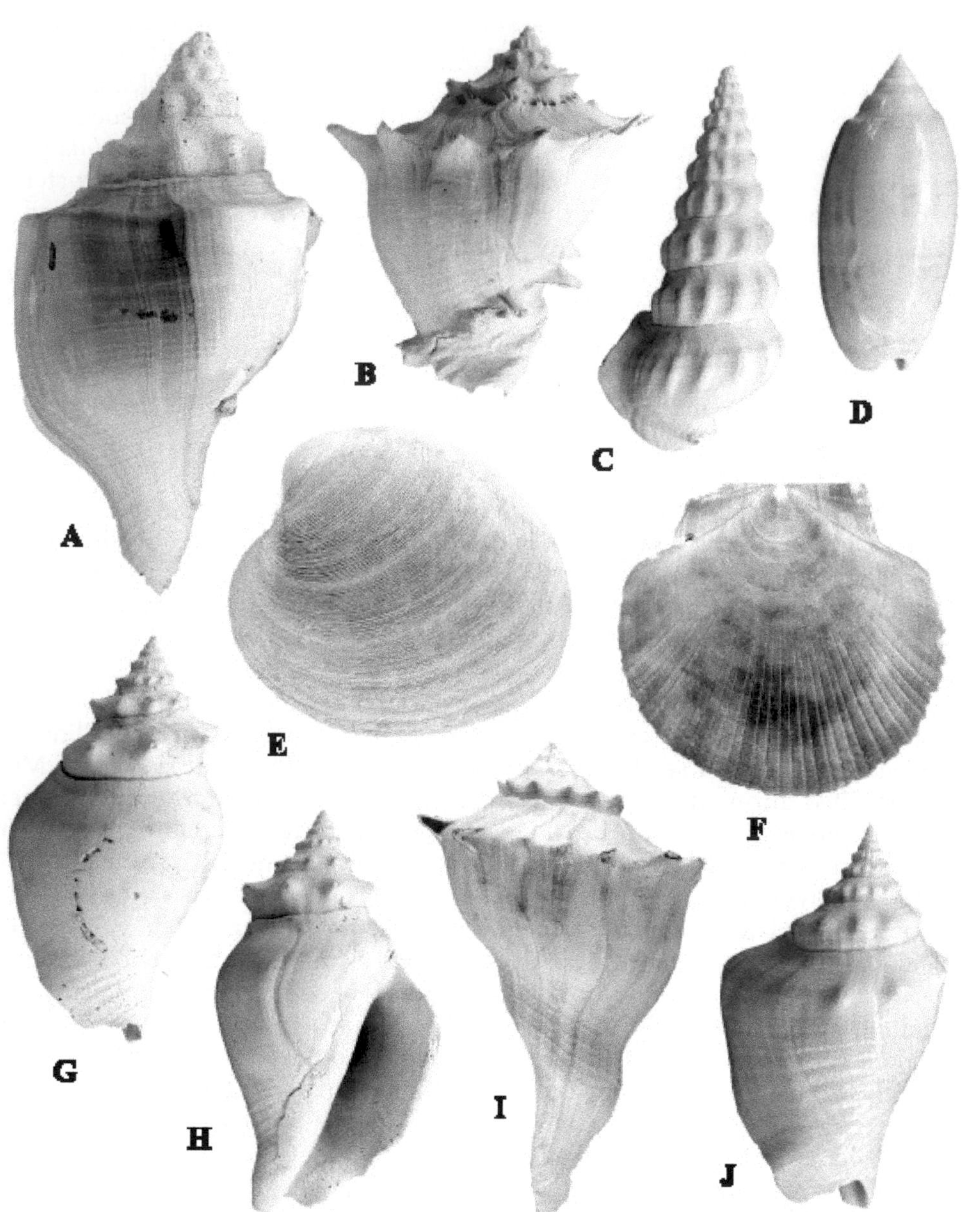

Plate 92. Index fossils and characteristic organisms of the *Pyrazisinus gravesae* Community (mangrove forests and intertidal mud flats during early Sangamonian time) and the *Strombus alatus kendrewi* Community (shallow lagoon carbonate sand flats and *Thalassia* beds during early Sangamonian time) of the Lake Worth Subsea. Included are: A= *Turbinella wheeleri* Petuch, 1994, length 258 mm; B= *Melongena (Rexmela) corona subcoronata* Heilprin, 1886, length 77 mm; C= *Pyrazisinus gravesae* Petuch, 1994, length 55 mm; D= *Oliva (Strephona) sayana sarasotaensis* Petuch and Sargent, 1986, length 41 mm; E= *Mercenaria campechiensis* (Gmelin, 1791), length 96 mm; F= *Pecten ziczac* (Linnaeus, 1758), length 58 mm; G,H= *Strombus alatus kendrewi* Petuch, new subspecies, holotype, length 83 mm (see Systematic Appendix); I= *Sinistrofulgur perversum okeechobeensis* Petuch, 1994, length 113 mm; J= *Strombus lindae dowlingorum* Petuch, new subspecies, holotype, length 55 mm (see Systematic Appendix). The specimens were collected in the Okaloacoochee Member of the Fort Thompson Formation, at either a construction dig at Parkland, Broward County, Florida (Figures A, G, and H), a road fill dig along State Road 80, near Belle Glade, Palm Beach County, Florida (Figures D, E, and F), at the Ferrell-Mattson pit, Okeechobee, Okeechobee County, Florida (Figures B,C, and I), or at the Boynton sand pit, US 441, Boynton Beach, Palm Beach County, Florida (Figure J).

lagoon *Thalassia* beds) and the *Pyrazisinus ultimus* Community (mangrove forests and intertidal mud flats). The faunas of the late Sangamonian Key Largo reefs are poorly known, with the fossil corals being exposed only in a few man-made channel cuts and small quarries. From these, it can be seen that the Key Largo reef platforms were composed primarily of large rounded, massive coral heads, mostly of the genera *Montastrea* and *Diploria*. The Miami oolite banks, although exposed at several natural outcrops and better-studied, were essentially sterile and unfossiliferous, much like the Recent Bahama Banks oolite deserts. Only the Fort Thompson communities will be discussed in the following sections. Of special interest was the fact that several older Belle Glade molluscan lineages managed to survive the Illinoian extinction event and persisted into Okaloacoochee time. These relicts, which include the *Strombus lindae* complex and *Turbinella*, lived only a short time, becoming extinct by the mid-Sangamonian. The relictual genus *Pyrazisinus* survived longer, becoming extinct at the beginning of the Wisconsinan Glacial Stage. During Wisconsinan time, the entire Okeechobean basin was emergent and the lowest elevation areas along the northern and eastern sides were filled by another, smaller fresh water lake, Lake Flirt (source of the latest Pleistocene-Holocene Lake Flirt Marl (actually calcarenite)). By the beginning of the Holocene, this last Okeechobean lake had dried up and the Lake Flirt calcarenites indurated into a soft limestone that covered most of the basin. This newly-formed limestone acted as a confining unit, preventing any water from infiltrating below it. Impounded water accumulated on top of the Lake Flirt limestone and calcarenite and, by 7000 years B.P., the modern Everglades marshlands first appeared.

The Pyrazisinus gravesae Community

During the early Sangamonian, the areas west of the Tomeu Islands and Palm Beach Archipelago and along the southern tip of the Floridian Peninsula (in the region of Recent Lake Okeechobee) housed extensive mangrove jungles and organic-rich mud flats. Here, a simple, low-diversity community existed, with a single species of potamidid gastropod forming immense monoculture aggregations. As has been seen in all similar Okeechobean mangrove-associated faunas, the genus *Pyrazisinus* (*P. gravesae* Petuch, 1994; namesake of the community; Plate 92, C) dominated the mud flats, feeding on surficial algal and bacterial films. These occurred together with large aggregations of the algivore neritid *Neritina reclivata* (Say, 1822). Feeding on the potamidids and neritids was the melongenid *Melongena (Rexmela) corona subcoronata* Heilprin, 1886 (Plate 92, B), which also occurred in large numbers. As in Recent mangrove forests, the littorinid gastropod *Littorinopsis angulifera* (Lamarck, 1822) lived on the mangrove leaves and branches. In muddy, brackish water areas adjacent to the mangrove forests, such as at the mouth of the proto-Kissimmee River and in the dying Lake Pahayokee during the earliest Sangamonian salt water incursions, the mactrid bivalve *Rangia cuneata* (Sowerby, 1831) formed immense monoculture beds.

The Strombus alatus kendrewi Community

Along the Tomeu Islands, the Miami and Palm Beach Archipelagos, and the southern tip of the peninsula, extensive shallow water (0.5-5 m) sand banks developed and these were the predominant environment of the early Sangamonian Lake Worth Subsea. These broad shoals were covered with an intricate patchwork of interfingering open sand bottom areas, calcareous algal thickets, and dense sea grass and *Thalassia* beds. Here, large aggregations of strombid gastropods dominated the open sand patches adjacent to the sea grass beds, and included *Strombus alatus kendrewi* Petuch, new subspecies (Plate 92, G and

Plate 93. Index fossils and characteristic organisms of the *Chione elevata* Community (shallow lagoon *Thalassia* beds during late Sangamonian time) of the Lake Worth Subsea. Included are: A= *Melongena melongena* (Linnaeus, 1758), length 109 mm; B= *Strombus alatus* Gmelin, 1791, length 80 mm; C= *Modulus pacei* Petuch, 1987, length 16 mm; D= *Columbella rusticoides* Heilprin, 1886, length 15 mm; E= *Cerithium muscarum* Say, 1832, length 24 mm; F= *Chione elevata* (Say, 1822), length 23 mm; G= *Sinistrofulgur sinistrum* (Hollister, 1958), length 212 mm; H= *Melongena (Rexmela) corona* (Gmelin, 1791), length 82 mm; I= *Solenosteira* s.l. *cancellarius* (Conrad, 1846), length 29 mm; J= *Gradiconus anabathrum patglicksteinae* (Petuch, 1987), length 38 mm; K= *Vokesinotus perrugatus* (Conrad, 1836), length 22 mm. The specimens were collected in the Coffee Mill Hammock Member of the Fort Thompson Formation, at either the Ferrell-Mattson pit, Okeechobee, Okeechobee County (Figures A, B, G, H, and I) or at a construction dig along Lake Worth Road, near the Florida Turnpike, Lake Worth, Palm Beach County, Florida (Figures C, D, E, F, J, and K).

H; see Systematic Appendix), *S. lindae dowlingorum* Petuch, new subspecies (Plate 92, J; see Systematic Appendix), *Macrostrombus costatus griffini* (Petuch, 1994), and *Eustrombus gigas verrilli* (McGinty, 1946). Several bivalves also occurred abundantly on these sand patches, including the venerids *Mercenaria campechiensis* (Gmelin, 1791) (Plate 92, E), Macrocallista nimbosa (Lightfoot, 1786), and Chione elevata (Say, 1822), the cardiids *Dinocardium robustum* (Lightfoot, 1786) and *Trachycardium muricatum* (Linnaeus, 1758), and the pectinid *Pecten ziczac* (Linnaeus, 1758) (Plate 92, F). Feeding on these bivalves and strombids were several large molluscivorous gastropods, some of which included the busyconids *Sinistrofulgur perversum okeechobeensis* Petuch, 1994 (Plate 92, I), and *S. sinistrum* subspecies, the naticid *Neverita duplicata* (Say, 1822), and the fasciolariids *Triplofusus giganteus* (Kiener, 1840) and *Fasciolaria tulipa* (Linnaeus, 1758). Also occurring with the molluscivores were several other carnivorous gastropods, including the general carnivore/scavenger olivid *Oliva (Strephona) sayana sarasotaensis* Petuch and Sargent, 1986 (Plate 92, D; also living in deep water off Recent western Florida), the echinoid-feeding cassid *Cassis spinella* Clench, 1944, and vermivores such as the conids *Spuriconus spurius* (Gmelin, 1791), *Jaspidiconus jaspideus* (Gmelin, 1791), *J. pfluegeri* Petuch, new species (see Systematic Appendix and Chapter 10), and *Gradiconus anabathrum* (Crosse, 1865). On the scattered Turtle Grass beds, the venerid bivalve *Anomalocardia aubreyana* (Orbigny, 1842) formed large interstitial aggregations within the root mats and algivorous gastropods such as the cerithiid *Cerithium muscarum* Say, 1832 and the neritid *Smaragdia viridis weyssei* (Russel, 1940) occurred in large numbers on the grass blades. These small gastropods and bivalves served as the main prey items for the busyconid *Fulguropsis spiratum* (Lamarck, 1816), the muricids *Phyllonotus pomum* (Gmelin, 1791) and *Vokesinotus perrugatus* (Conrad, 1846), the fasciolariid *Fasciolaria (Cinctura) hunteria* (Perry, 1811), and the melongenid *Melongena (Rexmela) bispinosa* (Philippi, 1844). Feeding on enteropneust and polychaete worms in the *Thalassia* beds was the large turbinellid *Turbinella wheeleri* Petuch, 1994 (Plate 92, A), the last-living member of its genus in Florida.

The Chione elevata Community

During the late Sangamonian, sea levels rose to the highest stand of the entire Pleistocene (Krantz, 1991). By this time, the Okeechobean Subsea basin was almost entirely infilled so this high sea level had little effect on the bathymetry of the Lake Worth Subsea. Although the western half of the paleosea basin was now dry land, occupied by the large, wide Immokalee Island, the entire eastern half still contained extensive sand shoal systems. These shallow (1-10 m) Coffee Mill Hammock banks resembled those from the earlier Okaloacoochee time, being covered with interfingering environments of open sand bottoms, calcareous algal thickets, and Turtle Grass beds. On the open sand areas, the venerid bivalve *Chione elevata* (Say, 1822) (namesake of the community; Plate 93, F) formed dense pavement-like beds. These served as the main food resource for several molluscivorous gastropods, some of which included the busyconid *Sinistrofulgur sinistrum* (Hollister, 1958) (Plate 93, G), the fasciolariids *Fasciolaria tulipa* (Linnaeus, 1758) and *Triplofusus giganteus* (Kiener, 1840), the melongenid *Melongena melongena* (Linnaeus, 1758) (Plate 93, A), and the muricids *Phyllonotus pomum* (Gmelin, 1791), *Urosalpinx cinereus* (Say, 1822), *Vokesinotus perrugatus* (Conrad, 1846), *Chicoreus dilectus* (A. Adams, 1855), *Eupleura sulcidentata* Dall, 1890, and *Eupleura tampaensis* (Conrad, 1846). Living along with these molluscivores were several other carnivorous gastropods, including vermivores such as the conids *Spuriconus spurius* (Gmelin, 1791) and *Gradiconus anabathrum patglicksteinae* (Petuch, 1987) (Plate 93, J), general carnivores/scavengers such as the haminoeid *Haminoea taylorae* Petuch, 1987 and the buccinids *Solenosteira* s.l. *cancellarius* (Conrad, 1846)

Plate 94. Index fossils and characteristic organisms of the *Pyrazisinus ultimus* Community (mangrove forests and intertidal mud flats during late Sangamonian time) of the Lake Worth Subsea. Included are: A= *Melongena (Rexmela) corona winnerae* Petuch, new subspecies, paratype, length 150 mm (see Systematic Appendix and Chapter 10); B= *Pyrazisinus ultimus* Petuch, new species, holotype, length 53 mm (see Systematic Appendix); C= *Cerithidea beattyi lakeworthensis* Petuch, new subspecies, holotype, length 11 mm (see Systematic Appendix); D= *Cerithidea scalariformis palmbeachensis* Petuch, new subspecies, holotype, length 23 mm (see Systematic Appendix); E= *Anomalocardia* cf. *hendriana* Mansfield, 1939, length 12 mm; F= *Cerithidea costata* (daCosta, 1778), length 16 mm; G= *Melampus coffeus* (Linnaeus, 1758), length 12 mm; H= *Sabal* palmetto, petrified wood, length 252 mm; I= *Batillaria minima* (Gmelin, 1791) form *degenerata*, length 16 mm; J= *Melampus bidentatus* Say, 1822, length 9 mm; K= *Melampus monilis* (Bruguiere, 1789), length 10 mm. The specimens were collected in the Coffee Mill Hammock Member of the Fort Thompson Formation, at either a construction dig along Okeechobee Road, near the Florida Turnpike, West Palm Beach, Palm Beach County, Florida (Figure A), a construction dig along US 441 and Forest Hill Boulevard, Wellington, Palm Beach County, Florida (Figure B), or along Lake Worth Road, near the Florida Turnpike, Lake Worth, Palm Beach County, Florida (Figures C, D, E, F, G, H, I, J, and K).

(Plate 93, I) and *Gemophos tinctus* (Conrad, 1846), and the scavengers/carrion-feeder nassariid *Phrontis vibex* (Say, 1822). Large aggregations of strombid gastropods also occurred on these open sand bottoms, including *Strombus alatus* Gmelin, 1791 (Plate 93, B), *Macrostrombus costatus* (Gmelin, 1791), and *Eustrombus gigas* (Linnaeus, 1758). On the Turtle Grass beds, the venerid bivalve *Anomalocardia aubreyana* (Orbigny, 1842) lived in dense interstitial beds along with other bivalves such as the lucinids *Codakia orbicularis* (Linnaeus, 1758) and *Pseudomiltha floridana* (Conrad, 1833) and several small tellinids and mactrids. Living on the sea grass blades were several algivorous gastropods, including the cerithiids *Cerithium muscarum* Say, 1832 (Plate 93, E) and *C. algicola* C. Adams, 1845, and the modulids *Modulus pacei* Petuch, 1987 (Plate 93, C) and *M. calusa* Petuch, 1988. These bivalves and small gastropods served as the principal food resource for several prominent molluscivorous gastropods, including the melongenid *Melongena (Rexmela) corona* (Gmelin, 1791) (Plate 93, H), the fasciolariid *Fasciolaria (Cinctura) hunteria* (Perry, 1811), the muricid *Vokesinotus perrugatus* (Conrad, 1836) (Plate 93, K), and the columbellid *Columbella rusticoides* Heilprin, 1886 (Plate 93, D).

The Pyrazisinus ultimus Community

On the Tomeu Islands and Palm Beach Archipelago during the late Sangamonian, extensive mangrove forest and sand flat complexes developed along their sheltered western coasts. Here, the last species of the potamidid gastropod genus *Pyrazisinus* lived in large aggregations on the sand flats and among the mangrove roots (*P. ultimus* Petuch, new species, namesake of the community; Plate 94, B; see Systematic Appendix). This relictual species occurred together with dense beds of the small venerid bivalve *Anomalocardia* cf. *hendriana* Mansfield, 1939 (Plate 94, E) and the batillariid gastropod *Batillaria minima* (Gmelin, 1791) form *degenerata* (Plate 94, I). These served as the principal prey items of the large melongenid *Melongena (Rexmela) corona winnerae* Petuch, new subspecies (Plate 94, A; see Systematic Appendix) and numerous shore birds. Farther within the Lake Worth mangrove jungles, in the Black Mangrove (*Avicennia*) zone, several high intertidal algivore and detritivore gastropods lived on the mangrove breather roots and rotting leaf litter. Some of these included the potamidids *Cerithidea beattyi lakeworthensis* Petuch, new subspecies (Plate 94, C; see Systematic Appendix), *C. scalariformis palmbeachensis* Petuch, new subspecies (Plate 94, D), and *C. costata* (daCosta, 1778) (Plate 94, F), and the air-breathing ellobiids *Melampus bidentatus* Say, 1822 (Plate 94, J), *M. coffeus* (Linnaeus, 1758) (Plate 94, G), and *M. monilis* (Bruguiere, 1789) (Plate 94, K). On the drier parts of the islands adjacent to the mangroves, dense palmetto forests (*Sabal*) (Plate 94, H) dominated and their logs frequently washed onto the sand flats. These served as both a food source and a refuge for the detritivorous gastropods.

Communities and Environments of the Champlain Sea

The northernmost of the eastern North American paleoseas, the isostatically-produced Champlain Sea, existed only as long as the Pleistocene continental ice sheets depressed the St. Lawrence River area. At the Pleistocene-Holocene boundary and the melting of the ice sheets, isostatic rebound took place and the Champlain Sea drained off the continent, being replaced by fresh water rivers and lakes. During the Wisconsinan Glacial Stage, the Champlain Sea was at its maximum development (Figure 34) and was composed of three distinct areas; the main Champlain basin (the Recent St. Lawrence River Valley), the deep and narrow Burlington Fjord (named for Burlington, Vermont, and encompassing the Lake Champlain area), and the shallow Ontario Embayment (Recent eastern Lake Ontario). From the fossil beds along the St. Lawrence River, in eastern Lake

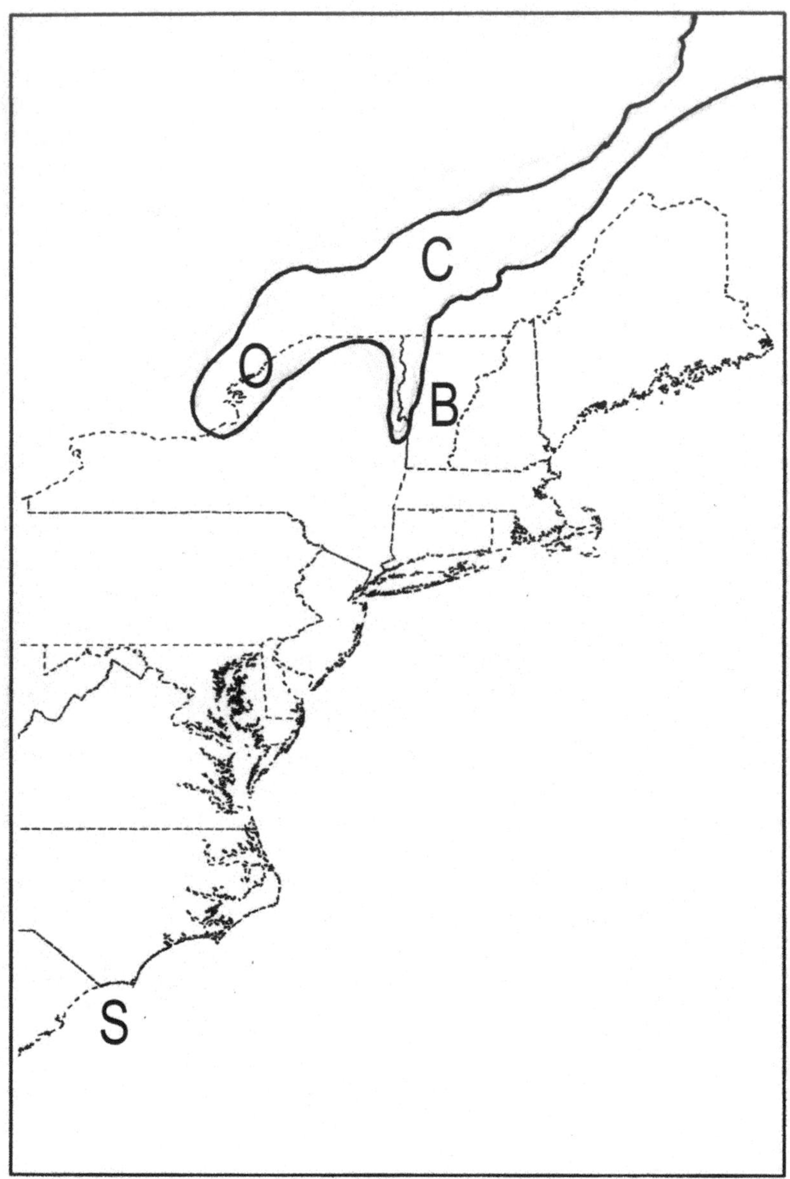

Figure 34. Approximate configuration of the St. Lawrence River Valley and eastern Lake Ontario during the Wisconsinan Pleistocene, superimposed upon the outline of the northeastern United States. Prominent features included: C= Champlain Sea, B= Burlington Fjord, O= Ontario Embayment, S= Socastee Lagoon System (for reference).

Ontario, and around Lake Champlain (the Riviere du Loup and Champlain Formations and the Prescott Beds), an impoverished high Arctic marine fauna is known to have existed within this short-lived paleosea (Richards, 1962). The Champlain ecosystems were represented by only two main community types: the *Macoma calcarea* Community (shallow muddy bays and intertidal mud flats) and the *Astarte laurentiana* Community (deep muddy and shell rubble bottom areas). Although the Champlain molluscan fauna was essentially the same as that of the Recent Gulf of St. Lawrence, some local endemic species were present, particularly the astartid bivalve *Astarte laurentiana* and dwarf subspecies of buccinid and neptuniid gastropods (Shimer, 1908).

The Macoma calcarea Community

Along the shoreline of the main Champlain Sea basin and in the shallow Ontario Embayment, in muddy intertidal areas and shallow (0-3 m) sand flats, extensive beds of three species of bivalves literally carpeted the sea floor. These included the macomids *Macoma calcarea* (Gmelin, 1791) (namesake of the community; Plate 95, K) and *M. balthica* (Linnaeus, 1758) (Plate 95, G), and the myid *Mya arenaria* Linnaeus, 1758 (Plate 95, C). These dense bivalve beds served as the principal food resource of several carnivorous gastropods, including the buccinids *Buccinum glaciale donovani* Gray, 1824 (Plate 95, A and B), *B. fragile* Verkruezen, 1878 (Plate 95, H and I), *B. undatum* Linnaeus, 1758, and *B. cyaneum* Bruguiere, 1789, and the naticid *Cryptonatica clausa* (Broderip and Sowerby, 1829) (Plate 95, E and F). In the intertidal areas along rocky shores and on sand flats, dense beds of the mytilid mussel *Mytilus edulis* Linnaeus, 1758 (Plate 95, J) accumulated in massive biohermal structures. Nestled between the mussels was the hiatellid bivalve *Hiatella arctica* (Linnaeus, 1767) (Plate 95, D) and several small gastropods, including algivores such as the trochids *Margarites olivacea* Brown, 1827 and *M. cinerea* (Couthouy, 1838) and the littorinid *Littorina saxatilis* (Olivi, 1792), and carnivores such as the columbellid *Mitrella holbolli* (Mueller, 1842). In the brackish water *Mytilus edulis* beds along the western edge of the Ontario Embayment (present-day Thousand Islands of Lake Ontario), a distinctive Arctic fish fauna was present (preserved in sandstone nodules), including Capelin smelts (*Mallotus*), Lumpfish (*Cyclopterus*), and Sticklebacks (*Gasterosteus*).

The Astarte laurentiana Community

In the deeper areas (3-100 m) of the main Champlain basin and within the deep Burlington Fjord, the small astartid bivalves *Astarte laurentiana* Lyell, 1845 (namesake of the community; Plate 96, K) and *A. banksii* (Leach, 1819) (Plate 96, J) formed densely packed beds on the open sea floor. Other larger bivalves also occurred with the abundant astartids, including the cardiid *Serripes groenlandicus* (Bruguiere, 1789), the myid *Mya truncata* Linnaeus, 1758 (Plate 96, E), and the pectinid *Chlamys islandicus* (Mueller, 1776) (which occurred in scattered large beds). Small aggregations of turritellid gastropods also were present on the bivalve beds, and were composed of only two species, *Tachyrhynchus acicula* (Stimpson, 1851) and *T. erosa* (Couthouy, 1838). The detritivore/suspension feeder *Trichotropis borealis costellata* Couthouy, 1838 (Plate 96, I) also occurred on these small turritellid beds. The bivalves, and the trichotropid and turritellids, were the principal prey items for a large array of molluscivorous gastropods, including the neptuniids *Neptunea despecta* subspecies (Plate 96, A and B) and *N.* species (Plate 96, F and G) (both dwarf forms and undescribed taxa), the buccinids *Plicifusus kroyeri* (Moeller, 1842), *Beringius turtoni* (Bean, 1834), *Buccinum totteni* Stimpson, 1865, *B. plectrum* Stimpson, 1865, *B. ciliatum* Fabricius, 1780, and *B. scalariformis* Beck, 1842, the naticid *Euspira pallida* (Broderip and Sowerby, 1829), the trophonine muricids *Boreotrophon clathratus* (Linnaeus, 1758) and *B.*

Plate 95. Index fossils and characteristic organisms of the *Macoma calcarea* Community (shallow muddy bays and intertidal mud flats) of the Champlain Sea. Included are: A,B= *Buccinum glaciale donovani* Gray, 1824, length 41 mm; C= *Mya arenaria* Linnaeus, 1758, length 41 mm; D= *Hiatella arctica* (Linnaeus, 1767), length 24 mm; E,F= *Cryptonatica clausa* (Broderip and Sowerby, 1829), length 26 mm; G= *Macoma balthica* (Linnaeus, 1758), length 16 mm; H,I= *Buccinum fragile* Verkruezen, 1878, length 44 mm; J= *Mytilus edulis* Linnaeus, 1758, length 41 mm; K= *Macoma calcarea* (Gmelin, 1791), length 28 mm. All specimens were collected in the Riviere du Loup Formation, in a sand quarry at St. Nicholas, Quebec, Canada.

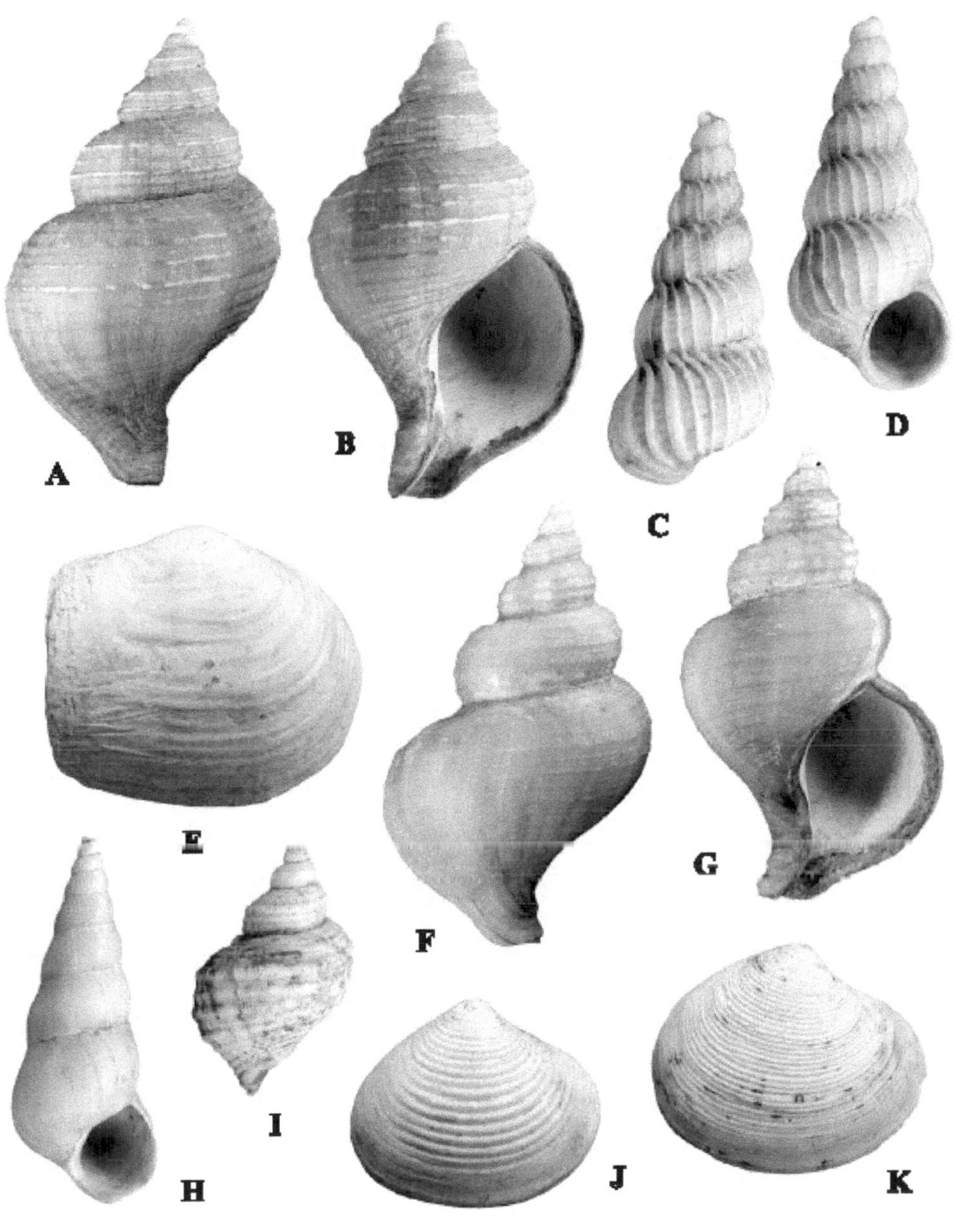

Plate 96. Index fossils and characteristic organisms of the *Astarte laurentiana* Community (deep muddy and shell rubble bottom areas) of the Champlain Sea. Included are: A,B= *Neptunea despecta* subspecies, length 53 mm; C,D= *Boreoscala greenlandicum* (Perry, 1811), length 38 mm; E= *Mya truncata* Linnaeus, 1758, length 42 mm; F,G= *Neptunea* species, length 56 mm; H= *Acirsa borealis* (Lyell, 1842), length 28 mm; I= *Trichotropis borealis costellata* Couthouy, 1838, length 10 mm; J= *Astarte banksii* (Leach, 1819), length 19 mm; K= *Astarte laurentiana* Lyell, 1845, length 20 mm. All specimens were collected in the Riviere du Loup Formation, in a sand quarry east of St. Nicholas, Quebec, Canada.

scalariforme (Gould, 1840), and the fasciolariid *Ptychotractus ligatus* (Mighels and Adams, 1842). Large numbers of zoantherians, such as *Metridium* and *Edwardsia*, also lived on these sand and shell rubble bottoms and these served as the prey of the large suctorial epitoniid gastropods *Boreoscala greenlandicum* (Perry, 1811) (Plate 96, C and D) and *Acirsa borealis* (Lyell, 1842) (Plate 96, H). The boreal polychaete fauna of the Champlain Sea was large and diverse, supporting a rich fauna of vermivorous gastropods that included a large radiation of the turrid genus *Oenopota*, with *O. elegans* (Moeller, 1842), *O. pleurotomaria* (Couthouy, 1838), *O. scalaris* (Moeller, 1842), *O. treveliana* (Turton, 1870), *O. violacea* (Mighels and Adams, 1842), and *O. harpularia* (Couthouy, 1838), and possibly the suctorial-feeding cancellariid *Admete viridula* (Fabricius, 1780). The Burlington Fjord also contained a rich crustacean fauna and this supported large resident populations of Beluga Whales (*Delphinapterus*).

Chapter 10. Biotic Patterns in Time and Space

Of all the patterns preserved in the fossil records of the eastern North American paleoseas, the cyclical turnover of their marine faunas is the most dramatic and obvious. This rhythm of constant biotic changes takes into account four interconnected subpatterns: extinctions of existing faunas and communities, evolutionary radiations of new taxa, faunal replacement and the formation of new communities, and the survival of relictual genera and species. In light of recent advances in coastal paleoceanography, including more detailed paleontological data and tighter stratigraphic control, these subpatterns can now, for the first time, be studied with a much higher resolution and with greater accuracy. When put into a well-defined time frame, these biotic patterns give unprecedented insight into the cyclicity and rapidity of catastrophic changes in marine climates throughout the late Cenozoic.

Patterns of Extinction in the Eastern North American Paleoseas

As demonstrated in the previous chapters, extinction events in paratropical and eutropical coastal marine communities are initiated by only two environmental factors that take place nearly simultaneously: a rapid decline in water temperature and an accompanying sea level drop. The former factor severely limits community structure by excluding physiologically-eutropical groups ("refridgeration" of Stanley, 1986) while the latter affects community structure by shrinking colonization space and reducing the number and types of available environments (Vermeij, 1978; Vermeij and Petuch, 1986). From earlier studies of Transmarian (Miocene) and Caloosahatchian (Plio-Pleistocene) gastropod faunas (Petuch, 1993, 1995), mass extinction events were seen to involve a two-part, bimodal pattern. In these cases, a severe regional extinction event was followed, within a geologically short period of time, by a second extinction event. During the first extinction time, which I here term the "wounding event," a large percentage of the invertebrate fauna (usually 15-30%) becomes extinct and the community structures are disrupted. If another catastrophic event does not occur within a geologically short period of time, the communities rebound and are repopulated by radiations of newly evolved descendants of survivors. On the other hand, if a second extinction event takes place before complete ecological order is restored, the entire system collapses, resulting in a mass extinction (with the disappearance of over 50% of the taxa at both the generic and species level). This second extinction event, which I termed the "coup de grace event" (Petuch, 1995), forms a couplet with the first, lesser extinction event. The time interval bounded by the extinction couplet varies from one-half to four million years, depending on the physiological make-up of the communities and their latitudinal and bathymetric distributions.

For the time frame covered in this book, five separate mass extinctions of marine molluscan faunas can be recognized as having occurred within the eastern North American paleoseas. Since these were major catastrophic events that resulted in dramatic faunal shifts and new ecological structures, I feel they are important enough to be given their own names. These include the *Suwannean Extinction* (mid-Chattian Oligocene), the *Chipolan Extinction* (late Burdigalian Miocene), the *Transmarian Extinction* (mid-Tortonian Miocene; previously described in Petuch, 1993), the *Caloosahatchian Extinction* (late Calabrian Pleistocene), and the *Evergladesian Extinction* (Illinoian Pleistocene). For each of these mass extinctions, a predecessor wounding event occurred and the five resulting couplets form a first-order cycle of extinctions. Numerous small localized extinctions, or extirpations, also took place on a regular basis during the couplet time and during the intervening times between the couplets. These form a second-order pattern of extinctions that is superimposed upon the first order system. These events are discussed in the following sections.

The Suwannean Extinction

This mid-Oligocene mass extinction (named for the Suwannee Formation which contains the typical fauna) was the coup-de-grace for an earlier wounding event that occurred at the Eocene-Oligocene boundary. During the intervening Rupelian time, several older Eocene gastropod genera and species lineages managed to survive and evolved into new complexes. Some of the more important of these survivors included *Cypraeorbis* (Cypraeidae), *Fusimitra* (Mitridae), *Pleurofusia* (Turridae), *Myristica* (Melongenidae), and *Clavolithes* (Turbinellidae). Later during the Suwannean Extinction, these, along with a large part of the Rupelian gastropod fauna, became either completely or regionally extinct. Also among the casualties of the Suwannean catastrophism was a series of cerithiid and potamidid radiations, with *Cestumcerithium*, *Prismacerithium*, *Semivertagus*, and *Telescopium* (with the last two genera surviving in the Recent Indo-Pacific Region), and other gastropod genera such as *Suwannescapha* (Cylichnidae) and *Terebrellina* (Terebridae). Several other prominent Rupelian mollusks became regionally extinct at this time, disappearing from eastern North America but persisting into the Recent in the Indo-Pacific Region. Some of these included the seraphsid gastropod *Terebellum* and the tridacnid bivalve *Hippopus*. During the intervening Rupelian time, several new, ecologically dominant groups appeared and these managed to survive the subsequent mass extinction event. Primary among these newly-evolved groups were *Orthaulax* (Strombidae) and *Pyrazisinus* (Potamididae). The *Telescopium*-dominated ecosystems of the Rupelian and early Chattian mud flats were especially hard-hit by the Suwannean Extinction. After that event and until the late Pleistocene, all eastern North American tropical mud flat and mangrove environments were dominated by *Pyrazisinus*.

The Chipolan Extinction

Following a period of climatic stability that occurred during the late Chattian Oligocene-early Aquitanian Miocene time interval, another major extinction event took place during the mid-Aquitanian (Planktonic Foraminiferal Zone N5). This wounding event removed several important and prominent gastropods from the Aquitanian ecosystems, some of which included "*Fusus*" (undescribed *Hemifusus*-like genus; Melongenidae), several complexes of *Tympanotonus*-like potamidids (undescribed genera), *Doxander* (Strombidae), and *Tritonopsis* (Thaididae). This first extinction event of the couplet, however, did not affect several other pre-Suwannean gastropod taxa, and these persisted on into the Burdigalian Miocene as ecologically-important and dominant groups. Some of these wounding event survivors included *Orthaulax* (Strombidae), *Falsilyria* (Volutidae), *Floradusta* and *Loxacypraea* (both Cypraeidae; which evolved during the interval between the Suwannean mass extinction and the mid-Aquitanian extinction), and *Viator* and *Homalacantha* (both Muricidae).

During the late Burdigalian Miocene (Planktonic Foraminiferal Zone N7), the second extinction event of the couplet, the coup-de-grace Chipolan Extinction (named for the Chipolan Subprovince of the Baitoan Molluscan Province), dramatically and permanently altered the composition of the American tropical marine ecosystems. At this time, the ecologically prominent strombid *Orthaulax* became extinct, along with old Eocene-derived groups such as *Falsilyria* (Volutidae) and *Spinifulgur* (Busyconidae), and Burdigalian chronoendemics such as *Floradusta, Loxacypraea* (both Cypraeidae) and *Psammostoma* (Fasciolariidae). Many other prominent Tethyan-derived groups became regionally extinct, dying out in eastern North America but persisting into the Recent in the tropical Indo-Pacific Region. Some of these groups included *Zoila* and *Lyncina* (both Cypraeidae), *Clavocerithium* and *Hemicerithium* (both Cerithiidae), *Globularia* (Ampullospiridae), *Terebralia* (Potamididae), and *Homalacantha* (Muricidae). During the Burdigalian stable climatic interval, between the two extinction events, radiations of several prominent eastern North American molluscan groups first evolved. Primary among these were the gastropods *Busycon* (Busyconidae) and *Triplofusus* (Fasciolariidae) and the bivalves *Nodipecten* (Pectinidae) and *Dinocardium* (Cardiidae), all of which survived the Chipolan Extinction, and all subsequent extinctions, to become dominant organisms in the Recent Carolinian ecosystems.

The Transmarian Extinction

The Transmarian Molluscan Province, extending along high latitude areas of Miocene eastern North America, was exceptionally vulnerable to climatic changes. Being paratropical in composition and physiologically stenothermal, the Transmarian faunas were extremely sensitive to both increased and decreased water temperatures. For these reasons, the mollusks of the Salisbury Sea are the best indicators of middle and late Miocene extinction events. From the fossil beds of the Choptank and St. Mary's Formations, two major extinction events can be seen to have occurred in quick succession. The first of these, during the mid-Serravallian at the end of Choptank deposition (Planktonic Foraminiferal Zone N13), was severe enough to be considered a wounding event. At this time, the spectacular ecphora fauna of the Calvert and Choptank Formations came to an abrupt end, with the total and immediate disappearance of all three ribbed (*Trioecphora*) and multiple-ribbed (*Chesathais*) species. Only two four-ribbed lineages (*Ecphora* and *Planecphora*) survived into the Tortonian (with *Planecphora* surviving farther south only in Florida and the Carolinas), and these were the progenitors of all subsequent late Miocene and Pliocene species. Also becoming extinct along with the ecphoras were several other classic Transmarian endemics, some of which included the gastropods *Calverturris* and *Transmariaturris* (both Turridae), *Cancellariella* (Cancellariidae), and *Patuxentrophon* and *Panamurex (Stephanosalpinx)* (both Muricidae).

During the latest Serravallian and early Tortonian, the marine climate of the Transmarian region warmed to near subtropical temperatures and remained relatively stable for over one million years. At this time, during the deposition of the St. Mary's Formation, an unprecedented explosion of evolution took place and an entirely new Transmarian fauna appeared. Besides containing radiations of older Transmarian genera such as *Conradconfusus* (Fasciolariidae), *Turrifulgur* and *Sycopsis* (Busyconidae), and *Mariaturricula* (Turridae), this fauna contained several distinctive new endemic groups, some of which included *Mariacassis* (Cassidae), *Pseudaptyxis* (Fasciolariidae), *Poliniciella* (Naticidae), *Mariasalpinx* (Muricidae), *Coronafulgur* (Busyconidae), and *Mariadrillia* and *Nodisurculina* (both Turridae). This new Transmarian fauna also contained endemic species offshoots of several tropical migrants, including *Gradiconus* (Conidae), *Cymatosyrinx* (Turridae), and *Mitra* s.l. (Mitridae), which appeared in the Salisbury Sea for the first time. By the beginning of mid-Tortonian time, in Planktonic Foraminiferal Zone N16, a second and much more severe extinction event took place. This coup-de-grace, the Transmarian Extinction (named for the Transmarian Province), was one of the greatest faunal decimations to take place during the entire Cenozoic. As an indicator of the severity of this extinction, almost 50% of the genera of St. Mary's buccinoidean, muricoidean, and conoidean gastropods became extinct at this time (Petuch, 1993). This percentage of gastropod extinction is even higher when other superfamilies, such as the naticoideans, cancellarioideans, and tonnoideans, are factored in. Some typical Transmarian genera did manage to survive the coup-de-grace event, however, and these persisted as relictual taxa within the Albemarle Sea to the south. These relicts, which included *Chesatrophon*, *Scalaspira*, and *Lirosoma* (all Muricidae), *Laevihastula* (Terebridae), and *Scaphellopsis* (Volutidae), lasted into the Piacenzian Pliocene. Other Transmarian endemics such as *Mariacolpus* (Turritellidae), *Bulliopsis* (Buccinidae), and *Coronafulgur* (*C. coronatus* complex, Busyconidae) persisted only into Messinian time in the Rappahannock Subsea (in the Eastover Formation), while others such as *Mariafusus* (Fasciolariidae) persisted longer, into the Zanclean Pliocene (in Yorktown Zone 1). In the south, in the Polk and Walton Subseas, the last North American member of the venerid bivalve genus *Clementia* became extinct during the Transmarian Extinction, while the busyconid genera *Coronafulgur* (*C. propecoronatum* complex) and *Turrifulgur* persisted on until the end of the Messinian. The Polk endemic cypraeid gastropods *Akleistostoma* and *Calusacypraea*, however, survived the coup-de-grace and persisted into the late Pliocene.

The Caloosahatchian Extinction

For almost seven million years, from the mid-Tortonian Miocene to the late Piacenzian Pliocene, the fused remnants of the southern Baitoan and northern Transmarian faunas underwent a continuous series of radiations that ultimately resulted in the evolution of the classic Caloosahatchian Molluscan Province fauna. During the Messinian Miocene and Zanclean Pliocene, the early Caloosahatchian fauna was impoverished and had a cool-water, marginally paratropical appearance. Only within the Charlotte and Murdock Subseas of the Okeechobean Sea did any eutropical components manage to survive and speciate. With the onset of the Piacenzian Pliocene, however, this low-diversity trend was reversed and paratropical and eutropical influences spread throughout the province. At this time, the sinistral gastropods *Sinistrofulgur* (Busyconidae) and *Contraconus* (Conidae) first appeared, simultaneously, and underwent species radiations in the southern subseas. Likewise, several older groups, such as the pectinid bivalve *Carolinapecten* (from the early Messinian) and the muricid gastropods *Ecphora* and *Planecphora* (late species groups, from the Serravallian Miocene), underwent species radiations and were dominant organisms in their respective ecosystems. *Ecphora,* itself, radiated into several new habitats, evolving contemporaneous offshoot groups such as *Globecphora* and *Latecphora*. Of particular interest during the Piacenzian was the evolutionary explosion of cypraeid gastropods in the Okeechobean Sea, resulting in large species clusters of the genera *Siphocypraea*, *Pahayokea*, *Calusacypraea*, *Myakkacypraea*, *Pseudadusta*, and *Akleistostoma*. Also confined to the Okeechobean Sea area were radiations of the genera *Echinofulgur* and *Tropochasca* (Melongenidae), *Pyrazisinus* (Potamididae), *Strombus* (with one exception in the Jackson Subsea) and *Macrostrombus* (both Strombidae), and *Hystrivasum* (with one exception in the Jackson Subsea) (Turbinellidae).

Besides these Okeechobean endemics, several widespread Caloosahatchian Province gastropod genera also underwent species radiations during Piacenzian time, some of which included *Trossulasalpinx*, *Pterorhytis*, *Vokesinotus*, and *Urosalpinx* (all Ocenebriniae-Muricidae), *Busycon*, *Pyruella*, *Fulguropsis*, *Busycotypus*, and *Brachysycon* (all Busyconidae), *Ptychosalpinx* and *Solenosteira* (*S. mengeana* complex) (both Buccinidae), *Terebraspira* (Fasciolariidae), *Scaphella* (Volutidae), and *Pleioptygma* (Pleioptygmatidae). An interesting and somewhat anomalous minor extinction event took place during the mid-Piacenzian, when the old, Transmarian-derived genera *Chesapecten* (Pectinidae) and *Ecphora* (along with its evolutionary offshoot, *Latecphora*) abruptly disappeared. Since water temperatures were at their warmest during this time, the extinction of these Transmarian relicts may have been due to ecological competition with more efficient and aggressive tropical groups. In this case, *Ecphora* and *Latecphora* may have been out-competed by the fast-moving and voracious muricids *Phyllonotus* and *Chicoreus*, and *Chesapecten* may have been overwhelmed by the immense beds of *Carolinapecten* and *Argopecten* species. This minor extinction presaged a major wounding event that took place in the late Piacenzian (end of Planktonic Foraminiferal Zone 21). The impact of this extinction episode was greatest in the Okeechobean Sea, where a large number of Turtle Grass, Mangrove, and mud flat-associated gastropods abruptly disappeared. The most prominent of these casualties was the large species radiation of cypraeid gastropods, including the genera *Akleistostoma*, *Calusacypraea*, *Myakkacypraea*, and *Pseudadusta*. Only two cypraeid genera, *Siphocypraea* and *Pahayokea*, survived this wounding event and persisted on into the early Pleistocene. Other prominent groups that became extinct at this time included *Tropochasca* (Melongenidae), *Eichwaldiella* (Turritellidae), *Eugeniconus* and *Virgiconus* (both Conidae; regionally extinct, surviving in the Recent Indo-Pacific), *Pachycrommium* (Ampullospiridae), and *Cymatophos* and *Trajana* (both Buccinidae; regionally extinct, surviving in the Panamic Province).

During the latest Piacenzian Pliocene and Calabrian Pleistocene, the survivors of the late Piacenzian wounding event began to reorganize their communities and underwent rapid pulses of evolution. Some of these reradiated into new genera and filled niches left vacant by the extinction of competing groups. This is particularly well-demonstrated in the case of the cypraeid subgenus *Okeechobea*, a neotenic offshoot of the sea grass bed-associated *Siphocypraea*. This inflated, bubble-like Calabrian complex moved into estuarine environments and filled the niches previously occupied by both *Calusacypraea* and *Myakkacypraea*. Several completely new groups also appeared at this time, including prominent genera such as *Liochlamys* (Fasciolariidae), *Porphyria* (Olividae), and *Rexmela* (Melongenidae). At the end of the Calabrian (end of Planktonic Foraminiferal Zone 22), a coup-de-grace event, the Caloosahatchian Extinction (named for the Caloosahatchian Molluscan Province), took place and this rivaled the Transmarian Extinction in its severity. The faunal compositions and trophic structures of all eastern North American marine communities were drastically altered by this extinction episode, so much so that the Caloosahatchian Molluscan Province ceased to exist as a biogeographic entity. Approximately 30% of the prominent and ecologically-important molluscan genera and 90% of the species-level taxa became extinct during this coup-de-grace, and included gastropods such as *Terebraspira* and *Liochlamys* (both Fasciolariidae), *Contraconus* and *Seminoleconus* (both Conidae), *Siphocypraea*, *Okeechobea*, and *Pahayokea* (all Cypraeidae), *Hystrivasum* (Vasinae-Turbinellidae), *Echinofulgur* (Melongenidae), *Cymatosyrinx* (Turridae), *Globinassa*, *Scalanassa*, and *Paranassa* (all Nassariidae), *Brachysycon* (Busyconidae), and *Trossulasalpinx*, *Subpterynotus* (regionally extinct, surviving in West Africa and southern Australia), *Pterorhytis* (regionally extinct, surviving off Ecuador), and *Acantholabia* (all Muricidae), and bivalves such as *Cunearca* and *Arcoptera* (Arcidae) and *Stralopecten* (Pectinidae). A few taxa did manage to survive the Caloosahatchian Extinction, and persisted into the middle Pleistocene as relicts within the warm, protected lagoonal environments of the Okeechobean Sea. These Caloosahatchian relictual elements included the gastropods *Malea* (*M. springi* complex) (Tonnidae), *Pyruella* (Busyconidae), *Cerithioclava* (Cerithiidae), *Porphyria* (Olividae), *Calusaconus* and *Ximeniconus* (both Conidae), and *Jenneria* (Ovulidae), and the bivalves *Carolinapecten* (Pectinidae) and Miltha (Lucinidae).

The Evergladesian Extinction

During the Aftonian Interglacial Stage of the mid-Pleistocene, only the Okeechobean Sea area contained eutropical and rapidly evolving faunas. At this time, the relictual taxa from the older Caloosahatchian faunas had integrated themselves into the newly evolved ecosystems of the enclosed Loxahatchee Subsea, resulting in chronologically hybridized assemblages. Even within this geologically short time, several new prominent endemic groups evolved, including the gastropods *Titanostrombus* (Strombidae), *Lindoliva* (Olividae), *Miccosukea* (Melongenidae), *Pseudozonaria* (Cypraeidae, *P. portelli* complex; a migrant from the Caribbean region), and the *Vasum floridanum* complex (Vasinae-Turbinellidae). By the beginning of the Kansan Glacial Stage, however, a major wounding event took place and this decimated the Loxahatchee faunas. The first extinction event of this couplet saw the disappearance of the Caloosahatchian relicts *Jenneria* (Ovulidae; regionally extinct, surviving in the Panamic Province), *Porphyria* (Olividae; regionally extinct, surviving in the Panamic Province), *Microcythara* (Columbellidae; regionally extinct, surviving in the Panamic Province), *Pyruella* (Busyconidae), *Malea* (Tonnidae; regionally extinct, surviving in the Panamic Province), *Carolinapecten* (Pectinidae), the Loxahatchee endemic *Titanostrombus* (Strombidae; regionally extinct, surviving in the Panamic and Brazilian Provinces), and the Caribbean migrant *Pseudozonaria* (Cypraeidae; regionally extinct, surviving in the Panamic Province). The now-impoverished mid-Pleistocene Okeechobean Sea fauna rebounded during the following Yarmouthian Interglacial Stage, evolving new species complexes in many of the remaining endemic genera. This evolutionary spurt was short-lived, however, as the Pleistocene climates plummeted to their coldest temperatures, and sea levels dropped to their lowest stands, during the Illinoian Glacial Stage (400,000 yrs. B.P.). At this time, the final coup-de-grace event of the Cenozoic took place; the Evergladesian Extinction (named for the fossil faunas of the Everglades area). Among the casualties of this severe extinction episode were gastropods such as *Lindoliva* (Olividae), *Miccosukea* (Melongenidae), *Cerithioclava* (Cerithiidae; regionally-extinct, surviving in the Recent western Caribbean), *Ximeniconus* (*X. perplexus* complex; regionally extinct, surviving in the southern Caribbean Sea and Panamic Province) and *Calusaconus* (both Conidae), and the bivalve *Miltha* (Lucinidae; regionally extinct, surviving in the Brazilian Province). The impoverished molluscan faunas and communities that remained after the Evergladesian Extinction resembled those of the Recent Carolinian Molluscan Province. Only one relictual genus survived all five extinction couplets throughout the Cenozoic; the potamidid gastropod *Pyrazisinus*. This highly-successful mangrove-associated group dominated its mud flat ecosystem during the late Pleistocene, and evolved at least two new species during the Sangamonian Interglacial Stage. With the initiation of the Wisconsinan Glacial Stage (75,000 yrs. B.P.), this amazing Florida endemic disappeared, becoming the last major casualty of the Pleistocene extinctions.

Geographical Heterochrony in Eastern North America

One of the more interesting patterns of community evolution and biogeography in the marine environment is seen in geographical heterochrony (Petuch, 1982). In this concept, all the communities and organisms of a biogeographical subdivision (usually a province or subprovince) are found not to be evolving at the same rate, but that some localized areas are falling behind chronologically, retaining relictual communities that date from older geological times (Petuch, 1988). The geographically-small areas that house these chronostatic communities come in two main types; *primary relict pockets*, where the relictual taxa have been retained at the species level, and *secondary relict pockets*, where the relictual taxa have been retained only at the genus level. As can be assumed, the older the relict pocket, the less likely it is that it will retain a primary relictual fauna. For this reason, all the primary relict pockets in the tropical western Atlantic contain faunas dating only from the mid-Pleistocene. Although having undergone a slower rate of evolution, the community structures and organisms of secondary relict pockets have changed enough to only superficially resemble their older and ancestral analogues (Petuch, 1980, 1982).

In the Recent tropical western Atlantic, only two areas are known to house relatively-intact mid Pleistocene primary relict pockets. These include the coasts of Florida and the eastern coasts of Honduras and Nicaragua (Petuch, 1988).The Floridian relict pockets are distributed in three small areas: the St. Lucie and Lake Worth Lagoon systems of Martin and Palm Beach Counties (the Lake Worth Primary Relict Pocket), the *Thalassia* environments of the Florida Keys and Florida Bay (the Florida Bay Primary Relict Pocket), and the edge of the continental shelf from Tampa north to Apalachicola (the Tampa Primary Relict Pocket). Each of these areas has its own distinctive composition of relictual taxa. The Honduran-Nicaraguan Primary Relict Pocket occurs in *Thalassia* environments on the offshore islands and banks, extending from the Caratasca Cays to the Miskito Cays.

The southeastern coast of Florida houses a mixed relictual fauna that combines species from both the Okaloacoochee and Coffee Mill Hammock Members of the Fort Thompson Formation. Here, only in the geographically small St. Lucie and Lake Worth estuaries, these relicts have managed to survive the Wisconsinan cold time. In the Lake Worth Lagoon, a relatively intact remnant of the *Pyrazisinus ultimus* Community still exists, minus, unfortunately, its potamidid namesake. The large Coffee Mill Hammock melongenid *Melongena (Rexmela) corona winnerae* Petuch, new subspecies (Plate 97, A and B) has recently been discovered to be living here, feeding on *Batillaria minima* (Gmelin, 1791) form *degenerata* and the large *Modulus pacei* Petuch, 1987 just as its Sangamonian ancestors did. Also occurring in Lake Worth, in slightly deeper tidal channels, is the Okaloacoochee cone shell *Jaspidiconus pfluegeri* Petuch, new species (Plate 97, F and I). Offshore of the Lake Worth and Palm Beach coasts (in 20-150 m depth), the Bermont endemic muricid *Vokesimurex bellegladeensis* (E. Vokes, 1963) and the triviid *Pusula lindajoyceae* Petuch, 1994 (Plate 97, K) have been discovered to have survived into the Recent, and these Aftonian and Yarmouthian species live with the Fort Thompson *Gradiconus anabathrum patglicksteinae* (Petuch, 1987) in this partially relictual ecosystem.

The Florida Bay Turtle Grass areas house not only an impoverished Sangamonian fauna, but also elements from the Yarmouthian Bermont Formation. Here, the abundant Fort Thompson and Bermont modulid *Modulus calusa* Petuch, 1988 (Plate 97, H and J) is equally abundant in the Recent. Occurring with the relict modulid, or in areas adjacent to the sea grass beds, are several other Bermont relicts, including the turrid *Neodrillia blacki* Petuch, new species (Plate 97, D), the columbellid *Zaphrona taylorae* Petuch, 1987, the muricid *Favartia pacei* Petuch, 1988, and the haminoeid bubble shell *Haminoea taylorae* Petuch, 1987. In the Tampa Pocket, in deeper water areas (50-150 m) off the northwestern coast of Florida, other Bermont relicts have been discovered to be extant. Some of these include the muricids *Chicoreus rachelcarsonae* Petuch, 1987 (Plate 97, E) and *Favartia lindae* Petuch, 1987 (Plate 97, C), and the ovulid *Cyphoma lindae* Petuch, 1987. The Okaloacoochee olivid *Oliva (Strephona) sayana sarasotaensis* Petuch and Sargent, 1986 is also living in these deeper water areas off northwestern Florida. Farther south in the western Caribbean, another partial Bermont relict fauna is now known to exist. Here, particularly in the Turtle Grass beds near the Caratasca Cays of Honduras, several classic Bermont species are extant, including the fasciolariid *Latirus jucundus* McGinty, 1940, the modulid *Modulus bermontianus* Petuch, 1994, the triviid *Niveria bermontiana* Petuch, 1994, and the cerithiid *Cerithioclava garciai* Houbrick, 1986. Underscoring the relictual nature of this primary relict

Plate 97. Floridian relict gastropods of Sangamonian, Aftonian, and Yarmouthian age. Included are: A,B= *Melongena (Rexmela) corona winnerae* Petuch, new subspecies, holotype, length 134 mm, Little Lake Worth Lagoon, Singer Island, Palm Beach, Florida; C= *Favartia lindae* Petuch, 1987, length 19 mm, 150 m depth off Cedar Key, Florida; D= *Neodrillia blacki* Petuch, new species, holotype, length 28 mm, Middle Torch Key, Florida Keys; E= *Chicoreus rachelcarsonae* Petuch, 1987, length 40 mm, 150 m depth off Cedar Key, Florida; F= *Jaspidiconus pfluegeri* Petuch, new species, paratype, length 31 mm, Okaloacoochee Member, Fort Thompson Formation, Belle Glade, Palm Beach County, Florida; G= *Melongena (Rexmela) corona sprucecreekensis* Tucker, 1994, length 148 mm, Spruce Creek, Volusia County, Florida (for comparison with *M. corona winnerae*); H= *Modulus calusa* Petuch, 1988, width 12 mm, Little Torch Key, Florida Keys; I= *Jaspidiconus pfluegeri* Petuch, new species, holotype, length 25 mm, Lake Worth Lagoon, Riviera Beach, Palm Beach County, Florida; J= *Modulus calusa* Petuch, 1988, width 12 mm, Okaloacoochee Member, Fort Thompson Formation, Belle Glade, Palm Beach County, Florida; K= *Pusula lindajoyceae* Petuch, 1994, 17 mm, 60 m depth off Palm Beach Island, Palm Beach, Florida (first record of a living *Pusula* in the Atlantic Ocean).

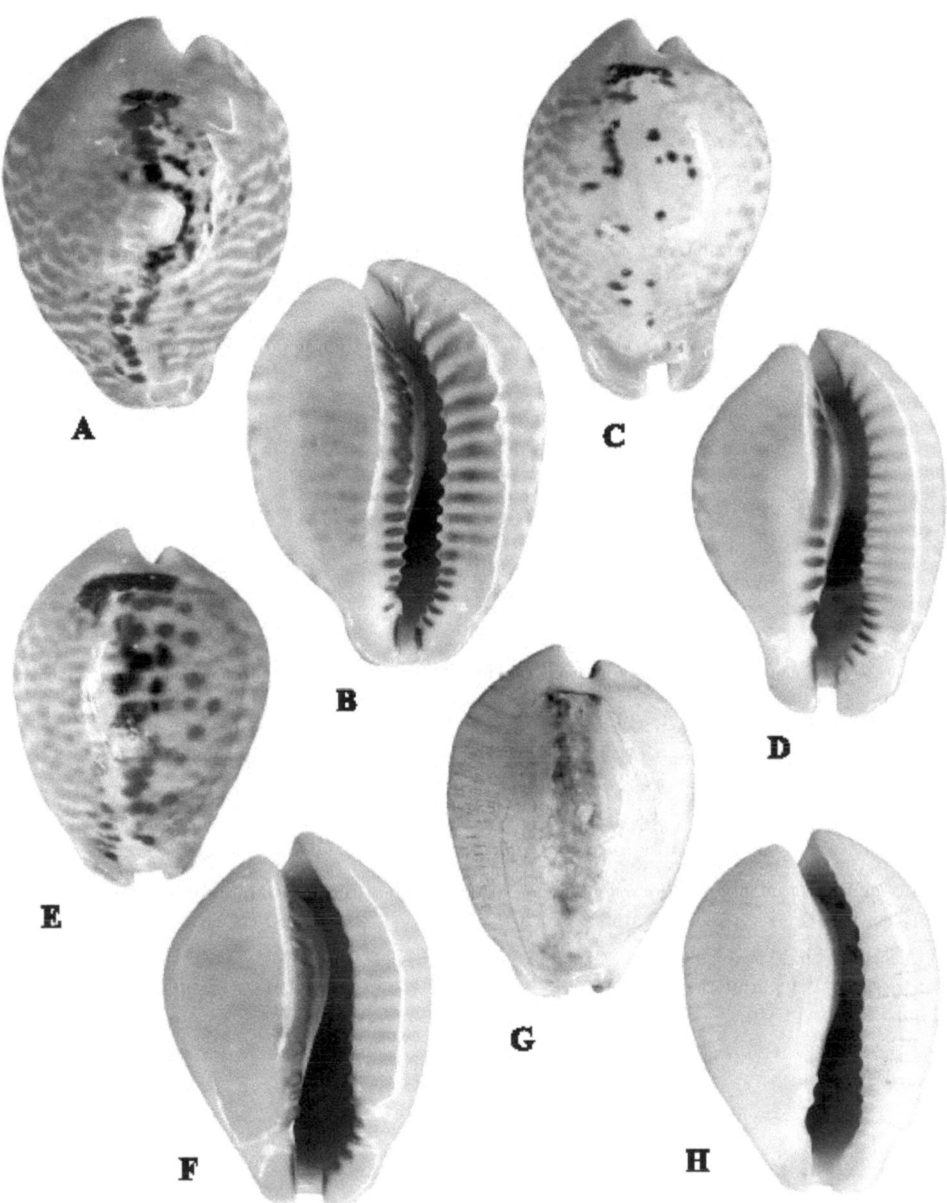

Plate 98. Colombian and Venezuelan relict cowries of the genus *Muracypraea*. Included are: A,B= *Muracypraea donmoorei* (Petuch, 1979), length 59 mm, dredged from 35 m depth off Cabo La Vela, Goajira Peninsula, Colombia; C,D= *Muracypraea tristensis* (Petuch, 1987), length 49 mm, dredged from 100 m depth east of the Monges Islands, Gulf of Venezuela, Venezuela; E,F= *Muracypraea mus* (Linnaeus, 1758), length 41 mm, in Turtle Grass bed, Amuay Bay, Paraguana Peninsula, Falcon State, Venezuela; G,H= *Muracypraea mus* (Linnaeus, 1758), fossil specimen, length 42 mm, un-named Sangamonian Pleistocene formation, Amuay Bay, Paraguana Peninsula, Falcon State, Venezuela.

pocket is the presence, in slightly deeper waters adjacent to the Cays, of the pleioptygmatid *Pleioptygma helenae* (Radwin and Bibbey, 1977), the last-living member of this Caloosahatchian genus. The Bermont-aged relict melongenid *Melongena (Rexmela) bispinosa* (Philippi, 1844) occurs just north of the Honduran area, where it is abundant in the mangrove jungles and mud flats of the Yucatan Peninsula of Mexico.

The only well-developed secondary relict pocket within the Caribbean Basin is found along the Colombian and Venezuelan coasts, from the mouth of the Magdalena River to the Golfo de Venezuela and across to the Golfo Triste. Here, the last-living assemblage of Baitoan and Gatunian taxa is now known to exist (Petuch, 1980, 1982, 1986, 1987). Of special interest along this coast is a living fauna of the relict cypraeid genus *Muracypraea*. This once-widespread Mio-Pliocene genus is represented today by only three species; the shallow water (0-10 m) *Muracypraea mus* (Linnaeus, 1758) (Plate 98, E, F, G, and H) and the deep water (20-100 m) *M. donmoorei* (Petuch, 1979) (Plate 98, A and B) (incorrectly synonymized with "*Cypraea bicornis*" Sowerby, a *nomen nudum*) and *M. tristensis* (Petuch, 1987) (Plate 98, C and D). *Muracypraea mus* prefers to live in intertidal depths on muddy Turtle Grass beds, where it feeds on the grass blades and on epibionts. This *Thalassia*-based ecology approximates the lifestyles of its extinct Caloosahatchian relatives, the genera *Akleistostoma*, *Pahayokea*, *Pseudadusta*, and *Siphocypraea*. The deep water *M. donmoorei* and *M. tristensis* live on offshore sponge banks, where they feed on both Euspongia and Calcarea, much like the living *Zoila* cowries of southern Australia. These deeper water *Muracypraea* species live together with an entire fauna of relictual Gatunian gastropod genera, some of which include the turrid *Paraborsonia*, the volutomitrid *Conomitra*, the muricid *Panamurex*, and the conids *Protoconus* (*P. consobrinus-granarius-archon* complex) and *Ximeniconus* (*X. puncticulatus-perplexus* complex). The Brazilian Province to the south also is known to have acted as a refugium for several Plio-Pleistocene relictual genera. Here, particularly along the Bahia coast, the Bermont strombid genus *Titanostrombus* and the Everglades Pseudoatoll and reef-associated fasciolariid genus *Pleuroploca* (*P. aurantiaca-lindae* complex) occur abundantly. Farther offshore in deeper water along the central Brazilian coast, an even older refugium has apparently existed since at least the late Miocene. Here, the Transmarian and Baitoan endemic genera *Sediliopsis* (Turridae), *Pseudaptyxis* (Fasciolariidae), and *Plochelaea* (subgenus *Plicoliva*) (Volutidae) have managed to survive to the Recent. Discoveries such as these have shown that many of the neritic ecosystems of the Recent western Atlantic are actually chronologically hybridized assemblages; being combinations of newly evolved late Pleistocene-Holocene faunas and faunal remnants from predecessor paleoseas.

Systematic Appendix

Descriptions of New Index Species and Genera

During the course of writing this book, I found that a number of stratigraphically and ecologically-important taxa were still undescribed and un-named. These included 37 species of gastropods, one species of bivalve, and 11 new genera and three new subgenera of gastropods. Since the completeness of my paleoecological, stratigraphic, and evolutionary analyses depended on these taxa having available, stable names, I have decided to describe them in the following systematic section. I feel this is the perfect place to describe these important new organisms; in context with their home paleoseas and the local paleoceanography, their resident communities and environments, and their regional biogeographical frameworks. Several of the new species are the namesakes of their communities and the need for formal scientific descriptions brings to mind the old adage from Confucius; "the beginning of wisdom is calling things by their right names ."

The following descriptions are arranged by standard molluscan systematic ordering. The type material of the new fossil species is deposited in the type collection of the Division of Paleontology, American Museum of Natural History, New York, and bears AMNH FI (fossil invertebrates) numbers. The type material of the new Recent species is deposited in the living mollusks type collection and bears AMNH numbers.

Mollusca
Gastropoda
Caenogastropoda
Cerithioidea
Potamididae
Cerithidea Swainson, 1840

Cerithidea beattyi lakeworthensis new subspecies
(Plate 94, C)

Description: Shell small for genus, elongated, turriculate; suture deeply indented, producing strongly concave whorls; body and spire whorls ornamented with 19-21 thin, sharp ribs; whorls sculptured with 2 large spiral cords around midbody; intersection of thin ribs and spiral cords producing low, faint raised bead; base of shell smooth, ornamented by single large cord at intersection with ribs; outer lip thickened and flaring in adults.
Type Material: Holotype; AMNH FI-50643, length 11 mm, width 5 mm; Paratypes; AMNH FI-50644, length 10 mm; 3 specimens, research collection of the author, lengths 10-12 mm, same locality as holotype.
Type Locality: The holotype was collected at a construction dig on Lake Worth Road, 1.3 km west of the Florida Turnpike, Lake Worth, Palm Beach County, Florida.
Stratigraphic Range: Confined to the Coffee Mill Hammock Member of the Fort Thompson Formation, Okeechobee Group, late Sangamonian Pleistocene.
Etymology: Named for Lake Worth, Florida, the type locality.
Discussion: The new taxon represents an extinct subspecies of *Cerithidea beattyi* Bequaert, 1942, a species now confined to the eastern Caribbean, from the Bahamas to Trinidad. The Pleistocene Floridian subspecies differs from the Recent Caribbean nominate form in consistently having more inflated, convex whorls, in having finer, thinner, and more numerous ribs, in having lower, less-developed spiral cords, and in having smaller, almost-obsolete beads at the intersections of the ribs and spiral cords.

Cerithidea scalariformis palmbeachensis new subspecies
(Plate 94, D)

Description: Shell of average size for genus, thin, elongated, protracted; whorls rounded and inflated; spire whorls with 16-18 low, coarse ribs per whorl; ribs become lower and less developed on last 3 whorls; base of shell (anterior end) ornamented with 6 strong spiral cords; spire and body whorls ornamented with 5 low spiral cords, with cords on shoulder near suture being smaller; intersection of ribs and spiral cords producing low, faint bead; outer lip thickened and flaring in adults.
Type Material: Holotype; AMNH FI-50645, length 23 mm, width 10 mm.
Type Locality: The holotype was collected at a construction dig on Lake Worth Road, 1.3 km west of the Florida Turnpike, Lake Worth, Palm Beach County, Florida.
Stratigraphic Range: Confined to the Coffee Mill Hammock Member of the Fort Thompson Formation, Okeechobee Group, late Sangamonian Pleistocene.
Etymology: Named for Palm Beach County, Florida.
Discussion: The new taxon represents an extinct subspecies of the wide-ranging western Atlantic (South Carolina-West Indies) *Cerithidea scalariformis* (Say, 1825). The new Pleistocene subspecies differs from its Recent descendant in having fewer, lower, and less-developed ribs on the spire whorls, in having smoother sculpture on the last two whorls, with poorly developed and almost obsolete ribs, and in having stronger and better-developed spiral cords around the spire and body whorls.

Pyrazisinus Heilprin, 1886

Pyrazisinus ultimus new species
(Plate 94, B)

Description: Shell of average size for genus, elongated and protracted; shoulder of whorls sharply-angled, flaring outward; subsutural area only slightly sloping, giving flattened appearance to entire shoulder area; spire whorls with 11-12 large, evenly-spaced flattened ribs; body whorl with 8-10 very large, prominent, knoblike ribs; both spire and body whorl ribs angled forward; spire whorls sculptured with 8 thin, flattened spiral threads; body whorl with 13-15 large, flattened spiral cords around anterior half; posterior half of body whorl with 8 thin, flattened spiral threads; siphonal canal short, stumpy, only partially (usually half) closed on adult specimens; outer lip of adults large, thickened, flaring.
Type Material: Holotype; AMNH FI-50647, length 53 mm; Paratypes; AMNH FI-50648, length 52 mm; 3 specimens in the research collection of the author, lengths 48-55 mm, same locality as holotype.
Type Locality: The holotype was collected at a construction dig along US 441 and Forest Hill Boulevard, Wellington, Palm Beach County, Florida.
Stratigraphic Range: Confined to the Coffee Mill Hammock Member of the Fort Thompson Formation, Okeechobee Group, late Sangamonian Pleistocene.
Etymolgy: Named in reference to the new species' being the last-living ("ultimus") member of its genus.
Discussion: *Pyrazisinus ultimus* is most similar to *P. gravesae* Petuch, 1994 (Plate 92, C) from the underlying and older Okaloacoochee Member of the Fort Thompson Formation, but differs in having a broader, wider shell with a much sharper and more pronounced shoulder angle, in having fewer and proportionally larger ribs, and in having a distinctly flattened subsutural area instead of a sloping, angled one.

Turritellidae
Turritellinae
Mariacolpus Petuch, 1988

Mariacolpus covepointensis new species
(Plate 32, J)

Description: Shell of average size for genus, elongated, thin and fragile; suture deeply indented, producing overhanging whorls; spire and body whorls ornamented with 5 thin main spiral cords; between main cords are 1 or 2 extremely fine spiral threads; spiral cord at whorl periphery largest, producing overhanging carina; subsutural area with 2-3 extremely thin spiral threads; body whorl often partially disconnected from previous whorls.
Type Material: Holotype; AMNH FI-50649, length 33 mm, width 11 mm; Paratypes; 3 specimens, research collection of the author, 31-34 mm, same locality as holotype.

Type Locality: The holotype was collected in Shattuck Zone 23, Little Cove Point Member, St. Mary's Formation, Chesapeake Group, at Little Cove Point, Calvert County, Maryland.
Stratigraphic Range: Confined to the Little Cove Point Member of the St. Mary's Formation, Chesapeake Group, latest Serravallian Miocene.
Etymology: Named for Little Cove Point, the type locality, where the new species is extremely abundant.
Discussion: This abundant and characteristic lower St. Mary's fossil has traditionally been assigned the taxon *Mariacolpus plebeia* (Say, 1824). Recently, however, Ward (1992) has demonstrated that the true *M. plebeia* (as "*Turritella plebeia* ") is actually from the much younger Eastover Formation of Virginia and is morphologically distinct from St. Mary's species. As a result of this discovery, it is now known that the *Mariacolpus* species found in the Little Cove Point and the Windmill Point Members are both in need of new names. The Little Cove Point species described here is similar to both the Eastover *M. plebeia* and Windmill Point *M. lindajoyceae* (described next) but is distinct in several subtle but consistent ways. From *M. lindajoyceae, M. covepointensis* differs in having proportionally much thinner and more threadlike spiral cords, in having more sharply overhanging peripheral edges on the last 3 or 4 whorls, and in having more numerous and finer secondary threads between the main cords, particularly in the subsutural area. From *M. plebeia, M. covepointensis* differs in having a much more pronounced and sharply angled periphery, in having more distinctly overhanging peripheral edges, and in having the peripheral cord being the largest as opposed to having the peripheral and central cords being equal in size as in *M. plebeia*.

Mariacolpus lindajoyceae new species
(Plate 36, H)

Description: Shell of average size for genus, elongated, thin and fragile; suture indented, producing rounded, convex whorls; peripheries of whorls slightly angled; whorls ornamented with 5 main spiral cords around mid-body; peripheral cord largest, producing slightly overhanging whorls; central main cords bounded on either side by bands of extremely fine spiral threads, usually 3-5 in number.
Type Material: Holotype; AMNH FI-50650, length 37 mm, width 11 mm; Paratypes; AMNH FI-50651, 5 specimens, lengths 35-38 mm, same locality as holotype; 5 specimens, research collection of the author, lengths 33-38 mm.
Type Locality: The holotype was collected in the Windmill Point Member of the St. Mary's Formation, Chesapeake Group, at Chancellor's Point, St. Mary's River, St. Mary's County, Maryland.
Stratigraphic Range: Confined to the Windmill Point Member of the St. Mary's Formation, Chesapeake Group, early Tortonian Miocene.
Etymology: Named for my wife, Linda Joyce Petuch, who collected the holotype at Chancellor's Point.
Discussion: Of the three closely-related St. Mary's and Eastover *Mariacolpus* species, *M. lindajoyceae* is closest to the younger *M. plebeia* from Virginia (last-living *Mariacolpus* species), particularly in having the same rounded whorls and lack of a distinct overhanging periphery. The new Windmill Point species differs from its Eastover descendant in having coarser and more evenly spaced spiral cords arranged in a group of five around the central area of each whorl. In *M. plebeia*, the cords are much finer and more numerous, with two of the central cords being stronger and better developed, giving each whorl a bicarinate appearance. The two strong cords are particularly well developed on the early whorls, a feature that is lacking in *M. lindajoyceae*. This bicarinate sculpture is carried to its fullest development in another Eastover species, *M. carinata* (Gardner, 1948).

Stromboidea
Strombidae
Strombus Linnaeus, 1758

Strombus alatus kendrewi new subspecies
(Plate 92, G and H)

Description: Shell of average size for genus, inflated, with high, protracted spire; body whorl smooth, polished, with rounded, unornamented shoulder; spire whorls with sharply angled shoulders, ornamented with large pointed knobs or short spikes, usually 8-10 per whorl; anterior end of shell ornamented with 10-12 low, thick cords.

Type Material: Holotype; AMNH FI-50652, length 83 mm, width 56 mm; Paratypes; 3 specimens, in the research collection of the author, lengths 80-86 mm, same locality as holotype.
Type Locality: The holotype was collected in the Okaloacoochee Member of the Fort Thompson Formation, Okeechobee Group, at a construction dig at Parkland, near Coral Springs, Broward County, Florida.
Stratigraphic Range: Confined to the Okaloacoochee Member of the Fort Thompson Formation, Okeechobee Group, early Sangamonian Pleistocene.
Etymology: Named for Mr. Eric Kendrew of Valrico, Florida, who collected the holotype.
Discussion: This new taxon represents a late Pleistocene subspecies of the Recent *Strombus alatus* Gmelin, 1791 and is the perfect morphological intermediate between the mid-Pleistocene *S. evergladesensis* Petuch, 1991 (Plate 90, A) and the Coffee Mill Hammock and living *S. alatus* (Plate 93, B). The new subspecies has the smooth, polished, inflated body whorl and lack of body whorl shoulder knobs of the Bermont *S. evergladesensis*, but yet has the spire of the Coffee Mill Hammock and Recent *S. alatus*, with the prominent, large, sharp shoulder knobs. Although occasional specimens of *S. evergladesensis* may show some shoulder knob development on the spire whorls, it is never as prominent as that of *S. alatus kendrewi*.

Strombus lindae dowlingorum new subspecies
(Plate 92, J)

Description: Shell small for genus, inflated, with high, protracted spire; body whorl shoulder angled, variably ornamented with either small knobs (like the holotype) or being completely smooth; on knobbed specimens, number of shoulder knobs on body whorl varies from 8 (as on holotype) to 10; spire whorls heavily ornamented with small, regularly spaced rounded knobs, averaging 12 per whorl; main part of body whorl sculptured with 10-12 low, wide spiral cords; anterior end heavily sculptured with 10 large raised spiral cords; outer lip of adults broad, flaring; color pattern, when preserved (as on holotype), composed of 2 broad bands, one around mid-body and one around anterior end.
Type Material: Holotype; AMNH FI-50653, length 55 mm, width 31 mm; Paratypes; 2 specimens, research collection of the author, lengths 53 mm and 57 mm.
Type Locality: The holotype was collected in the Okaloacoochee Member of the Fort Thompson Formation, Okeechobee Group, at the Boynton sand pit, along US 441 between Boynton Beach Boulevard and Lantana Road, west of Boynton Beach, Palm Beach County, Florida.
Stratigraphic Range: Confined to the Okaloacoochee Member of the Fort Thompson Formation, Okeechobee Group, early Sangamonian Pleistocene.
Etymology: Named for Mr. William and Mrs. Cathy Dowling of Lake Worth, Florida, who collected the holotype.
Discussion: The new taxon represents a late Pleistocene subspecies of the mid-Pleistocene *Strombus lindae* Petuch, 1991 (Plate 89, H). The new Okaloacoochee descendant differs from the nominate Bermont ancestor in having a broader, wider shell with a more flaring adult lip, in having a proportionally lower spire, and in having a smoother body whorl with reduced spiral cord sculpture. The members of the *Strombus lindae* species complex, with their small and heavily sculptured shells, represent a separate, dwarf offshoot of the Caloosahatchee *S. keatonorum* Petuch, 1994 (Plate 80, F). This Calabrian ancestor gave rise to both the *S. alatus* and *S. lindae* complexes, which co-occurred until the early Sangamonian. At that time, the *S. lindae* complex became extinct and only the *S. alatus* group survived to the Recent.

Strombus mardieae new species
(Plate 17, A, I, and J)

Description: Shell small for genus, proportionally narrow, with high, protracted spire; shoulder of body whorl sharply angled, ornamented with 10 large, pointed knobs; last 2 spire whorls with 10 large knobs per whorl; early whorls with 16-18 low, rounded knobs per whorl; body whorl heavily sculptured with 21-25 large, prominent, evenly spaced spiral cords; finer spiral threads present between cords; spire whorls sculptured with 6-7 large spiral cords; anterior end encircled with 8 small, low spiral cords; aperture proportionally narrow, with lip of adults not expanded, paralleling parietal area; posterior edge of lip almost perpendicular to shell axis, only slightly sloping.
Type Material: Holotype; AMNH FI-50654, length 66 mm, width 46 mm; Paratype; in the research collection of the author, length 37 mm, same locality as holotype.

Type Locality: The holotype was collected in the Chipola Formation, along the Chipola River, 0.5 km north of the mouth of Tenmile Creek, Calhoun County, Florida.
Stratigraphic Range: Confined to the Chipola Formation, Alum Bluff Group, Burdigalian Miocene.
Etymology: Named for Ms. Mardie Drolshagen Banks, Visual Communications Department, Florida Atlantic University, who patiently took and crafted every photograph used in this book.
Discussion: This new species represents the third known *Strombus* to be found in the Chipola Formation, occurring together with *S. chipolanus* Dall, 1890 (Plate 17, E and F) and *S. aldrichi* Dall, 1890 (Plate 17, G and H). Of these, *S. mardieae* is most similar to *S. chipolanus*, but differs in being a much more slender and elongated shell with a proportionally higher spire, in having a less-flaring, straight-edged lip, in having a proportionally-smaller and narrower aperture, and in having stronger and more prominent spiral cords on the body whorl and spire. The posterior end of the lip of *S. chipolanus* is angled anteriorly, sloping strongly downward from the shoulder, while that of *S. mardieae* is relatively straight and flat, with only a small angle, being essentially perpendicular to the shell axis. Of the three sympatric Chipola species, *S. mardieae* is the rarest, with only a few specimens having ever been collected, while *S. aldrichi* is the commonest, often accumulating in dense beds.

Cypraeoidea
Cypraeidae
Bernayinae
Akleistostoma Gardner, 1948

Akleistostoma bairdi new species
(Plate 71, E and F)

Description: Shell of average size for genus, inflated, elongately-pyriform, with protracted anterior end; base of shell slightly flattened; margins of shell thickened in adults but only moderately developed; apical sulcus wide, open, slotlike, with slight but noticeable clockwise twist; aperture proportionally narrow, widening at anterior end; edge of lip with 20-22 large teeth; columella with 17-18 thin, narrow teeth; columellar teeth often bifurcate or form smaller, partial teeth between main teeth; auricles poorly developed.
Type Material: Holotype; AMNH FI-50655, length 51 mm, width 32 mm; Paratypes; 3 specimens, in the research collection of the author, lengths 48-54 mm, same locality as the holotype.
Type Locality: The holotype was collected in the Fruitville Member (Sarasota Unit 3) equivalent of the Tamiami Formation, Okeechobee Group, from dredging along the Kissimmee River at Fort Basinger, Highlands County, Florida.
Stratigraphic Range: Confined to the Unit 3-equivalent Kissimmee beds of the Tamiami Formation, Okeechobee Group, Piacenzian Pliocene.
Etymology: Named for Dr. Donald Baird of Boynton Beach, Florida.
Discussion: This new cowrie forms the morphological intermediate between two Tamiami species; the older (Sarasota Unit 4 equivalent) *A. hughesi* (Olsson and Petit, 1964) (Plate 69, G and H) and the younger (Sarasota Unit 2 equivalent) *A. diegelae* (Petuch, 1994) (Plate 71, A and B). From the older *A. hughesi, A. bairdi* differs in being a much more slender, pyriform shell with less developed margins, in having a more open and twisted apical sulcus, and in having finer and more numerous columellar teeth. Although the elongated pyriform shape of *A. bairdi* was retained by the younger *A. diegelae,* that descendant species differed from the ancestral new species in being a much smaller shell with a narrower, slotlike apical sulcus, in having a narrower aperture, and in having fewer and much coarser columellar and labial teeth. The species complex of *A. hughesi, A. bairdi,* and *A. diegelae* all evolved from the older (Pinecrest equivalent) *A. transitoria* (Olsson and Petit, 1964) (Plate 69, C and D) and remained confined to the Kissimmee Embayment throughout their entire stratigraphic ranges. This tight-knit group appears to represent a separate lineage off the main *Akleistostoma* line and may represent a new subgenus.

Calusacypraea Petuch. 1996

Calusacypraea globulina new species
(Plate 5, F and G)

Description: Shell small for genus, extremely inflated, globiform, and fragile; base of shell rounded; margins not

developed; posterior end of lip projecting beyond spire; apical area open, with tip of spire visible within apical pit; aperture wide, open, flaring toward anterior end; lip with 22-23 small teeth; columella with 18-20 small teeth, some of which bifurcate; auricles poorly-developed, almost obsolete.
Type Material: Holotype; AMNH FI-50656, length 55 mm, width 35 mm; Paratypes; two specimens, in the research collection of the author, lengths 53 mm and 56 mm, same locality as holotype.
Type Locality: The holotype was collected in the Pinecrest Member (Sarasota Unit 7) of the Tamiami Formation, in Quality Aggregates, Inc. Phase 6 pit, Fruitville, Sarasota, Sarasota County, Florida.
Stratigraphic Range: Confined to the Pinecrest Member (Unit 7) of the Tamiami Formation, Okeechobee Group, Piacenzian Pliocene, and found only in the Myakka Lagoon area.
Etymology: "Little globe," in reference to the new species' extremely globular, inflated appearance.
Discussion: This newest species of *Calusacypraea* (s.s.) forms the morphological intermediate between two Tamiami estuarine cowries; the older (Buckingham Member, Unit 10) *C. duerri* Petuch, 1996 (Plate 52, J and K) and the younger (Fruitville Member, Unit 4) *C. tequesta* Petuch, 1996 (Plate 64, H and I). Of these two, *C. globulina* is closest to *C. duerri*, but differs in being consistently a smaller, broader, more globose, and inflated shell, in having a more rounded base, in lacking any marginal thickening (even on old adults), and in having coarser teeth. This last character is carried on and developed in the descendant species, *C. tequesta*, which has the largest and coarsest teeth of the genus.

Myakkacypraea new subgenus

Diagnosis: *Calusacypraea* species with more elongated and pyriform shells than seen in the nominate subgenus; apertures narrow, widening only at anterior end; apertural and columellar teeth small and fine; marginal calluses poorly developed or obsolete on old adults; posterior end of lip extending well beyond spire line; apical pit well developed, with tip of spire projecting within.
Type Species: *Calusacypraea (Myakkacypraea) myakka* Petuch, new species, described herein, Unit 4, Fruitville Member, Tamiami Formation, Okeechobee Group.
Other Species In Myakkacypraea: *C. (Myakkacypraea) briani* Petuch, 1996 (Unit 7, Pinecrest Member); *C. (Myakkacypraea) kelleyi* Petuch, 1999 (Unit 3, Fruitville Member); *C. (Myakkacypraea) schnireli* Petuch, new species, described herein (Unit 2, Fruitville Member).
Etymology: Named as a combination of "Myakka," for the Myakka Lagoon System, and "*Cypraea*."
Discussion: *Myakkacypraea* represents a separate lineage of *Calusacypraea* that broke off from *C. duerri* at the end of Buckingham time, and co-occurred with *Calusacypraea (Calusacypraea)* species until the late Piacenzian. Members of the new subgenus differ from members of the nominate subgenus in having elongated, distinctly pyriform shells that contrast with the rounded, globose shells of typical *Calusacypraea* species. The apertures of the two groups also differ markedly, with those of *Calusacypraea* s.s. being wide and flaring and those of *Myakkacypraea* being narrow and flaring only at the anterior end. The new subgenus has only been found in the Myakka Lagoon System deposits of the Sarasota area.

Calusacypraea (Myakkacypraea) myakka new species
(Plate 64, D and E)

Description: General shell characters as for description of subgenus; shell elongated, wide at posterior end and tapering gradually to anterior end; aperture proportionally narrow; lip with 22-24 small, pointed teeth; columella with 22-25 thin, almost obsolete teeth; posterior edge of lip projecting greatly beyond spire; auricles only slightly developed.
Type Material: Holotype; AMNH FI-50657, length 59 mm, width 33 mm; Paratypes; 3 specimens, in the research collection of the author, lengths 57-60 mm, same locality as holotype.
Type Locality: The holotype was collected in the Fruitville Member (Unit 4, "Black Layer") of the Tamiami Formation, Okeechobee Group, in the APAC pit, Fruitville, Sarasota, Sarasota County, Florida.
Stratigraphic Range: Confined to the Fruitville Member (Unit 4) of the Tamiami Formation, Okeechobee Group, Piacenzian Pliocene, and found only in the Myakka Lagoon area.
Etymology: Named for the Myakka Lagoon System of the Tamiami Subsea, Okeechobean Sea.
Discussion: The new species is only similar to *C. (Myakkacypraea) schnireli* new species (described next) from Unit 2 of the Fruitville Member, but differs in being a much larger, narrower, and more elongated shell with a more

cylindrical appearance. The posterior lip projection of *C. (Myakkacypraea) myakka* is also proportionally larger and better-developed than that of *C. (Myakkacypraea) schnireli.*

Calusacypraea (Myakkacypraea) schnireli new species
(Plate 67, I and J)

Description: General shell characters as for description of subgenus; shell small for subgenus, squat and distinctly pear-shaped, wide across shoulder and tapering rapidly to narrow anterior end; aperture proportionally narrow, widening only slightly at anterior end; lip with 15-17 large, coarse teeth; columella with 9-11 small, nearly obsolete teeth, with teeth being absent on the posterior one-third; apical pit shallow; posterior projection of lip well developed.
Type Material: Holotype; AMNH FI-50658, length 41 mm, width 26 mm; Paratypes; 2 specimens, in the research collection of the author, lengths 40 and 42 mm, same locality as holotype.
Type Locality: The holotype was collected in the upper part of the Fruitville Member (Unit 2) of the Tamiami Formation, Okeechobee Group, late Piacenzian Pliocene, in the Quality Aggregates, Inc. Phase 9 pit, Fruitville, Sarasota, Sarasota County, Florida.
Stratigraphic Range: Confined to the upper part of the Fruitville Member (Unit 2) of the Tamiami Formation, Okeechobee Group, latest Piacenzian Pliocene, and found only in the Myakka Lagoon area.
Etymology: Named for Mr. Brian Schnirel of Wellington, Florida, who collected the holotype.
Discussion: The morphological differences between this new species and its closest relative, the ancestral *C. (Myakkacypraea) myakka,* were discussed under the previous description.

Pseudadusta new genus

Diagnosis: Shells of average size for family, pyriform in shape, inflated, solid and heavy; posterior and anterior ends projecting, well-developed; shell base rounded; margins thickened, well-developed, often heavily callused; apical sulci simple, slotlike, almost straight or only slightly curved; apertures very narrow throughout, with no anterior widening; labial and columellar dentition extremely well-developed, with prominent coarse teeth.
Type Species: *Pseudadusta lindae* (Petuch, 1986) (as "*Cypraea lindae*") (Plate 62, A and B), Golden Gate Member of the Tamiami Formation, Okeechobee Group.
Other Species In Pseudadusta. *P. hertweckorum* (Petuch, 1991), Pinecrest Member (Unit 7); *P. ketteri* (Petuch, 1994), Fruitville Member (Unit 3); *P. kalafuti* (Petuch, 1994), Fruitville Member (Unit 3); P. *metae* (Petuch, 1994), Fruitville Member (Unit 4); *P. marilynae* (Petuch, 1994), upper part of the Fruitville Member (Unit 2); nine other undescribed species from various members of the Tamiami Formation, including the Buckingham Member (Unit 10), the Fruitville Member (Units 2, 3, and 4), and the Golden Gate Member.
Etymology: Named for the superficial resemblance to the Indo-Pacific erroniinine genus *Adusta* (with species such as *A. adusta, A. onyx, A. nymphae,* and *A. melanesiae*).
Discussion: This new, previously unrecognized genus of bernayinine cowries evolved from *Akleistostoma* Gardner, 1948 during the Zanclean Pliocene. Although sharing the same rounded, globose shape and having the same type of marginal callus development as *Akleistostoma, Pseudadusta* differs from its progenitor in having a very narrow aperture instead of an open one, in having narrower posterior and anterior canals, and in having a deeper and narrower apical sulcus. The new genus is also much more geographically limited, being found only in the Myakka Lagoon System and coral reef tracts of the Tamiami Subsea of the Okeechobean Sea. *Akleistostoma,* on the other hand, had a far greater range, with species being found all the way from Virginia to the northern Gulf of Mexico. *Pseudadusta* appears abruptly in the basal Piacenzian (Buckingham time) and existed only to the late Piacenzian (Fruitville Unit 2 time).

Siphocypraea Heilprin, 1886
Okeechobea new subgenus

Diagnosis: Neotenic *Siphocypraea* cowries that are convergent on *Calusacypraea,* having highly inflated, fragile, bulliform shells with juvenile appearances; apical sulci open and flaring, extruded, producing deeply depressed apical pit; tip of spire showing within apical pit; posterior extension of outer lip extremely well developed, extending beyond line of spire; aperture uniformly narrow; labial teeth small and undeveloped; columellar teeth

long and thin; auricles poorly developed, almost obsolete.
Type Species: *Siphocypraea (Okeechobea) brantleyi* Petuch, new species, described herein, Bee Branch Member of the Caloosahatchee Formation, Okeechobee Group.
Other Species In Okeechobea: *S. (Okeechobea) philemoni* Fehse, 1997, Ft. Denaud Member of the Caloosahatchee Formation (Plate 79, F and G); an elongated undescribed species from the uppermost bed of the Ft. Denaud Member of the Caloosahatchee Formation; a small, highly inflated undescribed species from the Ayer's Landing Member of the Caloosahatchee Formation.
Etymology: Named for the Okeechobean Sea, the home paleosea of the new subgenus.
Discussion: This newly discovered lineage of *Siphocypraea* was well on its way to becoming a full genus, but was killed-off by the late Calabrian and Nebraskan cold times. *Okeechobea* appears to have been a neotenic offshoot of *Siphocypraea*, just as *Calusacypraea* was a neotenic offshoot of *Akleistostoma* (Petuch, 1996). By the end of Caloosahatchee time, this group had evolved a number of bizarre forms, with uncoiled and flaring apical sulci and exaggerated posterior labial extensions. The members of the new subgenus differ from members of the nominate subgenus by having an extension on the posterior of the lip, causing it to project beyond the spire line, in having more open and flaring apical sulci, and in having reduced labial and columellar dentition. *Okeechobea* existed only during the Calabrian, being restricted to the Ft. Denaud, Bee Branch, and Ayer's Landing Members of the Caloosahatchee Formation of southern Florida.

Siphocypraea (Okeechobea) brantleyi new species
(Plate 82, A and B)

Description: General shell characters as for description of subgenus; shell highly inflated, rounded, globose, thin and fragile; posterior of lip greatly extended, projecting beyond line of spire; posterior labial projection involute, with columellar pillar exposed within; apical sulcus open, uncoiled, with tip of spire exposed within; lip with 19-21 small, rounded teeth; columella with 17-18 thin, elongated teeth.
Type Material: Holotype; AMNH FI-50659, length 56 mm, width 34 mm; Paratypes; 3 specimens, in the research collection of the author, lengths 52-59 mm, same locality as the holotype.
Type Locality: The holotype was collected in the Bee Branch Member of the Caloosahatchee Formation, Okeechobee Group, in the Brantley pit, south of Arcadia, De Soto County, Florida.
Stratigraphic Range: Confined to the Bee Branch Member of the Caloosahatchee Formation, Okeechobee Group, mid-Calabrian Pleistocene.
Etymology: Named for Mr. D. L. Brantley of Arcadia, Florida, who allowed me to conduct research on his property and who is an inspired amateur paleontologist.
Discussion: Of the described *Okeechobea* species, *S. (Okeechobea) brantleyi* is closest to the older *S. (Okeechobea) philemoni* Fehse, 1997 from the underlying Ft. Denaud Member (Plate 79, F and G). The new Bee Branch species differs from its Ft. Denaud ancestor in being a smaller, much more inflated and truncated shell with a much more produced and exaggerated posterior labial extension, and in having much smaller, finer, and less developed labial and columellar teeth.

Pahayokea new subgenus

Diagnosis: Small, squat, stocky *Siphocypraea* species with rounded, inflated shells and well-developed posterior and anterior tips; shell bases flattened, only slightly rounded; shell margins generally well-developed, often heavily callused and sometimes flanged; coiling of apical sulci variable, with some species having sulci with one complete turn and others having sulci with only partial coiling; labial and columellar dentition coarse and prominent, with the labial teeth often being bifurcated.
Type Species: *Siphocypraea (Pahayokea) penningtonorum* Petuch, 1994, upper Fruitville Member (Unit 2) equivalent in the Kissimmee Embayment, Tamiami Formation, Okeechobee Group.
Other Species In Pahayokea: *S. (Pahayokea) kissimmeensis* Petuch, 1994, Fruitville Member (Unit 4) equivalent in the Kissimmee Embayment, Tamiami Formation; *S. (Pahayokea) basingerensis* Petuch, new species, described herein, Fruitville Member (Unit 3) equivalent in the Kissimmee Embayment, Tamiami Formation; *S. (Pahayokea) gabrielleae* Petuch, new species, upper Fruitville Member (Unit 2) equivalent in the Kissimmee Embayment, Tamiami Formation; *S. (Pahayokea) rucksorum* Petuch, new species, described herein, upper Fruitville Member (Unit 2) in the Kissimmee Embayment, Tamiami Formation; *S. (Pahayokea) josiai* Fehse, 1997, Ft. Denaud Member

of the Caloosahatchee Formation; *S. (Pahayokea) aspenae* Petuch, new species, described herein, Bee Branch Member of the Caloosahatchee Formation; an undescribed species from the Ayer's Landing Member of the Caloosahatchee Formation.
Etymology: The subgenus name is a latinized version of "Pa-hay-okee," "River of Grass" in Choctaw-Seminole, in reference to the Everglades.
Discussion: This distinctive group of endemic Everglades cowries represents an offshoot of the *Siphocypraea* trippeana lineage that diverged during Fruitville Member (Unit 4) time. The new subgenus differs from *Siphocypraea* s.s. in having much fatter, broader shells with well developed and often flaring marginal calluses. Although members of this group resemble *Akleistostoma* in shell shape and general appearance, they retain typical *Siphocypraea* characteristics, such as having narrow apertures and coiled apical sulci. This last character is more variable in *Pahayokea* than in the nominate subgenus, with some species have only slightly coiled, slotlike sulci (such as *S. aspenae*) while others have tightly coiled sulci (such as *S. gabrielleae*). A unique feature of the group is the presence of raised lines, arranged in a reticulate pattern, on the dorsum of some individuals, giving the shell a cracked glass appearance. This reticulate dorsal pattern is seen in several species but is most prevalent on *S. (Pahayokea) penningtonorum* and *S. (Pahayokea) rucksorum*. Throughout its short stratigraphic range, *Pahayokea* was restricted to the Okeechobean Sea area, being found only in the upper beds of the Tamiami Formation (Fruitville Member equivalent in the Kissimmee Embayment) and in the Caloosahatchee Formation.

Siphocypraea (Pahayokea) aspenae new species
(82, G and H)

Description: General shell characters as for description of subgenus; shell cylindrical with marginal calluses only moderately developed; apical sulcus relatively straight, narrow, with only slight curve; lip with 24-26 strong, pointed teeth; columella with 21-23 long, thin teeth; auricles proportionally small, only slightly flared.
Type Material: Holotype; AMNH FI-50660, length 59 mm, width 35 mm; Paratypes; 3 specimens, in the research collection of the author, lengths 57-63 mm, same locality as holotype.
Type Locality: The holotype was collected in the Bee Branch
ember of the Caloosahatchee Formation, in the Brantley pit, south of Arcadia, DeSoto County, Florida.
Stratigraphic Range: Restricted to the Bee Branch Member of the Caloosahatchee Formation, Okeechobee Group, mid-Calabrian Pleistocene.
Etymology: Named for Ms. Aspen Brantley of Arcadia, Florida, daughter of Mr. D. L. Brantley, owner of the Brantley pit.
Discussion: Of the known *Pahayokea* species, *S. (Pahayokea) aspenae* is most similar to the older *S. (Pahayokea)* josiai Fehse, 1997 (Plate 79, K and L) from the underlying Ft. Denaud Member. The new descendant species differs from its ancestor in having a more cylindrical, less pyriform shell, in having more numerous and finer apertural and columellar teeth, and in having a straight, slotlike, almost uncoiled apical sulcus.

Siphocypraea (Pahayokea) basingerensis new species
(Plate 71, C and D)

Description: General shell characters as for description of subgenus; shell heavy, thickened, oval or pyriform in outline, squat and stocky; marginal calluses greatly developed, expanded, flaring, giving shell distinctly flattened appearance; lip with 20-21 large, pointed teeth; columella with 15-16 long, thin teeth; small secondary teeth sometimes present between primary teeth; apical sulcus narrow, only partially curved, roughly one-half turn; auricles well developed, flangelike.
Type Material: Holotype; AMNH FI-50661, length 58 mm, width 35 mm; Paratypes; 2 specimens, in the research collection of the author, lengths 54 and 56 mm, same locality as the holotype.
Type Locality: The holotype was collected in the upper Tamiami Formation (Fruitville Unit 3 equivalent), from dredgings of the Kissimmee River at Fort Basinger, Highlands County, Florida.
Stratigraphic Range: Confined to the Fruitville Member (Unit 3) equivalent of the Tamiami Formation Okeechobee Group, in the Kissimmee Embayment, Piacenzian Pliocene.
Etymology: Named for Ft. Basinger, Highlands County, Florida, the type locality.
Discussion: Of the known *Pahayokea* species, *S. (Pahayokea) basingerensis* is most similar to the older *S. (Pahayokea) kissimmeensis* Petuch, 1994 (Plate 69, A and B) from the underlying Unit 4 (Fruitville Member) equivalent beds in

the Kissimmee Embayment. The new species differs from its ancestor in being a smaller, squatter, more ovately pyriform shell with more rounded, less-flattened margins, in having fewer and coarser teeth, and in having a more-coiled apical sulcus.

Siphocypraea (Pahayokea) gabrielleae new species
(Plate 71, G, H, and I)

Description: General shell characters as for description of subgenus; shell broadly oval in shape; marginal calluses greatly developed, flaring, expanded, producing thin, flangelike marginal edge and distinctly flattened appearance; base of shell flattened; lip with 21-23 large teeth; teeth often bifurcated and with smaller secondary teeth often being present between primary teeth; four anteriormost labial teeth being largest; columella with 22-24 long, thin teeth, with some being bifurcated; apical sulcus highly coiled, completing one full turn; apical sulcus partially covered by overhanging edge of left posterior marginal callus; auricles very well developed, flattened, flaring, flangelike.
Type Material: Holotype; AMNH FI-50662, length 65 mm, width 47 mm; Paratypes; 3 specimens, in the research collection of the author, lengths 63-68 mm, same locality as the holotype.
Type Locality: The holotype was collected in the uppermost bed of the Tamiami Formation (Fruitville Member Unit 2 equivalent), at the bottom of the Rucks pit, Fort Drum, Okeechobee County, Florida.
Stratigraphic Range: Confined to Unit 2 (Fruitville Member) equivalent, Tamiami Formation, Okeechobee Group, in the Kissimmee Embayment, latest Piacenzian Pliocene.
Etymology: Named for Ms. Gabrielle LaRose of Ft. Drum, Florida, who collected the holotype (and ruined her cellular phone in the process).
Discussion: Of the known *Pahayokea* species, this interesting cowrie is the most flattened and has the broadest and most flaring marginal calluses. In this aspect, *S. (Pahayokea) gabriellae* is reminiscent of the flattened *Zoila rosselli* from Recent southwestern Australia. The new species is most similar to *S. (Pahayokea) penningtonorum* Petuch, 1994 (Plate 70, F and G) from the basal part of the Unit 2 Tamiami Formation equivalent, but differs in being a much broader, more oval and flattened shell, in having much better developed and flattened marginal calluses, and in having a much more coiled apical sulcus (one complete turn) that is overhung by the left posterior marginal callus. The extremely flattened shape and partially covered apical sulcus of the new species may have been adaptations for living in very muddy environments with soft, flocculent substrates.

Siphocypraea (Pahayokea) rucksorum new species
(Plate 71, J, K, and L)

Description: General shell characters as for subgenus; shell large for subgenus, ovately-pyriform, inflated, with high, domed dorsum; marginal calluses very thick, well developed, distinctly rounded, often ornamented with low, rounded wartlike protuberances and undulations; surface of dorsum characteristically covered with reticulate pattern of raised lines (faint on holotype, stronger on some individuals); lip with 21-23 large, pointed teeth; smaller secondary teeth often present between primary teeth; columella with 19-21 thin, long teeth, some of which bifurcate; apical sulcus narrow, open, with one complete turn.
Type Material: Holotype, AMNH FI-50663, length 63 mm, width 47 mm; Paratypes; 2 specimens, in the research collection of the author, lengths 67 and 71 mm, same locality as holotype.
Type Locality: The holotype was collected in the uppermost bed of the Tamiami Formation (Fruitville Member Unit 2 equivalent), at the bottom of the Rucks pit, Fort Drum, Okeechobee County, Florida.
Stratigraphic Range: Confined to Unit 2 (Fruitville Member) equivalent, Tamiami Formation, Okeechobee Group, in the Kissimmee Embayment, latest Piacenzian Pliocene.
Etymology: Named for the Rucks Family of Okeechobee and Fort Drum, Florida, in appreciation of their allowing me to conduct field work in their Fort Drum quarry.
Discussion: The new species has the general shell shape and rounded marginal callosities of the older *S. (Pahayokea) basingerensis*, but differs in being a much larger and more inflated shell and in having a much more coiled apical sulcus. *Siphocypraea (Pahayokea) rucksorum* is also similar to the contemporaneous, but probably not directly sympatric, *S. (Pahayokea) gabrielleae*, but differs in being a much more globose, inflated, and domed shell, in lacking sharp-edged, flattened margins and having rounded marginal calluses that are ornamented with low protuberances, in having coarser apertural teeth, and in lacking the posterior marginal edge that overhangs the

apical sulcus. With its rounded, inflated shell, *S. rucksorum* probably lived in Turtle Grass beds while its contemporary congener, *S. gabrielleae* with its flattened shell, probably lived on nearby open mud bottoms and mud flats.

Cypraeorbinae
Floradusta new genus

Diagnosis: Cypraeorbinine cowries with larger shells than average, pyriform, inflated but often somewhat flattened, thin and fragile; extremities projecting; spires covered with small, dimple-like callus; apertures uniformly narrow; teeth simple, proportionally large and coarse; fossula absent.
Type Species: *Floradusta heilprini* (Dall, 1890) (originally as "*Cypraea heilprini*") (Plate 19, B and C), Chipola Formation, Alum Bluff Group, Burdigalian Miocene of northern Florida.
Other Species In Floradusta: *F. ballista* (Dall, 1915), Tampa Formation, late Chattian Oligocene-early Aquitanian Miocene; *F. emilyae* (Dolin, 1991), Chipola Formation, Burdigalian Miocene; *F. praelatior* (Dolin, 1991), Chipola Formation, Burdigalian Miocene; *F. shirleyae* (Dolin, 1991), Chipola Formation, Burdigalian Miocene; *F. alumensis* (Ingram, 1948), Chipola Formation, Burdigalian Miocene; *F. willcoxi* (Dall, 1890), Chipola Formation, Burdigalian Miocene.
Etymology: Named as a combination of "Florida" and the cowrie genus *Adusta*, in reference to the new taxons' superficial resemblance to that Indo-Pacific erroniinine genus.
Discussion: Although having the same general shape and size as members of the genus *Adusta* of the Indo-Pacific (Erroniinae), this new group of endemic Floridian cowries still retains many primitive characters found in other American groups of the primarily-Paleogene subfamily Cypraeorbinae. Some of these plesiomorphic characters include simple dentition, the lack of a fossula, and a dimpled callus over the spire. *Floradusta* represents the last wave of cypraeorbinine evolution in North America and replaced *Cypraeorbis* in the early Miocene. By the end of Burdigalian time, this group of endemic, over-sized cypraeorbinines became extinct.

Loxacypraea new genus

Diagnosis: Cypraeorbinine cowries with larger shells than average, heavy, thickened, broadly oval, highly inflated; extremities blunt, only moderately projecting; margins extremely thickened and callused, giving adults of some species flattened appearance, marginal calluses often covering dorsum of shell, producing faint dorsal furrow or dorsal band; aperture uniformly narrow; dentition simple, proportionally large and coarse; fossula absent
Type Species: *Loxacypraea chilona* (Dall, 1900) (originally as "*Cypraea chilona*") (Plate 3, B and C), Chipola Formation, Alum Bluff Group, Burdigalian Miocene of northern Florida.
Other Species In Loxacypraea: *L. tumulus* (Heilprin, 1886), Tampa Formation, late Chattian Oligocene-early Aquitanian Miocene; *L. arlettae* (Dolin, 1991) (Plate 18, A and B), Chipola Formation, Burdigalian Miocene; *L. hertleini* (Ingram, 1948) (Plate 21, M and N), Chipola Formation, Burdigalian Miocene; *L. apalachicolae* (Ingram, 1948) (Plate 21, K and L), Chipola Formation, Burdigalian Miocene; *L.* species (possibly as many as 3), Culebra Formation, Burdigalian Miocene of Panama; *L. fresnoensis* (Anderson, 1905), Vaqueros and Coalinga Formations, Burdigalian Miocene of southern California.
Etymology: Named as a combination of "loxa," Choctaw-Seminole for "turtle," and *Cypraea*, in reference to the new taxons' broad, inflated, turtle-like appearance.
Discussion: Along with *Floradusta*, *Loxacypraea* represents the last American evolutionary radiation of the subfamily Cypraeorbinae. The new genus, like its more slender, pyriform sister group, retains the primitive cypraeorbinine characters of lacking a fossula and having simple, coarse apertural dentition. *Loxacypraea* differs from *Floradusta* in having larger, heavier, and broader shells with much better-developed marginal calluses. Unlike *Floradusta*, which was confined to Florida throughout its stratigraphic range, *Loxacypraea* was much more widely distributed, being found in Panama and southern California. Interestingly, the new genus coexisted with the bernayinine cowrie genus *Muracypraea* in Panama, where it may have survived until the Tortonian Miocene.

Cypraeinae
Luria Jousseaume, 1884

Luria voleki new species
(Plate 88, A and B)

Description: Shell small for genus, elongately ovoid, subcylindrical, inflated; extremities blunt, rounded; margins thickened, forming well-developed lateral and spire calluses; base rounded, only slightly flattened along edges; aperture uniformly narrow, arcuate at posterior end; edge of labrum with 32-34 very fine, thin, narrow teeth (34 on holotype); columellar edge with 20-22 thin, narrow teeth (22 on holotype); fossula well developed, with 8-10 small teeth (10 on holotype).
Type Material: Holotype; AMNH FI-50664, length 33 mm, width 20 mm; Paratypes; 2 specimens, in the research collection of the author, lengths 35 and 38 mm, same locality as holotype.
Type Locality: The holotype was collected in the Holey Land Member equivalent of the Bermont Formation, in the Palm Beach Aggregates pit, Loxahatchee, Palm Beach County, Florida.
Stratigraphic Range: Confined to the lower part of the Bermont Formation (Holey Land Member and equivalents), Okeechobee Group, Aftonian Pleistocene.
Etymology: Named for Mr. Edward Volek of West Palm Beach, Florida, Chief of Security at Palm Beach Aggregates, Inc., who collected the holotype.
Discussion: This rare and seldom-seen Floridian cowrie was previously referred to the taxon *Cypraea (Luria) morinis* Ingram, 1939 (Petuch, 1994). That species (with its synonyms *L. rutschi* Schilder, 1939) and *L. catiana* (Weisbord, 1972)), however, is now known to be a widespread Pliocene form from the Limon Formation of Costa Rica, the Jacmel Formation of Haiti, the upper part of the Gatun Formation of Panama, and the Mare Formation of Venezuela. *Luria voleki* differs from the older *L. morinis* in being a broader, wider shell, in having much more thickened margins, in having finer and more numerous teeth (av. 33 labial and 21 columellar on *L. voleki*, and av. 28 labial and 20 columellar on *L. morinis*), and in having a more arcuate aperture. The new species is also similar to the Recent *L. cinerea* (Gmelin, 1791) but differs in being more elongated and cylindrical and in having much finer and more numerous teeth (av. 33 labial and 21 columellar on *L. voleki* and av. 23 labial and 17 columellar on *L. cinerea*).

Macrocypraea Schilder, 1930

Macrocypraea joannae new species
(Plate 88, K and L)

Description: Shell small for subgenus, elongate, subcylindrical, with pointed, projecting extremities; aperture uniformly narrow, widening slightly toward anterior end; lip with 32-34 thin, elongated, pointed teeth; columella with 27-28 thin, long teeth; anteriormost columellar teeth more rounded and knoblike; fossula very well-developed, flangelike, with 11-13 elongated teeth; auricles proportionally small, pointed, flattened; color pattern, when preserved, composed of numerous small brown dots on dorsum and sides and thin brown lines that extend from labial and columellar teeth.
Type Material: Holotype; AMNH FI-50665, length 54 mm, width 31 mm; Paratypes; 2 specimens, in the research collection of the author, lengths 58 mm and 64 mm, same locality as the holotype.
Type Locality: The holotype was collected in the Holey Land Member equivalent, lower part of the Bermont Formation, in the Mecca orange grove fill pit, westernmost Palm Beach Lakes Boulevard, rural North Palm Beach, Palm Beach County, Florida.
Stratigraphic Range: Confined to the Holey Land Member and equivalents, lower Bermont Formation, Okeechobee Group, Aftonian Pleistocene.
Etymology: Named for Mrs. Joanna Arline of Lake Park, Florida, avid naturalist and fossil collector.
Discussion: This new species is most similar to, and is the direct Pleistocene ancestor of, the Recent *Macrocypraea zebra* (Linnaeus, 1758). The new species differs in being a more inflated and less-cylindrical shell, in having more numerous and finer teeth, and in having a color pattern composed of small dark dots and not large pale spots and rings. In this last feature, *M. joannae* is similar to the Recent *M. cervus* (Linnaeus, 1771), but differs in having the colors reversed, with dark spots on a light-colored shell and not light spots on a dark-colored shell.

Tonnoidea
Cassidae

Cassis Scopoli, 1777

Cassis jameshoubricki new species
(Plate 85, B and C)

Description: Shell of average size for genus, broadly triangular in outline, heavy, solid; spire flattened, only projecting on early whorls; shoulder sharply angled, ornamented with 6-8 large, widely separated, spinelike knobs; shoulder knob at mid-dorsum largest and most prominent, giving shell distinct triangular appearance when viewed from spire; body whorl sculptured with 22-25 very low, poorly developed, almost obsolete spiral cords and 2 large, wide, elevated, very prominent spiral cords, one anterior of midbody line and one posterior of midbody line; prominent midbody cords ornamented with 5-7 very large, widely separated knobs; spire whorls and subsutural area of body whorl ornamented with 5-7 thin, poorly developed spiral cords and very numerous thin, radiating irregular riblets; intersections of spiral cords and radiating riblets produce small beads, giving spire pebbly appearance; aperture proportionally narrow; parietal shield very broad and extended, flat across posterior and sharply angled at shoulder; entire peristomal shield broadly flattener, distinctly triangular; edge of columella with 18-24 large, narrow, irregular teeth; inner edge of lip with 12-14 large, widely separated teeth, with the central teeth being bifurcated.
Type Material: Holotype; AMNH FI-50666, length 183 mm, width 144 mm; Paratypes; 2 specimens, in the research collection of the author, 172 mm and 187 mm, same locality as holotype.
Type Locality: The holotype was collected in the Holey Land Member equivalent, lower Bermont Formation, in the Palm Beach Aggregates, Inc. quarry, Loxahatchee, Palm Beach County, Florida.
Stratigraphic Range: Confined to the Holey Land Member and equivalents, lower Bermont Formation, Okeechobee Group, Aftonian Pleistocene.
Etymology: Named for Mr. James Houbrick of Loxahatchee, Florida, noted local natural history artist and art teacher, who collected the holotype.
Discussion: With its triangular body cross-section and wide, sharply angled parietal shield, *Cassis jameshoubricki* does not resemble any known congener in the Florida Plio-Pleistocene fossil record. The new species most closely resembles the Recent *C. tuberosa* (Linnaeus, 1758) but differs in being a smoother, less sculptured shell, in lacking the characteristic fine, clothlike reticulated ornamentation of *C. tuberosa*, and in having proportionally larger and more numerous knobs on the shoulder and body whorl cords. *Cassis jameshoubricki* is most probably the direct ancestor of the modern, widespread *C. tuberosa*.

Muricoidea
Muricidae
Ocenebrinae
Ecphora Conrad, 1843

Ecphora conoyensis new species
(Plate 31, D and E)

Description: Shell small for genus, thick and compact; whorls rounded; spire proportionally low; body whorl ornamented with 4 large, wide, rounded ribs; fifth low, faint rib present around base of siphonal canal; large spiral threads and tiny cords present between ribs and between anteriormost rib and faint fifth rib; spire whorls with 2 rounded ribs and numerous spiral threads; siphonal canal narrow and elongated; umbilicus proportionally narrow; aperture wide and flaring, almost circular; interior of aperture ornamented with 12-15 large, prominent raised cords.
Type Material: Holotype; AMNH FI-50667, length 52 mm, width 43 mm; Paratype; in the research collection of the author, length 55 mm, same locality as holotype.
Type Locality: The holotype was collected in the Conoy Member (Shattuck Zone 20) of the St. Mary's Formation, 2 km north of Little Cove Point, Calvert County, Maryland.
Stratigraphic Range: Confined to the Conoy Member (Shattuck Zone 20), lower St. Mary's Formation, Chesapeake Group, latest Serravallian Miocene.
Etymology: Named for the Conoy Member of the St. Mary's Formation.
Discussion: The new St. Mary's species represents the morphological intermediate between the thick-ribbed

Ecphora williamsi Ward and Gilinsky, 1988 (Plate 30, A and B) from the Boston Cliffs Member (Shattuck Zone 19) of the Choptank Formation and the large *E. asheri* Petuch, 1988 (Plate 31, A and B) from Shattuck Zone 21 of the Little Cove Point Member, St. Mary's Formation. *Ecphora conoyensis* retains the small size, stocky thick shell, and narrow umbilicus of *E. williamsi* but has the flattened ribs of *E. asheri*. From *E. williamsi*, the Choptank ancestor, *E. conoyensis* differs in having flattened ribs instead of rounded ribs and in having a proportionally larger, more flaring aperture. From *E. asheri*, the middle St. Mary's descendant, the new species differs in being a smaller, less inflated shell and in having coarse ribs inside the aperture. The small, faint fifth rib of *E. conoyensis* presages the large, equal-sized fifth rib of *E. asheri*.

Ecphora mattinglyi new species
(Plate 24, F and G)

Description: Shell of average size for genus, distinctly cylindrical, with high protracted spire and proportionally long siphonal canal; shoulder angled, subsutural area flattened; body whorl ornamented with 4 large, thick, highly elevated spiral ribs; ribs flattened on edges, slightly "T"-shaped in cross-section; rib on shoulder raised, bladelike, producing slightly canaliculated spire whorls; spire whorls with 3 large spiral ribs; siphonal canal ornamented with 10 low, faint, closely-packed spiral threads; umbilicus proportionally large, open, flaring.
Type Material: Holotype; AMNH FI-50675, length 56 mm, width 46 mm; Paratype; in the research collection of the author, length 88 mm, Shattuck Zone 14, Calvert Beach Member, Calvert Formation at Scientist's Cliffs, Calvert County, Maryland.
Type Locality: The holotype was collected in the Calvert Beach Member (Shattuck Zone 14), 3 km south of Chesapeake Beach, Calvert County, Maryland (in a slump block from the highest layer in the cliff face).
Stratigraphic Range: Confined to Shattuck Zone 14 of the Calvert Beach Member, Calvert Formation, Chesapeake Group, early Langhian Miocene.
Etymology: Named for Mr. Edward Mattingly of Owings, Maryland, who collected the holotype at Chesapeake Beach.
Discussion: This new ecphora is the oldest-known member of the *Ecphora meganae* species complex, a four-ribbed group that is prevalent in the subsequent Choptank Formation of the Patuxent Subsea. Evolving from the Zone 10 Plum Point *E. wardi* Petuch, 1988, the Zone 14 *E. mattinglyi* gave rise to *E. sandgatesensis* Petuch, 1989 in Zone 16 time, *E. meganae* Ward and Gilinsky, 1988 in Zone 17 time, and an undescribed species in Zone 19. After that time, the entire species complex became extinct. Of these known species, *E. mattinglyi* is most similar to the Zone 16 *E. sandgatesensis* (Plate 27, F and G) but differs in being a larger shell with a more inflated body whorl, in having a proportionally higher and more exerted spire, in having a wider and more flaring umbilicus, and in having a more rounded, thicker, less upturned, and less bladelike shoulder rib.

Planecphora new genus

Diagnosis: Ecphoras of average size, with inflated body whorls, wide shoulders, and proportionally low spires; body whorls with 4 thin, bladelike ribs; ribs always rounded on edges, never with grooves, thickenings, or flattenings.
Type Species: *Planecphora choptankensis* (Petuch, 1988) (originally as "*Ecphora*" *choptankensis*) (Plate 29, G and H), Shattuck Zone 17, St. Leonard Member, Choptank Formation, Chesapeake Group, Maryland, Serravallian Miocene.
Other Species In Planecphora: *P. turneri* (Petuch, 1992), Shattuck Zone 14, Calvert Beach Member, Calvert Formation, Chesapeake Group, late Langhian Miocene; *P. vokesi* (Petuch, 1989), Shattuck Zone 16, Drumcliff Member, Choptank Formation, Chesapeake Group, early Serravallian Miocene; *P. delicata* (Petuch, 1989), Shattuck Zone 19, Boston Cliffs Member, Choptank Formation, Chesapeake Group; *P. hertweckorum* (Petuch, 1988), Murdock Station Formation, Hawthorn Group, southern Florida, Zanclean Pliocene; *P. mansfieldi* (Petuch, 1989), Yorktown Formation (Rushmere Member), Chesapeake Group, Virginia and North Carolina; Tamiami Formation (Buckingham Member), Okeechobee Group, southern Florida; Jackson Bluff Formation ("*Ecphora* Zone"), Alum Bluff Group, northern Florida.
Etymology: A combination of "flat" (planus) and *Ecphora*, in reference to the new taxons' flattened, thin ribs.
Discussion: The new genus represents a distinct offshoot of four-ribbed ecphoras that occurred in the late Langhian Miocene. *Planecphora* evolved from *Ecphora wardi* Petuch, 1989 and co-occurred with its sister group

(*Ecphora* s.s.) throughout the Miocene, the Zanclean Pliocene, and into the early Piacenzian Pliocene. Throughout this time, the *Planecphora* lineage consistently differed from *Ecphora* in having extremely thin, bladelike ribs that lacked any expansion or ornamentation on their edges. In *Ecphora*, the edges of the ribs are always flared or "T"-shaped in cross-section and have at least one or two incised grooves (*E. meganae* and *E. quadricostata* complexes). Some *Ecphora* species, such as those in the Miocene *E. gardnerae* complex, have very wide "T"-shaped flanges on their rib edges and presage the extreme "I"-beam rib structure of the Pliocene genus *Latecphora*. Although prominent in the Langhian and early Serravallian ecosystems of the Salisbury Sea, *Planecphora* disappears from Maryland at the end of Choptank time. During the late Serravallian, Tortonian, and Messinian Miocene, and the Zanclean Pliocene, the new genus survived farther south in the Okeechobean Sea (as the *P. hertweckorum* complex). In the warming time at the beginning of the Piacenzian Pliocene, *Planecphora* reinvaded the Atlantic Coastal Plain and northern Gulf of Mexico regions as the widespread *P. mansfieldi*. After early Piacenzian time (Rushmere Member, Yorktown Formation-Buckingham Member, Tamiami Formation), however, the genus became extinct and was outlived by its sister group (*Ecphora*), which persisted into the later Piacenzian.

Buccinoidea
Fasciolariidae
Fasciolariinae
Pleuroploca Fischer, 1884

Pleuroploca lindae new species
(Plate 61, D and K)

Description: Shell of average size for genus, with high, protracted, scalariform spire and long, narrow siphonal canal; shoulder sharply angled, subcarinate; subsutural area flattened, only slightly sloping to shoulder angle; body whorl sculptured with 14-15 large spiral cords; numerous secondary and tertiary spiral threads present between each pair of cords; 2 cords around shoulder, 2 around midbody, and one around base of siphonal canal larger than others and heavily ornamented with large, elongated beads, giving shell rough appearance; shoulder coronated, with 12-15 large pointed knoblike beads, corresponding to other beaded cords; spire sculptured with 2 large beaded cords and numerous smaller spiral threads; siphonal canal sculptured with 12 large spiral cords; aperture proportionally small, with 15 large cords in interior; protoconch proportionally large, rounded, composed of 1.5 whorls.
Type Material: Holotype; AMNH FI-50669, length (juvenile) 48 mm, width 20 mm; Paratypes; AMNH FI-50670, length (incomplete adult) 83 mm, same locality as holotype; 2 specimens, in the research collection of the author, lengths (incomplete) 72 and 88 mm, same locality as holotype.
Type Locality: The holotype was collected in the Golden Gate Member of the Tamiami Formation, in Florida Rock Industries Naples Quarry (old Mule Pen Quarry), East Naples, Collier County, Florida.
Stratigraphic Range: Confined to the Golden Gate Member (coral reef limestone facies) of the Tamiami Formation, Okeechobee Group, late Piacenzian Pliocene.
Etymology: Named for my wife, Linda Joyce Petuch.
Discussion: The discovery of this new species represents the first record of the genus *Pleuroploca* (type: *P. trapezium*) in the North American fossil record. Today, the genus is restricted to Brazil in the Atlantic Ocean, the Eastern Pacific (Panamic Province), and the tropical Indo-Pacific. Of all the taxa from these varied regions, *P. lindae* is most similar to *P. aurantiaca* (Lamarck, 1816) from Brazil, but differs in being a more slender shell with a much higher, more protracted spire, in having a more slender and elongated siphonal canal, in having more numerous and smaller shoulder knobs, and in being a more strongly sculptured shell with larger and more numerous beads on the spiral cords.

Peristerniinae
Latirus Montfort, 1810

Latirus duerri new species
(Plate 67, D)

Description: Shell of average size for genus, elongated, with high, protracted spire; whorls inflated, with rounded shoulders; siphonal canal proportionally short, stumpy, open; body whorl ornamented with 10 large spiral cords, with anteriormost 7 cords being strongest; small secondary cords present between 3 primary cords around midbody and shoulder; whorls with 9-10 low, wide, undulating ribs; siphonal canal ornamented with 6 strong spiral cords, with smaller secondary cords between some primary cords; spire whorls with 4 strong spiral cords, with smaller secondary cords present between pair of primary cords; aperture oval, with 7-8 large ribs on interior; columella with 3 large teeth; pseudoumbilicus narrow, almost closed.

Type Material: Holotype; AMNH FI-50668, length 52 mm, width 20 mm; Paratype; in the research collection of the author, length 56 mm, same locality as the holotype.

Type Locality: The holotype was collected in the uppermost bed of the Tamiami Formation (Unit 2, Fruitville Member), in the Quality Aggregates Phase 9 pit, University Drive, Sarasota, Sarasota County, Florida.

Stratigraphic Range: Confined to Unit 2 of the Fruitville Member, Tamiami Formation, Okeechobee Group, latest Piacenzian Pliocene.

Etymology: Named for Mr. Richard Duerr of Okeechobee, Florida, who collected the holotype.

Discussion: With its large size and strong, fine spiral sculpture, *Latirus duerri* is similar to only one Floridian fossil species, *L. maxwelli* Pilsbry, 1939 from the Holey Land Member of the Bermont Formation (Aftonian Pleistocene). The new late Tamiami species differs from the Bermont descendant species in being a more slender shell with distinctly more rounded and nonangled whorls, and in having a coarser shell sculpture with fewer and stronger spiral cords. A large, undescribed species has recently been found in the Caloosahatchee Formation (Calabrian Pleistocene), and this seems to form the morphological intermediate between *L. duerri* and *L. maxwelli*.

Melongenidae
Melongena Schumacher, 1817
Rexmela Olsson and Harbison, 1953

Melongena (Rexmela) corona winnerae new subspecies
(Plate 94, A; Plate 97, A and B)

Description: Shell large for species and genus, inflated, broad across shoulder, with shell width being almost equal to shell length; shoulder sharply angled, ornamented with 8-10 proportionally large, open spikelike spines; spire proportionally low, with spire height being somewhat variable between individuals; subsutural area flattened, almost planar; edge of suture ornamented with numerous large scales; aperture extremely wide and flaring, oval; siphonal canal very short and stumpy, open; body whorl-siphonal canal juncture ornamented with row of large open spines; body whorl and siphonal canal ornamented with 21-26 faint, low spiral cords, with cords becoming stronger toward anterior end; color pattern (on living speciemens) consisting of yellow-white base color overlain by 3 wide dark bluish-brown bands, one below shoulder spines, one around midbody, and one around siphonal canal; smaller subsidiary bands sometimes present between 3 main bands.

Type Material: Holotype; AMNH 308067, length 134 mm, width 96 mm; Paratypes; AMNH 308068, length 126 mm, same locality as holotype; length, 127 mm (Plate 94, A), Okaloacoochee Member, Fort Thompson Formation, in the researh collection of the author; length 150 mm, in the research collection of the author, same locality as holotype.

Type Locality: The holotype was collected alive in Little Lake Worth Lagoon, off Singer Island near Little Munyon Island, Riviera Beach, Palm Beach County, Florida.

Stratigraphic Range: From the Okaloacoochee Member of the Fort Thompson Formation, Okeechobee Group (early Sangamonian Pleistocene) to the Recent. The new subspecies ranges, in the Recent, from the St. Lucie estuaries of St. Lucie and Martin Counties, Florida to the estuaries of Lake Worth, Palm Beach County, Florida.

Etymology: Named for Ms. Bea Winner of West Palm Beach, Florida, noted local naturalist, who discovered this relictual subspecies and brought it to the attention of the author.

Discussion: This new giant subspecies of *Melongena (Rexmela) corona* (Gmelin, 1791) represents a relict of the Pleistocene Lake Worth Lagoon System fauna that has managed to survive into the Recent. As presently understood, the *M. (Rexmela) corona* complex contains six living subspecies, all of which are geographically or ecologically separated. These Recent subspecies include: *sprucecreekensis* Tucker, 1994 (Plate 97, G), inhabiting salt water creeks and creek mouths from St. Johns to Volusia Counties, northeastern Florida; *altispira* Pilsbry and Vanatta, 1934, inhabiting open lagoons from St. Johns to Palm Beach Counties, the entire eastern coast of Florida; *winnerae* Petuch, new subspecies, inhabiting salt water creeks, rivers, and creek mouths from southern St. Lucie County to Lake Worth, Palm Beach County, southeastern Florida; *bicolor* (Say, 1827), from southern Biscayne Bay, Dade County, all of Florida Bay, Collier and Monroe Counties, to Key West, Florida Keys, Monroe County; nominate subspecies *corona* (Gmelin, 1791), from Marco Island, Collier County to Cedar Key, Levy County, western Florida; *johnstonei* Clench and Turner, 1956, from Cedar Key, across the entire Florida Panhandle, to Mobile Bay, Alabama. The eastern Floridian subspecies are genetically isolated from the other populations, as no melongenids occur along southeastern Florida in a stretch of coastline extending from Lake Worth, Palm Beach County to southern Biscayne Bay, Dade County. This melongenid-free area, with its carbonate, coralline environments, must have existed since at least the Sangamonian. Along eastern Florida, the melongenids have broken into two main ecological habitats; salt water creeks and lagoons that extend inland and open lagoons behind barrier islands. The new subspecies *winnerae* and the northern subspecies *sprucecreekensis* inhabit the salt water creek and creek mouth environments, where they feed on oysters. The subspecies *altispira*, on the other hand, prefers Turtle Grass beds and mangrove areas in open lagoons, where it feeds (like the Florida Keys *bicolor*) primarily on small cerithiid and batillariid gastropods. In Little Lake Worth, the type locality of the new subspecies, winnerae lives on mud flats in tidal channels between mangrove islands and in creeks and canals that flow into the lake. In these habitats, the salinities and temperatures fluctuate wildly over the course of a year and this large subspecies has evolved the ability to survive these environmental extremes. This ecological plasticity has allowed the subspecies to survive the Wisconsinan cold time.

With its large size and single row of shoulder spines, the new subspecies is morphologically closest to *M. corona sprucecreekensis* , but differs in being a much wider, broader shell with a much more inflated body whorl, in having the shoulder spines projecting outward and not upward, and in having a proportionally shorter and wider siphonal canal. From the nominate subspecies, *M. corona winnerae* differs in being a much larger and more noticeably inflated shell, in having fewer and more spikelike shoulder spines, and in having a wider and more sloping subsutural area.

Busyconidae
Busyconine
Busycon Roeding, 1798

Busycon diegelae new species
(Plate 3, A)

Description. Shell of average size for genus, pyriform, with inflated body whorl and long, narrow siphonal canal; shoulder sharply angled, subcarinate, ornamented with 10-12 large, flattened, bladelike knobs; subsutural area greatly flattened, sloping abruptly upward to suture; spire proportionally high, stepped, scalariform; body whorl sculptured with 12-15 very faint, low, thin spiral cords; siphonal canal ornamented with 12 large; prominent spiral cords; spire whorls and subsutural area of body whorl ornamented with 3 large, prominent spiral cords; smaller secondary threadlike cords sometimes present between primary cords; aperture large, flaring; protoconch proportionally large, exerted, mammilate, composed of 2 whorls.
Type Material: Holotype; AMNH FI-50671, length 163 mm, width 92 mm; Paratype; length (incomplete) 89 mm, in the research collection of the author, same locality as holotype.
Type Locality: The holotype was collected in the Chipola Formation, along the Chipola River 0.7 km south of the mouth of Tenmile Creek, Calhoun County, Florida.
Stratigraphic Range: Confined to the Chipola Formation, Alum Bluff Group, Burdigalian Miocene of the Florida Panhandle.
Etymology: Named for Ms. Phyllis Diegel of West Palm Beach, Florida, noted amateur paleontologist and frequent field assistant to the author.
Discussion: At 163 mm length, this new busyconid is the largest carnivorous gastropod to be found in the Chipola Formation. *Busycon diegelae* is closest to the contemporaneous (or possibly slightly younger) *B. burnsi* Dall, 1890 (Plate 21, F), but differs in being a much larger and more inflated shell, in being proportionally broader across the shoulder with a wider subsutural area (even in juvenile specimens), in having larger and fewer shoulder knobs, in having shoulder knobs that are angled outward and not upward (as in *B. burnsi*), and in having a smoother, less-sculptured body whorl that lacks strong, prominent raised cords. With its large size and broad shape, it is obvious that *B. diegelae* is the direct ancestor of *B. montforti* Aldrich, 1907 from the Shoal River Formation. The new species differs from its mid-Miocene descendant in having a higher, more stepped spire and

in having a more slender, narrower siphonal canal.

Lindafulgur new genus

Diagnosis: Busyconines of average size for subfamily, with proportionally short, vase-shaped body whorls and extremely long, well-developed siphonal canals; siphonal canals generally at least 1.5 to 2 times length of body whorl; juncture of siphonal canal and body whorl gradational, without distinct indentation; spires generally flattened, but with some species having elevated, scalariform spires; shoulders sharply angled, carinated, ornamented with numerous open, pointed, spinelike knobs; body whorls and spire whorls characteristically heavily sculptures with numerous strong, prominent spiral cords, giving shell rough texture; apertures proportionally small, oval, with interiors heavily sculptured with strong cords; protoconchs large, rounded, domelike, composed of 2 whorls.

Type Species: *Lindafulgur lindajoyceae* (Petuch, 1994) (originally as *"Busycon* (*Busycoarctum*)*" lindajoyceae*) (Plate 53, I), Buckingham Member of the Tamiami Formation, Okeechobee Group, early Piacenzian Pliocene of southern Florida.

Other Species In Lindafulgur: *L. alencasterae* (Perriat, 1963), Agueguexquite Formation, early Miocene (late Burdigalian-Langhian ? or possibly early Pliocene) of Veracruz State, Mexico; *L. miamiensis* (Petuch, 1991), Golden Gate Member, Tamiami Formation, Okeechobee Group of southern Florida; *L. candelabrum* (Lamarck, 1816), Recent Gulf of Mexico, from Campeche Banks to Contoy Island, Yucatan, Mexico; *L. lyonsi* (Petuch, 1987), Recent Gulf of Mexico, from 100-200 m depth off northwestern Florida.

Etymology: Named as a combination of "Linda" and *"fulgur"* (the old name for busyconids), in recognition of the support given by two Lindas to two busycon workers; myself and M.G. Harasewych. These include our wives, Linda J. Petuch and Linda Harasewych.

Discussion: Throughout its entire geological range (early Miocene to the Recent), *Lindafulgur* has been restricted to the Gulf of Mexico region and represents a eutropical offshoot of the main busyconine line. *Lindafulgur* is most similar to *Busycoarctum* Hollister, 1958 (another genus that has always been restricted to the Gulf of Mexico region), but differs in having larger shells with larger and more prominent shoulder spines, in having more elongated body whorls, and in having siphonal canals that grade directly into the body whorl and do not have a sharply-defined constriction at the body whorl-siphonal canal junction.

Spinifulgur Petuch, 1994

Spinifulgur armiger new species
(Plate 18, I)

Description: Shell of average size for genus, with vase-shaped body whorl and very long, slender siphonal canal; body whorl-siphonal canal juncture highly constricted; shoulder wide, sharply angled, subcarinated, ornamented with 8-10 very large, prominent, pointed spinelike knobs; spire flattened, only exerted on early whorls; subsutural area flattened, extending upward only at edge of suture; suture undulating, following contours of spines on previous whorls; spire whorls ornamented with 5 spiral cords; base of body whorl marked with wide, smooth, unsculptured band; area between band and shoulder sculptured with 8-10 thin, faint spiral threads; siphonal canal ornamented with 21-24 strong, prominent spiral cords; aperture proportionally small, oval, with 18-20 large cords on interior.

Type Material: Holotype; AMNH FI-50672, length 42 mm, width 23 mm; Paratype; length 43 mm, in the research collection of the author, same locality as the holotype.

Type Locality: The holotype was collected in the Chipola Formation, along the Chipola River 0.7 km south of the mouth of Tenmile Creek, Calhoun County, Florida.

Stratigraphic Range: Confined to the Chipola Formation, Alum Bluff Group, Burdigalian Miocene of the Florida Panhandle.

Etymology: "Armiger," a mace or war club (Latin), in reference to the new species' long siphonal canal and large shoulder spines.

Discussion: The new species is most similar to the contemporaneous and sympatric *Spinifulgur epispiniger* (Gardner, 1944) (Plate 3, H) but differs in consistently having a lower, flattened spire, in having a wider, broader shoulder and body whorl, and in having fewer and much larger and better-developed shoulder spines. Both *S.*

armiger and *S. epispiniger* were the last-living members of their genus, which ranged from the Eocene to the end of the Chipola Formation (Burdigalian Miocene).

Turrifulgur Petuch, 1988

Turrifulgur covepointensis new species
(Plate 32, I)

Description: Shell of average size for genus, slender, elongated, with high, protracted, scalariform spire; shoulder sharply angled, carinated; subsutural area greatly angled, sloping; body whorl and shoulder whorls sculpture with numerous very fine, closely packed spiral threads; base of body whorl with wide, smooth, unsculptured band; siphonal canal very long, slender, almost of equal length to body whorl; siphonal canal heavily sculptured with 28-30 large, prominent spiral cords; aperture proportionally large, oval, blending in with open siphonal canal.
Type Material: Holotype; AMNH FI-50673, length 65 mm, width 31 mm; Paratype; length 72 mm, in the research collection of the author, same locality as the holotype.
Type Locality: The holotype was collected in Shattuck Zone 23 of the Little Cove Point Member, St. Mary's Formation, at Little Cove Point, Calvert County, Maryland.
Stratigraphic Range: Confined to the Little Cove Point Member (Shattuck Zone 23) of the St. Mary's Formation, Chesapeake Group, latest Serravallian of Maryland.
Etymology: Named for Little Cove Point, the type locality.
Discussion: This new Little Cove Point busyconine species is most similar to its descendant, *Turrifulgur fusiforme* (Conrad, 1840) (Plate 33, C) from the younger Windmill Point Member of the St. Mary's Formation. *Turrifulgur covepointensis* differs from *T. fusiforme* in consistently being a much more slender and elongated shell with a narrower body whorl, in having a much higher, more protracted spire with distinctly scalariform whorls, and in being a more sculptured species, with fine spiral threads covering most of the body and spire whorls.

Busycotypinae
Coronafulgur Petuch, new genus

Diagnosis: Busycotypine busyconids of average size for subfamily, with elongated shells, vase shaped body whorls, and long, proportionally narrow siphonal canals; siphonal canals characteristically blend into body whorls with no distinct constriction; spire height variable, with some species having elevated, scalariform spires and some species with low, only slightly stepped spires; shoulders sharply angled, characteristically heavily ornamented with numerous small, sharply pointed spinelike knobs; sutural areas bounded by narrow, deeply incised sulcus; apertures proportionally large, flaring, elongated.
Type Species: *"Fulgur" coronatus* Conrad, 1841, Windmill Point Member, St. Mary's Formation, Chesapeake Group, early Tortonian Miocene of Maryland.
Other Species In Coronafulgur: *C. scalaspira* (Conrad, 1863), Kirkwood Formation, early Langhian Miocene of New Jersey and Delaware; *C. calvertensis* (Petuch, 1989), Plum Point Member (Shattuck Zone 10), Calvert Formation, Langhian Miocene of Maryland; *C. choptankensis* (Petuch, 1989), Boston Cliffs Member (Shattuck Zone 19), Choptank Formation, Serravallian Miocene of Maryland; *C. kendrewi* Petuch, n.sp. (described next), Cobham Bay Member, Eastover Formation, early Messinian Miocene of Virginia and North Carolina; *C. propecoronatum* (Mansfield, 1935), Red Bay Formation, Messinian Miocene of northwestern Florida.
Etymology: Named as a combination of "corona" (crown) and "*fulgur*" (the old name for busyconids), in reference to the prominent shoulder spikes of the new genus.
Discussion: Members of this new genus of busyconines were traditionally placed in the genus *Busycotypus* Wenz, 1943 by most recent workers (including Ward, 1992 and myself). As more evolutionary and stratigraphic data has come to light, it can now be seen that *Coronafulgur* represents an older, ancestral lineage that gave rise to *Busycotypus*. This split occurred just prior to the deposition of the Little Cove Point Member (probably during Conoy Member time) of the St. Mary's Formation, when the oldest and most primitive species, *Busycotypus* un-named species (*B. rugosum* complex), suddenly appeared and occurred together with *Coronafulgur chesapeakensis*. The two genera coexisted until the end of the Messinian Miocene. At that time, *Coronafulgur* became extinct while *Busycotypus* continued on into the Pliocene, Pleistocene, and Recent. True, morphologically advanced

Busycotypus (as typified by *B. incile, B. concinnum, B. scotti,* and *B. canaliculatum*) does not appear until the late Zanclean and earliest Piacenzian Pliocene, with the Choctaw-Okeechobean Seas *B. libertiense* being the oldest described species.

Coronafulgur differs from its offshoot, *Busycotypus*, in having smaller, less inflated, more elongated, and vase-shaped shells, in having proportionally larger and more prominent spikelike shoulder knobs, in having narrower and less-developed sutural canals, and in lacking the distinct constriction at the body whorl-siphonal canal juncture.

Coronafulgur kendrewi new species
(Plate 39, A and B)

Description: Shell small for genus, slender, elongated, with high, protracted, scalariform spire, and long, slender siphonal canal; body whorl vase-shaped, with only slight constriction at body whorl-siphonal canal juncture; body whorl ornamented with very numerous, extremely fine, closely-packed spiral threads, giving shell silky appearance; shoulder sharply angled, ornamented with 18-22 proportionally-large, pointed knobs; suture bordered with large, wide, deep channel; edge of sutural channel bordered by low spiral rib; subsutural area flattened, only slightly sloping; aperture proportionally large, open, flaring.
Type Material: Holotype; AMNH FI-50679, length 146 mm, width 72 mm; Paratypes; AMNH FI-50646, length 95 mm, same locality as holotype; in the research collection of the author, length 123 mm, same locality as holotype; length 72 mm, Kendrew collection, Valrico, Florida.
Type Locality: The holotype was collected in the Cobham Bay Member of the Eastover Formation, 5 km southeast of Bowler's Wharf, Essex County, Virginia, in an exposure along the Rappahannock River.
Stratigraphic Range: Confined to the Cobham Bay Member of the Eastover Formation, Chesapeake Group, latest Tortonian-earliest Messinian Miocene.
Etymology: Named for Mr. Eric Kendrew of Valrico, Florida, who assisted the author in collecting along the Rappahannock River.
Discussion: This new Eastover Formation busyconid is most similar to the older *Coronafulgur coronatum* (Conrad, 1840) (Plate 33, A and B) from the Windmill Point Member of the St. Mary's Formation of the Salisbury Sea. The new species differs from its St. Mary's ancestor, however, in being a much more slender and elongated shell with a less pyriform outline, in having a less distinct and more gradual body whorl-siphonal canal juncture, in having a much more elevated, protracted, and scalariform spire, and in having proportionally smaller and more numerous shoulder knobs (18-22 per whorl on *C. kendrewi*, 14-16 on *C. coronatum*). *Coronafulgur kendrewi* was the largest molluscivorous gastropod and the principal predator in the *Mulinia rappahannockensis* Community of the Rappahannock Subsea.

Pyruella Petuch, 1982

Pyruella tomeui new species
(Plate 86, G)

Description: Shell large for genus, pyriform, inflated, wide across shoulder; spire elevated, scalariform; shoulder sharply angled, carinated; subsutural areas of body and spire whorls flattened, only slightly angled; suture adherent; siphonal canal elongated, with distinct constriction at body whorl-siphonal canal juncture; body whorl ornamented with 28-32 large raised spiral cords; 1-3 fine spiral threads present between pair of spiral cords, giving shell rough appearance; flattened subsutural areas ornamented with 6-8 large spiral cords, with smaller spiral threads intercollated in between; 16-18 large spiral cords on siphonal canal, with smaller secondary cords present between some pairs of primary cords; shoulder angle of body whorl smooth, with single large carinalike cord; spire whorls ornamented with 4 large spiral cords; shoulder angle of spire whorls heavily coronated with 22-24 small, evenly-spaced elongated beads; aperture large, flaring, with 30 large cords on interior; protoconch large, bulbous, rounded, composed of 2 whorls.
Type Material: Holotype; AMNH FI-50674, length 89 mm, width 47 mm; Paratypes; 3 specimens, lengths 85-91 mm, in the research collection of the author, same locality as holotype.
Type Locality; The holotype was collected in the Holey Land Member equivalent ("Loxahatchee beds") of the lower Bermont Formation, in the Palm Beach Aggregates, Inc. quarry, Loxahatchee, Palm Beach County, Florida.

Stratigraphic Range: Confined to the lower member (Holey Land Member equivalent) of the Bermont Formation, Okeechobee Group, Aftonian Pleistocene of southern Florida.
Etymology: Named for Mr. Enrique Tomeu, President of Palm Beach Aggregates, Inc., Loxahatchee, Florida, in recognition of his support of local geological research.
Discussion: *Pyruella tomeui* is the last-living member of its genus, and its discovery was viewed with surprise when the first specimens were collected in the lower Bermont beds at Loxahatchee. The genus *Pyruella* was previously thought (Vermeij and Petuch, 1986; Petuch, 1995) to have died-out in the earliest Pleistocene, at the end of Caloosahatchee and Waccamaw time. The presence of a *Pyruella* species in the lower Bermont Formation demonstrates that the genus survived the Caloosahatchian Extinction event and persisted into the early mid-Pleistocene, at least to late Aftonian time. The new species most closely resembles the older Caloosahatchee Formation *P. eismonti* Petuch, 1991, especially in size and sculpture pattern, but differs in being a less inflated, more cylindrical and straight-sided shell, in having much flatter, less-sloping spire whorls and subsutural areas, in having a more prominent shoulder carina, and in having a lower spire. This new species, and the genus as a whole, became extinct during the Kansan Glacial Stage, in the interval between the deposition of the two members of the Bermont Formation.

Buccinidae
Photiinae
Cymatophos Pilsbry and Olsson, 1941

Cymatophos kellumi new species
(Plate 13, H)

Description: Shell small for genus, elongated, slender, with high protracted spire; shoulder rounded, only slightly angled; subsutural area sloping, highly angled; shoulder and most of body whorl ornamented with 7-8 large, rounded, elongated, longitudinal ribs, producing undulating surface on body and spire whorls; siphonal canal short, stumpy, open, bordered with large fascicular ridge; body and spire whorls sculptured with numerous closely packed, low, small spiral threads; aperture proportionally small, oval, with 16-18 small cords along inner edge of lip; columella with 1-3 very small plications.
Type Material: Holotype; AMNH FI-50676, length 46 mm, width 21 mm; Paratype; length 44 mm, in the research collection of the author, same locality as holotype.
Type Locality: The holotype was collected in the Haywood Landing Member of the Belgrade Formation ("Silverdale Marl"), in the Siverdale Marl Company quarry, Silverdale, Onslow County, North Carolina.
Stratigraphic Range: Restricted to the Haywood Landing Member of the Belgrade Formation, latest Chattian Oligocene-earliest Aquitanian Miocene of North Carolina.
Etymology: The new taxon honors Dr. L. B. Kellum, who conducted the pioneer research on the Silverdale faunas in the 1920's.
Discussion: This late Oligocene-Aquitanian Miocene species is the oldest-known member of its genus, indicating that the Silverdale Subsea may have been the center of origin for the genus *Cymatophos*. The new species is similar to *C. tuberaensis* (Anderson, 1929) from the early Miocene Tubera Formation of Colombia, but differs in being a smaller, more elongated shell, in having a less angled, more sloping shoulder, and in having weaker, finer spiral sculpture.

Volutoidea
Pleioptygmatidae

Pleioptygma apalachicolae new species
(Plate 21, I and J)

Description: Shell small for genus, elongated, fusiform, with protracted spire; shoulder rounded, almost obsolete; body whorl sculptured with 21-23 large spiral cords, with cords becoming wider apart toward posterior; spire whorls ornamented with 8 large spiral cords; siphonal canal proportionally short, open, broad, blending in with body whorl; siphonal canal ornamented with 17-18 large spiral cords, with cords being larger than those on body whorl; aperture long, narrow; columella with 4 large, rounded plications.

Type Material: Holotype; AMNH FI-50677, length (incomplete) 64 mm, width 23 mm; Paratype; length 66 mm, in the research collection of the author, same locality as holotype.
Type Locality: The holotype was collected in the Chipola Formation, basal bed at Alum Bluff, Liberty County, Florida, along the Apalachicola River.
Stratigraphic Range: Confined to the uppermost bed of the Chipola Formation, Alum Bluff Group, Burdigalian Miocene of the Florida Panhandle.
Etymology: Named for the Apalachicola River of the Florida Panhandle, along which the holotype was collected.
Discussion: Along with *Pleioptygma prodroma* (Gardner, 1937) (Plate 3, I and J), *P. apalachicolae* is the second-known species of pleioptygmatid giant miters to be found in the Chipola Formation. The new species differs from *P. prodroma* in being a much more slender and elongated shell, in having a more rounded, less broad, and almost obsolete shoulder, and in being a much more sculptured species, with a coarse ornamentation of large spiral cords covering the entire shell. *Pleioptygma prodroma* is a much smoother shell, with spiral cord sculpturing being present only around the sutural area and around the siphonal canal. Each species appears to have been the progenitor of one of the two pleioptygmatid species lines seen in the Floridian Pliocene; with *P. prodroma* giving rise to the smooth-shelled *P. carolinensis-lineolata* complex and *P. apalachicolae* giving rise to the corded, heavily-sculptured *P. lindae-ronaldsmithi* complex.

Conoidea
Conidae
Calusaconus new genus

Diagnosis: Shells small for family, squat, stumpy, turnip-shaped, with widest part of shell being just below shoulder; spires high, elevated, subscalariform; shoulders rounded; spire whorls slightly canaliculate; body whorl sculpture varies with species, with some being smooth, some partially sculptured, and others being fully sculptured with heavy spiral cords; apertures narrow; color pattern (when preserved) composed of large, irregular blotches or rectangular spots arranged in wide spiral bands, usually 8-10 on body whorl; spire whorl color pattern composed of large, evenly spaced rectangular flammules.
Type Species: *Calusaconus evergladesensis* (Petuch, 1991) (Plate 89, E) (originally "*Conus* (*Lithoconus*)" *evergladesensis*), upper beds of the Bermont Formation, Okeechobee Group, Yarmouthian Pleistocene of southern Florida.
Other Species In Calusaconus: *C.* un-named species, Buckingham Member, Tamiami Formation, early Piacenzian Pliocene; *C.* un-named species, Golden Gate Member, Tamiami Formation, late Piacenzian Pliocene; *C. spuroides* (Olsson and Harbison, 1953), Caloosahatchee Formation, Calabrian Pleistocene; *C. tomeui* Petuch, new species (described herein), Holey Land Member and equivalents, Bermont Formation, Aftonian Pleistocene.
Etymology: Named as a combination of "Calusa" (honoring the Calusa Indian Tribe of precolumbian Florida, and the genus *Conus*.
Discussion: Throughout its entire geochronological range, *Calusaconus* was confined to the Okeechobean Sea region of southern Florida, where it was a common organism in most of the late Pliocene-mid-Pleistocene marine communities. This endemic Floridian group of small, turnip-shaped cones appears to have been an offshoot of the much more widespread western Atlantic genus *Spuriconus* (type *S. spurius*; described herein). The new genus differs from the members of the typical "alphabet cone" complex in being much smaller, more truncated, and more pyriform shells with proportionally higher spires. Both genera, however, share the same color pattern of large blotches or spots arranged in wide bands.

Calusaconus tomeui new species
(Plate 85, H)

Description: General shell characters as for description of genus; posterior half of body whorl smooth, shiny, unsculptured; anterior half of body whorl and siphonal canal heavily sculptured with 18-20 large spiral cords; early spire whorls elevated, protracted; later spire whorls slightly canaliculated; color pattern composed of 6-8 wide bands of large irregular blotches.
Type Material: Holotype; AMNH FI-50678, length 34 mm, width 19 mm; Paratypes; 3 specimens, in the research collection of the author, lengths 34-37 mm, same locality as holotype.
Type Locality: The holotype was collected in the Holey Land Member equivalent ("Loxahatchee beds") of the Bermont Formation, in the Palm Beach Aggregates, Inc. quarry, Loxahatchee, Palm Beach County, Florida.

Stratigraphic Range: Confined to the Holey Land Member and equivalents, Bermont Formation, Okeechobee Group, Aftonian Pleistocene of southern Florida.
Discussion: In size, shape, and color pattern, the new Holey Land species is most similar to its descendant species, *C. evergladesensis* (type of the genus) from the upper beds of the Bermont Formation (equivalent to the type Bermont). *Calusaconus tomeui* differs from its descendant in being a less sculptured shell, with spiral cords being present only on the anterior half of the body whorl. The new species forms the perfect morphological intermediate between the older, completely smooth *C. spuroides* from the Caloosahatchee Formation and the completely sculptured *C. evergladesensis* from the upper Bermont Formation.

Cariboconus new genus

Diagnosis: Shells very small for family, averaging less than 15 mm in length; shells varying from being short and stocky to long and slender; spires flattened, almost completely planar; shoulders and spires smooth and uncoronated; body whorls variably sculptured, with some species being smooth and polished and others being ornamented with spiral threads and cords; apertures narrow; protoconchs proportionally large, mammilate; shell colors (on living species) varying from white to yellow to bright orange-red, with lighter midbody band; some species with large brown or orange longitudinal flammules; periostracum (on living species) thin, smooth, and silky; animals (on living species) pale red or pinkish-red.
Type Species: *Cariboconus flammeacolor* (Petuch, 1992) (originally "*Conus*" *flammeacolor*), Recent coral reef systems off Honduras and Nicaragua and the Gorda and Rosalind Banks of the central Caribbean.
Other Species In Cariboconus: North American Fossil Species; *C. harbisonae* (Petuch, 1994), Caloosahatchee Formation, Calabrian Pleistocene of Florida; *C. griffini* (Petuch, 1990) (Plate 87, D), Bermont Formation, Aftonian Pleistocene of southern Florida; Living Species; *C. bessei* (Petuch, 1992), coral banks off eastern Honduras; *C. brunneofilaris* (Petuch, 1990), San Blas Islands, Panama; *C. deynserorum* (Petuch, 1995), Banco Chinchorro Atoll, Quintana Roo, Mexico; *C. kalafuti* (daMotta, 1987), Roatan Island, Honduras; *C. kirkandersi* (Petuch, 1987), Isla Cozumel and Isla Mujeres, Mexico; *C. magnottei* (Petuch, 1987), Roatan Island, Honduras; *C. sahlbergi* (daMotta and Harland, 1986), western Bahamas and Cay Sal Bank.
Etymology: Named as a combination of "Carib" (honoring the Carib Indians of the Caribbean Basin) and the genus *Conus*.
Discussion: The new genus *Cariboconus* most closely resembles the endemic Caribbean-Brazilian genus *Magelliconus* daMotta, 1990, but differs in having smaller, proportionally narrower shells, in having smooth, uncoronated shoulders and spires, and in having proportionally larger protoconchs. *Cariboconus* is the western Atlantic ecological equivalent of the Indo-Pacific genus *Harmoniconus* daMotta, 1990 (type: *H. musicus* (Hwass, 1792)), as both genera have very small shells and live on coral rubble in the intertidal areas at the edges of coral reefs. The genus is described here to accommodate the two fossil species from the Caloosahatchee and Bermont Formations. Although known from the Pleistocene Floridian fossil record, *Cariboconus* is absent from Florida in the Recent, occurring only from the Bahamas and Cuba southward to the San Blas Islands of Panama.

Contraconus Olsson and Harbison, 1953

Contraconus petiti new species
(Plate 72, I and J)

Description: Shell of average size for genus, cylindrical, with straight sides; spire proportionally low, subpyramidal, distinctly stepped; spire whorls shallowly canaliculated; early whorls of spire faintly coronated with small beads; shoulder sharply angled, bordered by large, rounded, carina-like cord; body whorl smooth, faintly sculptured with numerous low, nearly obsolete spiral cords; spiral cords become larger and stronger on anterior third of shell, 24-26 large, prominent cords encircling siphonal canal; aperture proportionally wide, flaring at anterior end.
Type Material: Holotype; AMNH FI-50680, length 87 mm, width 46 mm; Paratype; length 73 mm, in the research collection of the author, same locality as holotype.
Type Locality: The holotype was collected in the Edenhouse Member of the Chowan River Formation, in the Lee Creek Texasgulf Mine, Aurora, Beaufort County, North Carolina.
Stratigraphic Range: Found in the Edenhouse Member of the Chowan River Formation, Chesapeake Group, the

lower Waccamaw Formation (Myrtle Beach, South Carolina), and the Sarasota Unit 1 and 2 equivalent beds in the Kissimmee Embayment (uppermost Tamiami Formation?) of southern Florida; all latest Piacenzian Pliocene.
Etymology: Named for Mr. Richard E. Petit of South Myrtle Beach, South Carolina, in recognition of his many contributions to the paleomalacology of the eastern United States.
Discussion: In general shape and proportions, this new species of left-handed cone is most similar to the older, mid-Piacenzian *C. adversarius* (Conrad, 1840) from the Yorktown, Duplin, and Tamiami Formations. *Contraconus petiti* differs from its ancestor in being a broader, stockier, and less elongated shell, in having distinctly canaliculate spire whorls and a stepped spire, and in having a thick, rounded, carinalike cord around the edge of the shoulder. A specimen of this new species from the lower Waccamaw Formation of Horry County, South Carolina, showing its banded color pattern, was illustrated by Olsson and Petit (1964, plate 79, figs. 1, 1a).

Jaspidiconus new genus

Diagnosis: Cone shells of small size for family, most averaging 25 mm or less but with few species averaging 45 mm or more; shells biconic, slender, with high protracted, often scalariform, spires; shoulders sharply angled, carinated, generally smooth but with few species being subcoronate, with low undulating knobs on shoulder carina; body whorls heavily sculptured with varying numbers of incised cords, producing grooved shell texture; some species smooth and polished, with grooves only around anterior tip; other species heavily sculptured with spiral rows of large pustules; base colors vivd, generally in shades of pink, orange, violet, or yellow, overlaid with variable amounts of amorphous flammules, bands, or rows of dots and dashes; protoconchs proportionally large, rounded, generally composed of 2 whorls; periostracum thin, translucent, silky; animals white or pale yellow, with variable amounts of black speckling.
Type Species: *Jaspidiconus jaspideus* (Gmelin, 1791) (originally "*Conus*" *jaspideus*), tropical western Atlantic, from southeastern Florida (Fort Lauderdale to the Florida Keys), throughout the Antilles Arc, south to the Grenadines, and from Veracruz, Mexico to Belize.
Other Species In Jaspidiconus: North American Fossil Species; *J. harveyensis* (Mansfield, 1930), Buckingham Member, Tamiami Formation and Jackson Bluff Formation, early Piacenzian Pliocene; *J. marymansfieldae* (Petuch, 1994), Pinecrest Member, Tamiami Formation, Piacenzian Pliocene; *J. susanae* (Petuch, 1994), Fruitville Member, Tamiami Formation, late Piacenzian Pliocene; *J. sarasotaensis* (Petuch, 1994), Fruitville Member (Unit 4), Tamiami Formation, late Piacenzian Pliocene; *J. hertwecki* (Petuch, 1988), Golden Gate Member, Tamiami Formation, Piacenzian Pliocene; *J. jaclynae* (Petuch, 1994), Kissimmee beds, Tamiami Formation, Piacenzian Pliocene; *J. laurenae* (Petuch, 1994), Kissimmee beds, Tamiami Formation, Piacenzian Pliocene; *J. wilsoni* (Petuch, 1994), Caloosahatchee Formation, Calabrian Pleistocene; *J. hyshugari* (Petuch, 1994), Holey Land Member, Bermont Formation, Aftonian Pleistocene; *J. maureenae* (Petuch, 1994), Holey Land Member, Bermont Formation, Aftonian Pleistocene: Living Species; *J. agassizii* (Dall, 1886), off Barbados and Grenadines; *J. branhamae* (Clench, 1953), central and eastern Bahamas; *J. duvali* (Bernardi, 1862), Puerto Rico and Lesser Antilles; *J. iansa* (Petuch, 1979), eastern Brazil; *J. mindanus* (Hwass, 1792), southeastern Florida to northern South America, Gulf of Mexico; *J. mindanus bermudensis* (Clench, 1942), Bermuda; *J. nodiferus* (Kiener, 1845), Greater Antilles; *J. pealii* (Green, 1830), Florida Keys and Florida Bay; *J. pfluegeri* Petuch, new species (described herein), St. Lucie and Lake Worth lagoon systems; *J. pusillus* (Lamarck, 1810), southeastern Brazil; *J. stearnsi* (Conrad, 1869), western Florida; *J. vanhyningi* (Rehder, 1944), southeastern Florida; *J. verrucosus* (Hwass, 1792), Bahamas and Greater Antilles; *J. verrucosus piraticus* (Clench, 1942). southeastern Florida and Florida Keys.
Etymology: Named as a combination of "jaspideus," jasper, and the genus *Conus*.
Discussion: Like *Cariboconus*, *Dauciconus*, *Magelliconus*, and *Purpuriconus*, *Jaspidiconus* is uniquely western Atlantic. Because of their ribbed sculpture pattern and sharp, carinated shoulders, members of *Jaspidiconus* were formerly placed in *Conasprella* Thiele, 1929 (typified by species such as *C. cancellatus*, *C. atractus*, and *C. pagodus*). That genus, however, is a deep water (50-300 m depths) group with larger, more elongated shells that are more heavily sculptured with strong spiral cords, are paler in color, being mostly white with brown flammules, and are more widely distributed, being found in all tropical and subtropical waters worldwide. The largest number of *Jaspidiconus* species found in one single area of the western Atlantic is seen in Florida, where seven species occur: with *J. stearnsi* (siliciclastic substrate shallow bays and sand flats) and *J. mindanus* (deeper water offshore) along the west coast; *J. pealii* (Turtle Grass beds), *J. verrucosus piraticus* (offshore coralline sediments) and *J. jaspideus* (carbonate sand) in the Florida Keys and Florida Bay; *J. vanhyningi*, (offshore coralline sediments), *J. verrucosus piraticus* (offshore coralline sediments), *J. jaspideus* (shallow water carbonate sediments), and *J. mindanus* (deeper water

offshore) along southeastern Florida; and with *J. pfluegeri* (siliciclastic substrate coastal lagoons and estuaries) and *J. mindanus* (deeper water offshore) from St. Lucie, Martin, and Palm Beach Counties (Lake Worth Lagoon only).

Jaspidiconus pfluegeri new species
(Plate 97, F and I)

Description: Shell of average size for genus, biconic, slightly pyriform, with high, protracted spire; shoulder sharply angled, carinated; spire whorls slightly canaliculated; body whorl shiny and polished; anterior half of body whorl sculptured with 10-12 large, deeply-impressed spiral sulci; posterior half of body whorl smooth, unsculptured; shell base color (in living specimens) typically purple or purplish-brown, with some specimens being tan (like holotype) or pinkish-lavender (like paratype); base color overlaid with 20-22 spiral rows of closely packed, alternating brown and white dots; edge of shoulder carina marked with large, evenly spaced dark brown spots; spire whorls same color as base color, marked with scattered widely spaced pale brown amorphous flammules; interior of aperture pale lavender; protoconch proportionally large, smooth, composed of 2 whorls, lavender in color.
Type Material: Holotype; AMNH 308069, length 25 mm, width 13 mm; Paratypes; AMNH 308070, length 24 mm; fossil specimen from the Okaloacoochee Member of the Fort Thompson Formation (Plate 97, F), in the research collection of the author, length 31 mm; 2 specimens, in the research collection of the author, lengths 23 mm and 24 mm, same locality as holotype.
Stratigraphic Range: From the early Sangamonian Pleistocene (Okaloacoochee Member of the Fort Thompson Formation, Okeechobee Group) to the Recent of southeastern Florida.
Etymology: Named for Frederick C. Pflueger III, Department of Chemistry and Biochemistry, Florida Atlantic University, who collected the type material in Lake Worth Lagoon.
Discussion: With its shiny, polished shell, bright purple, violet, or pinkish-tan color, and rows of vivid brown and white spots, the new species is most similar to *J. duvali* (Bernardi, 1862) from Puerto Rico and the Lesser Antilles. The endemic Floridian *J. pfluegeri* differs from its southern Caribbean congener, however, in having a larger shell, in being distinctly more pyriform and biconic in appearance, in having a more strongly carinated shoulder, and in having a stepped, scalariform spire. The new species is also similar to the widespread *J. jaspideus* (Gmelin, 1791), but differs in being a much larger shell, in having a proportionally higher, more protracted and scalariform spire, in being a smoother, less sculptured shell, and in being much more colorful, exhibiting tones of purple and pinkish-tan instead of white and brown. *Jaspidiconus pfluegeri* has also been collected in the Okaloacoochee Member of the Fort Thompson Formation, demonstrating that the new species is a relict of the once more-widely distributed Sangamonian fauna. Today, *J. pfluegeri* is restricted to muddy, silicate sand substrates in deeper tidal channels within the southeastern Floridian coastal lagoons, from the St. Lucie Inlet south to Lake Worth.

Eugeniconus daMotta, 1991

Eugeniconus irisae new species
(Plate 61, L)

Description: Shell of average size for genus, elongated, slender; shoulder sharply angled, smooth, subcarinate; spire flattened, only slightly raised on early whorls, with spire whorls shallowly canaliculate; shell smooth and shiny, with 10-12 small, low spiral cords encircling anterior tip; aperture uniformly narrow; protoconch exerted, distinctly mammilate; shell color pattern, when preserved, composed of dense network of tiny, triangular and heart-shaped markings covering entire body whorl.
Type Material: Holotype; AMNH FI-50680, length 31 mm, width 16 mm; Paratype; length 27 mm, in research collection of author, same locality as holotype.
Type Locality: The holotype was collected in the Golden Gate Member of the Tamiami Formation, in the Florida Rock Industries, Inc. Naples quarry (old Mule Pen Quarry), East Naples, Collier County, Florida.
Stratigraphic Range: Confined to the coral reef facies of the Golden Gate Member, Tamiami Formation, Okeechobee Group, late Piacenzian Pliocene.
Etymology: Named for Mrs. Iris Shellhorn of Golden Gate Estates, Naples, Florida.

Discussion: The new species is the second-known member of the Indo-Pacific coral reef-associated genus *Eugeniconus* (typified by species such as *E. nobilis*, *E. victor*, *E. marchionatus*, and *E. skinneri*) to be found in the Florida fossil record. *Eugeniconus irisae* occurred together on the Everglades Pseudoatoll reefs with the closely related *E. paranobilis* (Petuch, 1991) (Plate 61, C), with the finely triangle-patterned *E. irisae* belonging to the Recent *E. victor* (Broderip, 1842) species complex and the coarsely triangle-patterned *E. paranobilis* belonging to the Recent *E. nobilis* (Linnaeus, 1758) species complex. Besides differences in color pattern, the new species differed from *E. paranobilis* in being a more slender shell with a narrower shoulder, in having more distinctly canaliculate spire whorls, and in being a smoother shell, lacking the prominent fine spiral threads of *E. nobilis*.

Seminoleconus new genus

Diagnosis: Cone shells of small-to-average size, elongated, slender; shoulders sharply angled, smooth; spires high, protracted; body whorls smooth and polished, with 12-15 spiral cords around anterior tip; spire whorls variably ornamented, with older, primitive species having shoulders of all whorls edged with small beads and with more advanced species having beading confined to earliest whorls; apertures uniformly narrow; protoconchs proportionally large, rounded, composed of 2 whorls; color pattern (commonly preserved) composed of 8-12 spiral bands of large, evenly spaced rectangular spots.
Type Species: *Seminoleconus violetae* (Petuch, 1988) (originally "*Conus* (*Leptoconus*)" *violetae*), Buckingham Member of the Tamiami Formation, Okeechobee Group, early Piacenzian Pliocene of southern Florida.
Other Species In Seminoleconus: *S. trippae* (Petuch, 1991), Pinecrest Member, Tamiami Formation of southern Florida, Jackson Bluff Formation of northern Florida, Piacenzian Pliocene; *S.* un-named species, Fruitville Member, Tamiami Formation and equivalent Kissimmee beds, Chowan River Formation of North Carolina, late Piacenzian Pliocene; *S. diegelae* (Petuch, 1994), Caloosahatchee Formation of southern Florida, Nashua Formation of eastern Florida, James City Formation of North Carolina, latest Piacenzian Pliocene-Calabrian Pleistocene.
Etymology: Named as a combination of "Seminole" (honoring the Seminole Indian Tribe of Florida) and the genus *Conus*.
Discussion: This distinctive group of endemic Caloosahatchian Province cone shells is most similar to the genus *Gradiconus* daMotta, 1991, especially in shape, spire height, and having sharply angled shoulders. *Seminoleconus* differs from the widespread *Gradiconus*, however, in having the distinctive bands of large spots and in having beaded spire whorls. This last character is best-developed on *S. violetae*, where all the spire whorls are strongly beaded. On later species, such as *S. trippae* and *S. diegelae*, the spire beading is greatly reduced, being present only on the earliest few whorls.

Spuriconus new genus

Diagnosis: Cone shells of average-to-large size for family; shells distinctly conical, broad across shoulder, heavy, thickened; early whorls protracted and scalariform; later whorls becoming flattened, producing low-spired appearance on adult shells; shoulders angled but slightly rounded; spire whorls shallowly canaliculate; apertures narrow; body whorls smooth, often polished, with variable amounts of raised spiral sculpture; spiral sculpture species-specific, with some being smooth with silky texture and others having corded, rough-textured shells; shell color white or pale cream-white overlaid with numerous spiral bands of dark brown or orange-brown spots; spots often coalesce into large rectangular patches; spires colored with large, evenly-spaced flammules; protoconchs proportionally small, rounded; periostracum thick, heavy, opaque; operculum proportionally large, slender, filling one-third of aperture; animals cream colored with large amounts of brown mottlings and specklings; siphon black-banded.
Type Species: *Spuriconus spurius* (Gmelin, 1791) (originally "*Conus*" *spurius*), widespread Caribbean Sea, from southern Florida, Bahamas, to northern South America. Nominate subspecies not found in the Gulf of Mexico.
Other Species In Spuriconus: North American Fossil Species; *S. cracens* (S. Hoerle, 1976), Chipola Formation, Burdigalian Miocene; *S.* un-named species, Buckingham Member, Tamiami Formation, early Piacenzian Pliocene; *S. cherokus* (Olsson and Petit, 1964), Pinecrest Member, Tamiami Formation, Piacenzian Pliocene; *S. streami* (Petuch, 1994), Fruitville Member, Tamiami Formation, late Piacenzian Pliocene; *S. jeremyi* (Petuch, 1994), Golden Gate Member, Tamiami Formation, late Piacenzian Pliocene; *S. martinshugari* (Petuch, 1994), Golden Gate Member, Tamiami Formation, late Piacenzian Pliocene; (?) *S. yaquensis* (Gabb, 1873), upper Fruitville Member, Tamiami Formation and upper Gurabo Formation, Dominican Republic, late Piacenzian Pliocene; *S. brankampi*

(Hanna and Strong, 1949), Imperial Formation, Piacenzian Pliocene of Imperial County, California; *S. durhami* (Hanna and Strong, 1949), Imperial Formation, Piacenzian Pliocene of Imperial County, California; *S. jonesorum* (Petuch, 1994), upper Fruitville and Golden Gate Members, Tamiami Formation, latest Piacenzian Pliocene; *S. lemoni* (Petuch, 1990), Holey Land Member, Bermont Formation, Aftonian Pleistocene; *S. spengleri* (Petuch, 1991), Holey Land Member, Bermont Formation, Aftonian Pleistocene; *S. micanopy* (Petuch, 1994), upper beds, Bermont Formation, Yarmouthian Pleistocene; Living Species; *S. baylei* (Jousseaume, 1872) (=*S. arubaensis*), northern Colombia to Isla Margarita, Venezuela and Aruba; *S. lorenzianus* (Dillwyn, 1817), Belize to eastern Costa Rica; *S. phlogopus* (Tomlin, 1934), Caribbean Panama to Gulf of Venezuela; *S. spurius atlanticus* (Clench, 1942), Cape Hatteras, North Carolina to the Florida Keys and into the Gulf of Mexico to Cabo Catoche, Yucatan, Mexico.
Etymology: Named as a combination of "spurius," for the type species, and the genus *Conus*.
Discussion: Species belonging to this new genus had traditionally been placed in *Lithoconus* Moerch, 1852, primarily because of their similar spotted color patterns and their superficial resemblance to species such as *L. leopardus* and *L. litteratus*. The genus *Lithoconus* is now known to be confined to the Indo-Pacific region and is absent from both the Eastern Pacific and the entire Atlantic Ocean. *Spuriconus* differs from *Lithoconus* in having much smaller, broader shells, in having higher, more protracted, and often scalariform spires on the earlier whorls, and in having members that are heavily sculptured with strong spiral cords. The new genus is also similar to the endemic West African genus *Kalloconus* daMotta, 1991 (type: *K. pulcher* (Lightfoot, 1786), and the two genera may form an Atlantic sister group pair. Although sharing the same type of spotted pattern and large shell size, *Kalloconus* differs from *Spuriconus* in having broader, more inflated shells, proportionally wider apertures, and lower, flatter early spire whorls. Spuriconus is absent from the entire Brazilian Province.

Turridae
Clavinae
Neodrillia Bartsch, 1943

Neodrillia blacki new species
(Plate 97, D)

Description: Shell large for genus, elongated, with high protracted spire; adult specimens with 9 whorls; spire whorls with 12-13 prominent, thin, rounded axial ribs per whorl; body whorl with large, smooth, well-developed mid dorsal hump; remainder of body whorl, from mid-dorsal hump to edge of lip, relatively smooth, ornamented only with low, irregular axial undulations; entire shell surface covered with extremely fine, almost microscopic, closely-packed spiral threads, giving shell silky appearance; shoulder placed low on body whorl, sharply angled and prominent; subsutural area proportionally wide, sharply sloping, distinctly concave; axial ribs in subsutural area recurved posteriorly; suture impressed, causing low, beadlike formation on end of subsutural axial rib; aperture proportionally large, open, flaring; siphonal canal wide, open, truncated, with large stromboid notch; posterior sinus large, only slightly channeled, recurved posteriorly; edge of lip thin, smooth; parietal area smooth, with narrow, well-developed parietal shield; protoconch large, bulbous, rounded, smooth, composed of 2 whorls; shell color pure white overlaid with wide, diffuse band of pale salmon-orange; band darker on body whorl and penultimate whorl; interior of aperture pure white; operculum large, oval, smooth, yellow-tan in color.
Type Material: Holotype; AMNH 308071, length 28 mm, width 11 mm; Paratypes; (in the collections of the Department of Malacology, Field Museum of Natural History, Chicago, Illinois) FMNH 287808, length 25 mm, same locality as holotype; FMNH 287809, 2 specimens, lengths 25 mm and 27 mm, same locality as holotype; FMNH 287810, 7 specimens, lengths 17 mm to 26 mm, same locality as holotype; 2 fossil specimens, in the research collection of the author, Bermont Formation, lengths 24 mm and 27 mm.
Type Locality: In carbonate sand pocket, 1.5 m depth along eastern shore of Middle Torch Key, Pine Channel, Lower Florida Keys, Monroe County, Florida.
Stratigraphic Range: From the upper beds of the Bermont Formation, Okeechobee Group, Yarmouthian Pleistocene of the Everglades Basin, to the Recent, where the species is restricted to the southern Florida Keys.
Etymology: Named for Michael D. Black of West Palm Beach, Florida, avid Florida Keys naturalist who aided me in the collection of the type lot.
Discussion: *Neodrillia blacki* is a member of a relict mid-Pleistocene community that has survived into the Recent along the southern Florida Keys. Here, the new species occurs with several other Bermont relicts, including

Zafrona taylorae, Columbella rusticoides, Favartia pacei, and *Modulus calusa.* The new species is also commonly collected in the carbonate facies of the upper Bermont Formation in Palm Beach County, Florida, where it occurs with these same species (fossil specimen illustrated in Petuch, 1994, plate 99, fig. N, and incorrectly identified as "*Splendrillia pagodula*"). This large new Florida Keys endemic turrid is most similar to the common, widespread Caribbean *Neodrillia cydia* Bartsch, 1943, but differs in being a larger, narrower, more elongated shell with a proportionally much higher spire, in having a more sharply angled shoulder that is located lower on the body whorl, and having a wider and more sloping subsutural area. The most noticeable difference between the two species is seen in the form and number of axial ribs; those of *N. cydia* are wider, more rounded, and number 8-9 per whorl while those of *N. blacki* are thinner, sharper, and number 12-13 per whorl.

Bivalvia
Pterioida
Pectinoidea
Pectinidae
Pectininae
Carolinapecten Ward and Blackwelder, 1987

Carolinapecten jamieae new species
(Plate 86, C)

Description: shell of average size for genus, flattened, rounded in outline, slightly wider than long; auricles well-developed, proportionally large, rounded; upper (right) valve with 21-22 large, rounded ribs (round in profile); single, small, threadlike rib present between pair of large ribs; intercostal areas heavily sculptured with large scalelike growth increments; lower (left) valve with 21-22 large, flattened ribs; intercostal areas heavily sculptured with strong, prominent scalelike growth increments as on upper valve; auricles sculptured with 8 small, evenly spaced radiating ribs; color pattern (when preserved, as on holotype) composed of intermittent large flammules on each rib, all aligned in broad, undulating, concentric bands.
Type Material: Holotype; AMNH FI-50682, length 94 mm, width 101 mm; Paratype; length 90 mm, in the research collection of the author, same locality as the holotype.
Type Locality: The holotype was collected in the Holey Land Member equivalent ("Loxahatchee beds") of the lower Bermont Formation, in the Palm Beach Aggregates quarry, Loxahatchee, Palm Beach County, Florida.
Stratigraphic Range: Confined to the Holey Land Member, and equivalent units, of the Bermont Formation, Okeechobee Group, Aftonian Pleistocene of the Everglades region, Florida.
Etymology: Named for Ms, Jamie Smith, Department of Geology and Geography, Florida Atlantic University, who collected the holotype.
Discussion: The new species was the last-living member of the prominent Caloosahatchian Province genus *Carolinapecten*. Of the known species of the genus, which range from the Messinian Miocene to the Aftonian Pleistocene, *C. jamieae* is most similar to *C. solaroides* (Heilprin, 1886) (Plate 76, A) from the Calabrian Pleistocene Caloosahatchee, Nashua, and Waccamaw Formations, but differs in being a much smaller, more rounded shell, in having more numerous ribs (21-22 on *C. jamieae*, 19-20 on *C. solaroides*), in being a more ornate, sculptured shell with strong intercostals scales and subsidiary riblets, and in having more rounded ribs (when viewed in profile).

References

Briggs, J.C., *Marine Zoogeography*, McGraw-Hill, New York, 270 pp., 1974.

Briggs, J.C., *Global Biogeography*, Elsevier, Amsterdam, Netherlands, 452 pp., 1995.

Campbell, L.D., *Pliocene Molluscs From The Yorktown And Chowan River Formations In Virginia*, Publication **127**, Virginia Division of Mineral Resources, Charlottesville, 259 pp., 1993.

Case, E.C., Mammalia and Reptilia, Systematic Paleontology Section, in *The Miocene Deposits Of Maryland*, Clark, W.B., G.B. Shattuck, and W.H. Dall, Eds., Dept. of Natural Resources, Maryland Geological Survey, 3-70, 1904.

Collins, R.L., *Psammodulus*, a new middle Miocene modulid from the Isthmus of Tehuantepec, Mexico, *Nautilus* **47**, 127-130, 1934.

Cooke, C.W., New Vicksburg (Oligocene) Mollusks from Mexico, *Proc. U.S. Nat. Mus.* **73** (10), 1-11, 1928.

Cunningham, K.J., S.D. Locker, A.C. Hine, D. Bukry, J.A. Barron, and L.A. Guertin, Interplay of Late Cenozoic Siliciclastic and Carbonate Response on the Southeast Florida Platform. *Journ. of Sed. Res.* **73** *(1), 31-46, 2003.*

Dall, W.H., Contributions to the Tertiary fauna of Florida, with especial reference to the Miocene Silex Beds of Tampa and the Pliocene Beds of the Caloosahatchie River, Part 1, *Trans. Wagner Free Inst. Sci. Philadelphia* **3** (1), 1-200, 1890.

Dall, W.H., Contributions to the Tertiary fauna of Florida, with especial reference to the Miocene Silex Beds of Tampa and the Pliocene Beds of the Caloosahatchie River, Part 2, Trans. *Wagner Free Inst. Sci. Philadelphia*, **3** (2), 201-474, 1892.

Dall, W.H., On a brackish water Pliocene fauna of the southern Coastal Plain, *Proc. U.S. Nat. Mus.*, **46**, 225-237, 1913.

Dall, W.H., A monograph on the molluscan fauna of the *Orthaulax pugnax* Zone of the Oligocene of Tampa, Florida, *U.S. Nat. Mus. Bull.*, **90**, 1-173, 1915.

Dall, W.H., A contribution to the invertebrate fauna of the Oligocene Beds of Flint River, Georgia, *Proc. U.S. Nat. Mus.*, **51** (2162), 487-524, 1916.

DuBar, J.R., *Stratigraphy And Paleontology Of The Late Neogene Strata Of The Caloosahatchee River Area Of Southern Florida*, Geological Bulletin **40,** The Florida Geological Survey, Tallahassee, 267 pp., 1958.

DuBar, J.R., Summary of the Neogene Stratigraphy of southern Florida, in *Post-Miocene Stratigraphy, Central And Southern Atlantic Coastal Plain*, Oaks, R.Q. and J.R. DuBar., Eds., Utah State University Press, Provo, 206-231, 1974.

Eastman, C.R., Pisces, Systematic Paleontology Section, in *The Miocene Deposits Of Maryland*, Clark, W.B., G.B. Shattuck, and W.H. Dall, Eds., Dept. of Natural Resources, Maryland Geological Survey, 71-93, 1904.

Etheridge, R., Notes on the Mollusca collected by C. Barrington Brown from the Tertiary deposits of the Solimoes and Javary Rivers, Brazil, *Geol. Soc. London (Quart. Jour.)*, **35**, 82-88, 1924.

Figueras, A and J. Broggi, Nuevas especies de gastropodos marinos de la Formacion Camacho (Mioceno Superior de Uruguay), 1. con una resena geo-paleontologica de la Transgresion Paraense (=Entrerriana), *Com.*

Soc. Malac. Uruguay **6** (48), 257-285, 1985.

Frenguelli, J., Especies del genero *Conus* vivientes en el litoral Platense y fosilies en el Neozoico Superior Argentino-Uruguayo, *Not. Mus. Plata (Paleontologia)* **11**, 231-250, 1946.

Gibson, T.G., Stratigraphy of Miocene through lower Pleistocene Strata of the United States Central Atlantic Coastal Plain, in Ray, C.E., Ed., *Geology And Paleontology Of The Lee Creek Mine, North Carolina*, Smithsonian Institution press, Washington, D.C., 35-80, 1983.

Gibson-Smith, J., The genus *Voluta* (Mollusca: Gastropoda) in Venezuela, with the description of two new species, *Geos* **20**, 65-73, 1973.

Hanna, G.D., Paleontology of Coyote Mountain, Imperial County, California, *Proc. Cal. Acad. Sci.* **14** (18), 427-503, 1926.

Heilprin,A., Explorations on the west coast of Florida and in the Okeechobee wilderness, with special reference to the geology and zoology of the Floridian Peninsula, *Trans. Wagner Free Inst. Sci. Philadelphia* **1**, 365-506, 1886.

Hollister, S.C., New *Vasum* species of the subgenus *Hystrivasum*, *Bull. Amer. Paleo.* **58**, 289-304. 1971.

Hunter, M.E., Molluscan guide fossils in late Miocene sediments of southern Florida, *Trans. Gulf Coast Assoc. Geol. Soc.* **18**, 439-450, 1968.

Ingram, W.M., Type fossil Cypraeidae of North America, *Bull. Amer. Paleo.* **27** (104), 5-32, 1942.

Ingram, W.M., New fossil Cypraeidae from Colombia and Venezuela, *Bull. Amer. Paleo.* **31** (121), 1-12, 1947.

Jung, P., Miocene Mollusca from the Paraguana Peninsula, Venezuela, *Bull. Amer. Paleo.* **49** (223), 384-652, 1965.

Jung, P., Miocene and Pliocene mollusks from Trinidad, *Bull. Amer. Paleo.* **55** (247), 289-657, 1969.

Jung, P., Fossil mollusks from Carriacou, *Bull. Amer. Paleo.* **61** (269), 1-262, 1971.

Kellum, L.B., *Paleontology And Stratigraphy Of The Castle Hayne And Trent Marls In North Carolina*, U.S. Geological Survey Professional Paper **143,** 56 pp., 1926.

Kiegwin, L.D., Pliocene closing of the Isthmus of Panama, based on biostratigraphic evidence from nearby Pacific Ocean and Caribbean cores, *Geology* **6**, 630-634, 1978.

Krantz, D.E., A chronology of Pliocene sea level fluctuations, U.S. Atlantic Coastal Plain, *Quat. Sci. Reviews* **10,** 163-174, 1991.

Lyons, W.G., Caloosahatchee-age and younger molluscan assemblages at APAC mine, Sarasota County, Florida, in *The Plio-Pleistocene Stratigraphy And Paleontology Of Southern Florida*, Scott, T.M. and W.D. Allmon, Eds., Special publication **36**, Florida Geological Survey, Tallahassee, 133-165, 1992.

Magalhaes, J. and S. Mezzalira, Moluscos *Fosseis Do Brasil*, Biblioteca Cientifica Brasileira, Serie A-IV, Instituto Nacional do Livro, Rio de Janeiro, 283 pp., 1953.

MacNeil, F.S. and D.T. Dockery, *Lower Oligocene Gastropoda, Scaphopoda, And Cephalopoda Of The Vicksburg Group In Mississippi*, Mississippi Department of Natural Resources (Bureau of Geology), Bulletin **124**, 415 pp., 1984.

Mansfield, W.P., New fossil mollusks from the Miocene of Virginia and North Carolina, with a brief outline of the divisions of the Chesapeake Group, *Proc. U.S. Nat. Mus.* **74** (14), 1-11, 1928.

Mansfield, W.P., *Miocene Gastropods And Scaphopods Of The Choctawhatchee Formation Of Florida*, Bulletin **3**, Florida Geological Survey, Tallahassee, 185 pp., 1930.

Mansfield, W.P., *Miocene Pelecypods Of The Choctawhatchee Formation Of Florida*, Bulletin **8**, Florida Geological Survey, Tallahassee, 240 pp., 1932.

Mansfield, W.P., *New Miocene Gastropods And Scaphopods From Alaqua Creek Valley, Florida*, Geological Bulletin **12**, State of Florida Department of Conservation, Tallahassee, 75 pp., 1935.

Mansfield, W.P., *Mollusks Of The Tampa And Suwannee Limestones Of Florida*, Geological Bulletin **15**, State of Florida Department of Conservation, 334 pp., 1937.

Mansfield, W.P., *Notes On The Upper Tertiary And Pleistocene Mollusks Of Peninsular Florida*, Geological Bulletin **18**, State of Florida Department of Conservation, 75 pp., 1939.

Marks, J.G., Miocene stratigraphy and paleontology of southwestern Ecuador, *Bull. Amer. Paleo.* **33** (139), 7-162, 1951.

Martin, G.C., Mollusca (Cephalopoda, Gastropoda, Amphineura, Scaphopoda), Systematic Paleontology Section, in *The Miocene Deposits Of Maryland*, W.B. Clark, G. B. Shattuck, and W.H. Dall, Eds., Dept. of Natural Resources, Maryland Geological Survey, 131-404, 1904.

Maury, C.J., Scientific Survey of Porto Rico and the Virgin Islands, *New York Acad. Sci.*, **3** (1), 12-68, 1920.

Maury, C.J., *Fosseis Terciarios Do Brasil, Com Descricao De Novas Formas Cretaceas*, Monograph **4**, Servicio de Geologia e Mineralogia, Rio de Janeiro, 665 pp., 1924.

Maury, C.J., Fossil Invertebrata from northeastern Brazil, *Bull. Amer. Mus. Nat. Hist.* **67** (4), 123-179, 1934.

Missimer, T.M., Pliocene Stratigraphy of South Florida: Unresolved Issues of Facies Correlation in Time, in Harris, H.B., V.A. Zullo. and T.M. Scott, Eds., *The Third Baldhead Island Conference On Coastal Plains Geology* (Abstract Volume), University of North Carolina at Wilmington, pp. 36-44, 1992.

Nomland, J.O., The Etchegoin Pliocene of middle California, *Bull. Dept. Geol.* **10** (14), 191-254, University of California Publications, 1917.

Odum, E.P., *Fundamentals Of Ecology*, 3rd ed., W.B. Saunders, Philadelphia, 574 pp., 1971.

Olsson, A.A., *Neogene Mollusks From Northwestern Ecuador*, Paleontological Research Institute, Ithaca, New York, 256 pp., 1964.

Olsson, A.A., *Some Tertiary Mollusks From South Florida And The Caribbean*, Paleontological Research Institute, Ithaca, New York, 161 pp., 1967.

Olsson, A.A. and A. Harbison, *Pliocene Mollusca of southern Florida with special reference to those from north Saint Petersburg*, Mon. Acad. Nat. Sci. Philadelphia **8**, 1-361, 1953.

Olsson, A.A. and R.E. Petit, Some Neogene Mollusca from Florida and the Carolinas, *Bull. Amer. Paleo.* **47** (217), 556-561, 1964.

Olsson, A.A. and H.G. Richards, Some Tertiary fossils from the Goajira Peninsula of Colombia, *Notulae Naturae* **350**, 1-16, 1961.

Perrilliat, M.C., Moluscos de la Formacion Agueguexquite (Mioceno medio) del Istmo de Tehuantepec, Mexico, *Paleo. Mexicana* **14**, 1-45, 1963.

Petuch, E.J., *Strombus costatus* and *Morum dennisoni* collected off the North Carolina coast. *The Veliger* **15**, 32-35, 1972.

Petuch, E.J., An unusual molluscan assemblage from Venezuela, *The Veliger* **18**, 320-325, 1976.

Petuch, E.J., A new species of *Siphocypraea* from northern South America, with notes on the genus in the Caribbean, *Bull. Mar. Sci.* **29** (2), 216-225, 1979

Petuch, E.J., New gastropods from the Abrolhos Archipelago and reef complex, Brazil, *Proc. Biol. Soc. Washington* **92** (3), 510-526, 1979.

Petuch, E.J., A relict Caenogastropod fauna from northern South America, *Malacologia* **20** (2), 307-347, 1980.

Petuch, E.J., Notes on the molluscan paleontology of the Pinecrest Beds at Sarasota, Florida with the description of *Pyruella*, a stratigraphically important new genus, *Proc. Acad. Nat. Sci. Philadelphia* **134**, 12-30, 1982.

Petuch, E.J., Geographical heterochrony: contemporaneous co-existence of Neogene and Recent molluscan faunas in the Americas, *Palaeogeo., Palaeoclim., and Palaeoecol.* **37**, 277-312, 1982.

Petuch, E.J., The Pliocene reefs of Miami: their geomorphological significance in the evolution of the Atlantic Coastal Ridge, southeastern Florida, *Jour. Coast. Res.* **2** (4), 391-408, 1986.

Petuch, E.J., The Florida Everglades: A buried pseudoatoll?, *Jour. Coast. Res.* **3** (2), 189-200, 1998.

Petuch, E.J., *New Caribbean Molluscan Faunas*, The Coastal Education and Research Foundation, Charlottesville, Virginia, 154 pp., 1987.

Petuch, E.J., *Neogene History Of Tropical American Mollusks: Biogeography And Evolutionary Patterns Of Tropical Western Atlantic Mollusca*, The Coastal Education and Research Foundation, Charlottesville, Virginia, 217 pp., 1988.

Petuch, E.J., New species of *Ecphora* and ecphorine thaidids from the Miocene of Chesapeake Bay, Maryland, U.S.A., *Bull. Paleomalac.* **1** (1), 1-16, 1988.

Petuch, E.J., New gastropods from the Maryland Miocene, *Bull. Paleomalac.* **1** (4), 69-80, 1988.

Petuch, E.J., *Field Guide to the Ecphoras*, The Coastal Education and Research Foundation, Charlottesville, Virginia, 140 pp., 1989.

Petuch, E.J., New gastropods from the Bermont Formation (middle Pleistocene) of the Everglades Basin, *The Nautilus* **104** (3), 96-104, 1990.

Petuch, E.J., The Pliocene Pseudoatoll of southern Florida: a template for the evolution of the Everglades, in Allmon, W. and T. Scott, Eds., *Plio-Pleistocene Stratigraphy And Paleontology Of South Florida*, Southeastern Geological Society Guidebook **31**, 183-194, 1990.

Petuch, E.J., *New Gastropods from the Plio-Pleistocene of Southwestern Florida and the Everglades Basin.* Special Publication **1**, W.H. Dall Paleontological Research Center, Florida Atlantic University, 85 pp., 1991.

Petuch, E.J., The Pliocene Pseudoatoll of southern Florida and its associated gastropod fauna, in Scott, T. and W. Allmon, Eds., *Plio-Pleistocene Stratigraphy Of Southern Florida*, Special Publication **36**, Florida Geological Survey,

101-116, 1992.

Petuch, E.J., New Ecphoras from the Calvert Formation of Maryland (Langhian Miocene), *The Nautilus* **106** (2), 68-71, 1992.

Petuch, E.J., Patterns of diversity and extinction in Transmarian muricacean, buccinacean, and conacean gastropods, *The Nautilus* **106** (4), 155-173, 1993.

Petuch, E.J., *Atlas Of Florida Fossil Shells (Pliocene And Pleistocene Marine Gastropods)*, The Graves Museum of Archaeology and Natural History, Dania, Florida with Chicago Spectrum Press, Chicago, 394 pp., 1994.

Petuch, E.J., Molluscan diversity in the late Neogene of Florida: evidence for a two-staged mass extinction, *Science* (13 October) **270**, 275-277, 1995.

Petuch, E.J., *Coastal Paleoceanography Of Eastern North America*, Kendall-Hunt Pub. Co., Dubuque, 373 pp., 1997.

Petuch, E.J., A new gastropod fauna from an Oligocene back-reef lagoonal environment in west central Florida, *The Nautilus* **110** (4), 122-138, 1997.

Pilsbry, H.A., Revision of W.M. Gabb's Tertiary Mollusca of Santo Domingo, *Proc. Acad. Nat. Sci. Philadelphia* **73** (2), 305-435, 1922.

Pilsbry, H.A. and A. Harbison, Notes on the Miocene of southern New Jersey, *Proc. Acad. Nat. Sci. Philadelphia* **85**, 107-120, 1933.

Richards, H.G., Studies on the marine Pleistocene: Part 2, The marine Pleistocene mollusks of eastern North America, *Trans. Amer. Phil. Soc. (New Series)*, **1-141**, 1962.

Richards. H.G. and A. Harbison, Miocene invertebrate fauna of New Jersey, *Proc. Acad. Nat. Sci. Philadelphia* **94**, 167-250, 1942.

Rios, E.C., *Seashells Of Brazil*, Museu Oceanografico, Universidade do Rio Grande, Fundacao Cidade do Rio Grande, Brazil, 368 pp., 1994.

Rossbach, T.J. and J.G. Carter, Molluscan biostratigraphy of the lower River Bend Formation at the Martin-Marietta quarry, New Bern, North Carolina, *Jour. Paleo.* **65**, 80-118, 1991.

Scott, T.M., *The Lithostratigraphy Of The Hawthorn Group (Miocene) Of Florida*, Florida Geological Survey Bulletin **59**, 148 pp., 1988.

Scott, T.M., Coastal Plain stratigraphy: the dichotomy of biostratigraphy and lithostratigraphy — a philosophical approach to an old problem, in Scott, T. and W. Allmon, Eds., Plio-Pleistocene Stratigraphy And Paleontology *Of Southern Florida*, Special Publication **36**, Florida Geological Survey, 21-25, 1992.

Shattuck, G.B., Geographic and Geologic Relations, in *The Miocene Deposits Of Maryland*, Clark, W.B, G.B. Shattuck, and W.H. Dall, Eds., Department of Natural Resources, Maryland Geological Survey, lxv-clv, 1904.

Sloss, L.L., Sequences in the cratonic interior of North America, *Geol. Soc. Amer. Bull.*, **74**, 97-100, 1963.

Squires, R.L. and R.A. Demetrion, *Paleontology of the Eocene Bateque Formation, Baja California Sur, Mexico.* Cont. Science **434**, Nat. Hist. Mus. Los Angeles County, 1992.

Stanley, S.M., Anatomy of a regional mass extinction: Plio-Pleistocene decimation of the western Atlantic bivalve fauna, *Palaios* **1**, 17-36, 1986.

Valentine, J.W., *Evolutionary Paleoecology Of The Marine Biosphere*, Prentice-Hall, Inc., Englewood, New Jersey, 511 pp., 1973.

Vaughan, T.W., Fossil corals from Central America, Porto Rico, and Cuba, with an account of the American Tertiary, Pleistocene, and Recent Coral Reefs, *Bull. U.S. Nat. Mus.* **103**, 189-524, 1919.

Vermeij, G.J., Molluscs in mangrove swamps: physiognomy, diversity, and regional differences, *Syst. Zoo.* **22** (4), 609-624, 1973.

Vermeij, G.J., *Biogeography And Adaptation*: Patterns Of Marine Life, Harvard University Press, Cambridge, 332 pp., 1978.

Vermeij, G.J. and E.J. Petuch, Differential extinction in tropical American mollusks: endemism, architecture, and the Panama land bridge, *Malacologia* **27** (1), 29-41, 1986.

Vermeij, G.J. and E.H. Vokes, Cenozoic Muricidae of the western Atlantic Region. Part XII- the subfamily Ocenebrinae (in part), *Tulane Stud. Geol. Paleo.* **29** (3), 69-118, 1997.

Vermeij, G.J. and F.P. Wesselingh, Neogastropod mollusks from the Miocene of western Amazonia, with comments on marine to freshwater transitions in mollusks, *Jour. Paleo.* **76** (2), 265-270, 2002.

Vokes, E.H., The age of the Baitoa Formation, Dominican Republic, using Mollusca for correlation, *Tulane Stud. Geol. Paleo.* **15** (1-4), 105-116, 1979.

Vokes, E.H., A new species of *Turbinella* from the Pliocene of Mexico, with a revision of the geologic history of the line, *Tulane Stud. Geol. Paleo.* **18** (2), 47-52, 1984.

Vokes, E.H., Muricidae of the Esmeraldas Beds, northwestern Ecuador, *Tulane Stud. Geol. Paleo.* **21** (1,2), 1-50, 1988.

Vokes, E.H., An overview of the Chipola Formation, northwestern Florida, *Tulane Stud. Geol. Paleo.* **22** (1), 13-24, 1989.

Vokes, E.H., Cenozoic Muricidae of the western Atlantic Region. Part 8. Murex s.s., *Haustellum*, *Chicoreus*, and *Hexaplex*; Additions and Corrections, *Tulane Stud. Geol. Paleo.* **23** (1-3), 1-96, 1990.

Vokes, H.E., Notes on the fauna of the Chipola Formation-XXI; a new species of the genus *Lopha*, *Tulane Stud. Geol. Paleo.* **13**, 190-191, 1977.

Vonhoe, H.B., F.P. Wesselingh, and G.M. Ganssen, Reconstruction of the Miocene western Amazonian aquatic system using molluscan isotopic signatures, *Palaeogeo. Palaeoclim. Palaeoecol.* **141**, 85-93, 1998.

Waldrop, J.S. and D. Wilson, Late Cenozoic Stratigraphy of the Sarasota area, in Allmon, W. and T. Scott, Eds., *Plio-Pleistocene Stratigraphy And Paleontology Of South Florida*, Southeastern Geological Society Guidebook **31**, 195-227, 1990.

Ward, L.W., *Stratigraphy And Characteristic Mollusks Of The Pamunkey Group (Lower Tertiary) And The Old Church Formation Of The Chesapeake Group, Virginia Coastal Plain*, U.S. Geological Survey Professional Paper **1346**, 78 pp., 1985.

Ward, L.W., *Molluscan Biostratigraphy Of The Miocene, Middle Atlantic Coastal Plain Of North America*, Memoir 2, Virginia Museum of Natural History, Martinsville, 159 pp. 1992.

Ward, L.W., Mollusks from the lower Miocene Pollack Farm site, Kent County, Delaware: a preliminary analysis, in Benson, R.N., Ed., *Geology And Paleontology Of The Lower Miocene Pollack Farm Fossil Site, Delaware*, Special Publication **21**, Delaware Geological Survey, 59-131, 1998.

Ward, L.W. and B.W. Blackwelder, *Stratigraphic Revision Of Upper Miocene And Lower Pliocene Beds Of The Chesapeake Group, Middle Atlantic Coastal Plain*, U.S. Geological Survey Bulletin **1482-D**, 61 pp., 1980.

Ward, L.W. and B.W. Blackwelder, Late Pliocene and early Pleistocene Mollusca from the James City and Chowan River Formations at the Lee Creek Mine, in Ray, C., Ed., *Geology And Paleontology Of The Lee Creek Mine, North Carolina*, **II**, Smithsonian Contributions to Paleontology **61**, 113-283, 1987.

Ward, L.W. and N.L. Gilinsky, *Ecphora* from the Chesapeake Group of Maryland and Virginia, *Notulae Naturae* **469**, 1-21, 1988.

Ward, L.W. and P.F. Huddlestun, Age and stratigraphic correlation of the Raysor Formation, Late Pliocene, South Carolina, *Tulane Stud. Geol. Paleo.* **21** (1,2), 59-75, 1988.

Warmke, G. and Abbott, R.T., *Caribbean Seashells*, Livingston Publishing Co., Narberth, 348 pp., 1962.

Weisbord, N.E., Late Cenozoic gastropods from northern Venezuela, *Bull. Amer. Paleo.* **42** (193), 7-672, 1962.

Weisbord, N.E., Late Cenozoic corals of south Florida, *Bull. Amer. Paleo.* **66** (285), 259-544, 1974.

White, C.A., Contribuicoes a paleontologia do Brasil, *Arquiv. Mus. Nac. Brasil* **7**, 1-270, 1887.

Whitfield, R.P., *Mollusca And Crustacea Of The Miocene Formations Of New Jersey*, Monograph **24**, U.S. Geological Survey, 195 pp., 1894.

Wiedey, L.W., Notes on the Vaqueros and Temblor Formations of the California Miocene, with descriptions of new species, *Trans. San Diego Soc. Nat. Hist.* **5** (10), 95-182, 1928.

Woodring, W.P., *Miocene Mollusks From Bowden, Jamaica*, Publication **385**, Carnegie Institute of Washington, 564 pp., 1928.

Woodring, W.P., *Geology And Paleontology Of Canal Zone And Adjoining Parts Of Panama; Description Of Tertiary Mollusks (Gastropods: Vermetidae to Thaididae)*, U.S. Geological Survey Professional Paper **306-B**, pp. 147-239, 1959.

Woodring, W.P., *Geology And Paleontology Of Canal Zone And Adjoining Parts Of Panama; Description Of Tertiary Mollusks (Gastropods: Columbellidae to Volutidae)*, U.S. Geological Survey Professional Paper **306-C**, pp. 241-297, 1964.

Woodring, W.P., Endemism in middle Miocene Caribbean molluscan faunas, *Science* **148** (3672), 961-963, 1965.

Woodring, W.P., *Geology And Paleontology Of Canal Zone And Adjoining Parts Of Panama; Description Of Tertiary Mollusks (Gastropods: Rulimidae, Marginellidae to Helminthoglyptidae)*, U.S. Geological Survey Professional Paper **306-D**, pp. 299-452, 1970.

Woodring, W.P., The Miocene Caribbean Faunal Province and its subprovinces, *Verhand. Naturforsch. Gesell. Basel*, **84** (1), 209-213, 1974.

Zullo, V.A. and W.B. Harris, Sequence stratigraphy of the marine Pliocene and lower Pleistocene deposits of southwestern Florida: preliminary assessment, in Scott, T. and W. Allmon, Eds., *Plio-Pleistocene Stratigraphy And Paleontology Of Southern Florida*, Special Publication **36**, Fla. Geological Survey, Tallahassee, pp. 27-40, 1992.

Index

Milton Keynes UK
Ingram Content Group UK Ltd.
UKHW051947071024
449327UK00026B/2209